IET CONTROL, ROBOTICS AND SENSORS SERIES 103

Open Resonator Microwave Sensor Systems for Industrial Gauging

The IET International Book Series on Sensors

IET International Book Series on Sensing–Call for Authors

The use of sensors has increased dramatically in all industries. They are fundamental in a wide range of applications from communication to monitoring, remote operation, process control, precision and safety, and robotics and automation. These developments have brought new challenges such as demands for robustness and reliability in networks, security in the communications interface, and close management of energy consumption. This Book Series covers the research and applications of sensor technologies in the fields of ICTs, security, tracking, detection, monitoring, control and automation, robotics, machine learning, smart technologies, production and manufacturing, photonics, environment, energy, and transport.

Proposals for coherently integrated international multi-authored edited or co-authored handbooks and research monographs will be considered for this Book Series. Each proposal will be reviewed by the IET Book Series Editorial Board members with additional external reviews from independent reviewers. Please email your book proposal to: vmoliere@theiet.org or author_support@theiet.org.

Open Resonator Microwave Sensor Systems for Industrial Gauging

A practical design approach

Nathan Ida

The Institution of Engineering and Technology

Published by The Institution of Engineering and Technology, London, United Kingdom

The Institution of Engineering and Technology is registered as a Charity in England & Wales (no. 211014) and Scotland (no. SC038698).

First published 2018

The Institution of Engineering and Technology
Michael Faraday House
Six Hills Way, Stevenage
Herts, SG1 2AY, United Kingdom

www.theiet.org

British Library Cataloguing in Publication Data
A catalogue record for this product is available from the British Library

ISBN 978-1-78561-140-7 (hardback)
ISBN 978-1-78561-141-4 (PDF)

Typeset in India by MPS Limited
Printed in the UK by CPI Group (UK) Ltd, Croydon

Contents

Preface

1.1 General remarks on microwave systems

One of the goals of the present work is to present a coherent and entirely practical approach to the design of open resonator microwave sensors. After preliminary discussion of transmission lines and transmission line resonators and discussing the general issues of microwave measurements and resonance, we embark on a detailed design of a sensor system including dimensions, equipment, and results. The idea behind this approach stems from the fact that whenever one deals with microwaves, the "distance" between theory and practice is large and not easily bridged without years of experience or extensive experimentation. The description, which at times may seem lengthy, is intended to be inclusive so that by its end, there should be no doubt in the reader's mind on the whats and the hows of the design. It is genuinely hoped that this approach can reduce the angst of any potential designer and encourage the use of this powerful technique should the need arise. Nevertheless, it should be realized that microwave equipment design is a mixture of disciplines. In addition to understanding of the behavior of microwaves, the design of the mechanical components is equally important. It is not possible to achieve a good design with sloppy mechanical systems just as it is not possible to achieve proper operation with bad electromagnetics design. Although one may take the view that a sensing system is merely the implementation of a sensor, a source to feed the sensor, and a measurement system to read and analyze the results, this would be an oversimplification leading to an inadequate system. Every part of the system must be accurately designed and the components must fit together. For example, one can do everything right and still end up with a bad system by not paying attention to, say, impedance matching between the probes and the source. Or, one can again properly design the system but make a shield which is too thin that reacts to mechanical vibrations thereby introducing changes in the output that are not due to the measured quantity.

Microwave sensors in the industrial environment have been neglected in the past for a number of reasons, some valid, some less so. But it seems for the most part because of perceptions associated with microwaves and with the design of microwave systems. Whereas the military and communication industries have used microwaves on varied large scales, the use of microwaves in industry has been limited to microwave heating and limited use in sensing. A few of the reasons for this are as follows:

Microwave systems are expensive

Indeed, they usually are. There are many objective reasons why microwave systems are not inexpensive. First, there is the issue of design at high frequencies, which imposes certain requirements on the design, not to mention limited availability of trained designers. Cables for example must be properly shielded and what passes as an adequate shield at low frequencies may be totally inadequate at higher frequencies. Stable oscillators are more difficult to build, components that operate at microwave frequencies are fewer and more expensive and microwave integrated circuits are even rarer. Materials used in microwave components are also unique. Lossless dielectrics are rare and hence their production requires care. Conductors must be very good to limit losses. The use of silver and gold for conductors or for coating of conductors is very common. Then there are the issues of propagation of microwaves, which are best done with transmission lines or waveguides. Although these can be integrated, they are not usually off-the-shelf devices or components. Even printed circuit boards must be selected carefully as their properties at high frequencies are different than at lower frequencies. These issues, and many others, mean simply that one cannot decide that a microwave sensor is the best solution without thoroughly evaluating its costs and benefits. If it turns out that the microwave system envisioned is too costly or its cost-benefit ratio is inappropriate, one should look at alternatives. There are however applications in which microwave sensing and, in particular, microwave gauging are most appropriate both in terms of the physics involved and in terms of cost-benefit.

Microwave systems are sensitive and "fussy"

They can be but only to the extent that they are not properly designed or used. Many an engineer has found out the hard way that a microwave system is affected by unexpected sources leading to poor performance. This is not in fact a problem with the microwave idea but rather with its implementation. In fact, it is exactly this sensitivity to a variety of factors that is so appealing in microwave sensors. However, at the core of any microwave system, there are only three properties that can be sensed: permittivity, conductivity, and permeability. Any changes in these properties of materials and anything that causes changes in these properties will be sensed by a microwave sensor. It should then not be surprising that, for example, a cavity resonator designed to measure the permittivity of a plastic should also react to changes in humidity. Another example might be sensitivity of a sensor to personnel in its vicinity. The design must be able to cope with these issues by proper calibration if necessary, perhaps by compensation of the sensor's output or any other means that will eliminate the error due to external sources.

Hidden reasons for visible effects

One of the most frustrating issues with microwave sensing and in particular with microwave gauging is that the output of the sensor sometimes seems to be capricious—reacting as if it had a mind of its own. One gets a reading without being able to pinpoint the source. Of course, microwave systems just as any electronic system react to inputs in a specific and fully predictable fashion. Any effect seen is due to a physical condition, but in microwave, the sources are not always easy to identify. The source may be due to a bad cable or a leak in a waveguide, moisture in the system, or a myriad of other sources often mechanical in nature but not only. For example, a microwave may react to a radio or TV transmission or even to sources from power lines, not to mention mechanical issues such as vibrations. Careful measurements and/or simulation should be used to eliminate all undesired effects before a design can be called acceptable.

Complexity of microwave sources

A microwave source in its most general form is an oscillator operating either as a fixed frequency oscillator or perhaps in a range of frequencies. Amplitude stabilization may also be a part of that as can be a system of measuring both. That may not sound like much, but building such a source can be an arduous task. Similarly, connection of an external cable to a printed circuit board may require special attention. One cannot simply solder the cable and hope that all will be well. Proper matching to the board must be considered as must be the connector, length of wires, delays on various paths, materials involved, and so on. If a printed circuit board needs to be used, its design again calls for special techniques. One must remember that any conducting path is part of a transmission line with its propagation properties, losses, phase delays, and impedance. Every length of conductor is an antenna that will radiate into space and may interfere with the circuit itself or with other circuits. Special attention must be paid to issues of electromagnetic compatibility, power levels, and frequencies used as well as matching.

Expensive equipment: the use of a network analyzer

Anyone that has used a network analyzer must have been impressed with the performance of this exceptional instrument, its capabilities, its accuracy, its many uses, and, in particular, its cost. For a network analyzer is, necessarily, an expensive instrument. And it is expensive for all the good reasons—primarily its performance—and for all the bad reasons—primarily its limited use in laboratories. But, in spite of that, I suggest very much its use in the industrial environment as part of the sensor system proposed here. At first glance, it may seem that this will only make the system more expensive. However, when considering the time and cost needed to design a gauging sensor, its source and measuring system, the network analyzer, which includes all the necessary functions as well as the software needed to analyze the data and transmit it where it is needed, the cost is likely to be lower overall. Similarly, a single component system, as costly as it is, will be less expensive to maintain as it will not require more than an adept engineer that can set its parameters or modify some software and, in case of breakdown, replace it and send it for repair. There are of course some additional constraints imposed by the instrument. It will have to be protected from the environment, likely will need to be kept in a climatized space and may require periodic calibration. But these requirements are not that different from those required by industrial controllers and computers.

Safety of microwave radiation

Microwaves have been implicated in the popular press in health effects including cancer. None of this has been proven in reality but there are standards for exposure limits to microwaves both for the general public and for occupational environments. Microwave radiation is nonionizing throughout its range, but, as a matter of prudence, microwave designs should use the lowest intensity fields that will properly address the problem. Proper shielding of radiation is another requirement of the design.

Any time one designs a system rather than a component, the parameters to deal with are necessarily intertwined. There are constrained parameters of space, frequency, sensitivity, and the like. Then there are system parameters such as size, coverage, and cost. There are mechanical parameters imposed by the measurement environment, and finally, there are parameters that can be chosen to affect particular operational details of the system. In the design that follows, we look at all of these in details.

1.2 Measurement versus gauging

There is a rich and varied literature on microwaves and microwave measurements—from tutorials and textbooks to important monographs on circuits and measurements. The present book attempts to go beyond the general topic of microwave measurements and present a new topic—that of microwave gauging. The very name implies measurements, specifically accurate measurements. The premise is that in many cases, to detect something, or that a physical parameter such as thickness or permittivity can be measured is not sufficient. It has to be measured accurately and the measurement must be repeatable under reasonable conditions. Only then can one claim that true gauging takes place.

Microwave measurements are quite common and encompass the range of physical properties from simple dimensional measurements, general electric and magnetic properties of materials, to very specific applications such as measurement of moisture content, distance, and speed. Yet, the application of microwave measurements for a wide range of industrial processes has been slow to evolve and find acceptance for a number of reasons. One reason is objective—often, the cost of microwave equipment is high and hence the application must justify the cost. But there are two other reasons that come into play. One is the fact that most microwave measurements are seen as not sufficiently accurate for the task at hand and the second is the perceived complexity of such systems.

If microwave measurements are common, they are not simple. And one constant in all microwave measurements is that they are difficult to interpret uniquely. That is, a particular measurement may be influenced by other, sometimes secondary effects that skew the results. For example, measurement of transmission through a dielectric, in itself a very simple process, may be inaccurate because of high humidity in air, through which waves must propagate. In other instances, the material properties are only approximately known or measured at frequencies lower or higher, and again, the interpretation of results suffers. Then there is noise. Of all kinds.

It would seem that while microwaves are very well suited for many industrial applications, gauging, which by its very name implies accuracy, is not one of them. And yet, this is exactly what this work attempts to do: To show that with proper care, appropriate equipment, appropriate sensors, and sound calibration methods, gauging is not only possible but is accurate, repeatable, and effective. Like other gauging methods, microwave gauging is not inexpensive. The equipment is sophisticated, produced in low volumes and therefore expensive. Few if any of the sensors that may be used are off the shelf, again contributing to the overall cost.

On the other hand, microwave measurements and the sensors used are often trivially simple and therefore rugged, well suited to the industrial environment. Continuous production line gauging is practical, safe, and minimally intrusive. In addition, in some applications microwave sensing replaces radioactive sensing. This is the case in fabric thickness and density measurements.

The purpose of the present book is 2-fold. First, it attempts to dispel the notion that microwave measurements are inaccurate and lay the groundwork for meaningful, accurate, and industrially viable methods of gauging. Second, it tackles the issue of the measuring environment, sensors, and equipment to show, in detail, how these are made, the equipment needed, and specific procedures for practical gauging.

The presentation starts with the necessary theoretical subjects. Specifically, the ideas of microwave measurements, waveguides, and transmission lines, so necessary to the understanding of any microwave system, are first introduced, followed by their extension to waveguide and cavity resonators. In all of these, we emphasize the use of scattering parameters as these will then be used for practical measurements. This part of the book may seem lengthy but it is necessary for the proper understanding of the design and, in particular, for the measurement process.

1.3 Text contents

Chapter 1 introduces the subject of microwaves and places the sensors to be described later in perspective. Following a short section on the microwave range and a historical section, we discuss the advantages and disadvantages of microwave testing and gauging. The properties and energy associated with microwaves and material properties in the microwave domain are introduced.

Chapter 2 starts with basic theory of transmission lines followed by transmission line circuits and transmission line resonators. The theory is limited to those aspects of transmission lines that have a bearing on the work in this book. Similarly, we only discuss transmission line circuits that are relevant to the design of the sensors or to the measurement process including matching circuits, attenuators, and directional couplers. Properties of resonators and their uses are discussed with a view to the chapters to follow, where these resonators are implemented and evaluated. The discussion of resonators is particularly important since the core of this work is resonant microwave sensors based on transmission lines. Nevertheless, much of the discussion is general and applies to other types of resonators from simple RLC circuits to microwave cavity resonators. The chapter also includes the fundamentals of the Smith chart as a tool in design and display of transmission line properties.

Chapter 3 introduces the general concept of planar transmission lines with emphasis on striplines, and the stripline resonators are then introduced and their theory outlined. Stripline resonators and in particular the broadside-coupled resonator is emphasized as the structure used for the sensors in this work. A section on waveguides and cavity resonators is also included as an extension of the concept of transmission lines and because the analysis of the transmission line resonators to be introduced later uses methods borrowed from cavity resonators. The issues of calculation of changes in resonant frequency due to changes in material properties are addressed through introduction of the perturbation method. Since this work uses open stripline resonators with partial shielding, some of the aspects in this chapter overlap with that of cavity resonators. In particular, a section on cavity perturbation applies classical microwave cavity resonator ideas to the present work.

Chapter 4 introduces the general context of microwave measurements including reflection, transmission, refraction, resonance, and their relation to gauging. The case for resonant methods as particularly sensitive and applicable to gauging is made. Because the sensors presented here are designed to work with a network analyzer which provides both the driving and measurement capabilities, the chapter emphasizes the use of *S*-parameters as the means to measure the various quantities required. This approach is not only convenient but necessary to achieve the accuracy required from a gauging system, and, as an added benefit, it is modern and well suited for use in connected systems. The chapter also discusses the general methods for measurement of frequency and *Q*-factors in resonators, measurement of power, and the application of the cavity perturbation method introduced in Chapter 3 for analysis. The chapter concludes with the important issue of measurement of material properties.

Chapter 5 discusses the design of the resonator for two applications of interest. One is a wide latex soaked fabric moving on a production line with the sensors monitoring its thickness or, alternatively, the moisture content in the fabric for the purpose of defining the amount of latex on the fabric. The second is a rubber sheet moving on a calender (a cylindrical drum) where the interest is the thickness of the sheet. Selection of physical parameters, dimensions and operational parameters, and the simulations necessary are discussed. Alternative designs including multiple sensors and moving sensors for full coverage of the fabric are weighed and an appropriate design is reached. The details of design are given in full with alternatives and justification so that the reader has full accounting of what the design involves and what to expect from the final product. As is often the case in microwave systems, design is a mix of science and art. The design of stripline resonators is no exception, but, fortunately, some of the uncertainties can be eliminated by simulation. The use of simulation tools is emphasized at every step of the process. Although there are two sensors involved, some of the design parameters are common to both, and hence, the rubber thickness sensor is viewed as a modification of the fabric sensor. The modification also change some of the properties of the sensor and these are discussed as well.

Chapter 6: It is one thing to come up with a sensor but a whole different thing to come up with a sensor that satisfies the strict criteria required in a gauging application. The present chapter discusses the performance of the sensor, its sensitivity, accuracy, calibration, and other parameters such as drift, sensitivity to environmental changes, and long-time stability. The evaluation is based on both simulations and on extensive in-plant measurements under normal production conditions with a prototype sensor over extended periods of time. The process of optimization continues with particular emphasis on performance and the effects of various design parameters have on errors.

Chapter 7 gives full details of the implementation to the extent they are important in terms of performance, but it is also recognized that different applications will have different physical requirements. Particular attention is given to the mechanical system and its effects on the measurements. A full evaluation of the mechanical system is included. Testing follows the implementation. Calibration of the measurements is done with calibration standards, and compensation for environmental conditions is implemented based on the odd-mode resonant frequency for the fabric sensor.

Chapter 8: As was indicated above, one of the features of the current work is the incorporation of a network analyzer as part of the overall sensing system. Although a network analyzer is a very expensive piece of equipment often associated with laboratory work, its use in a system of this type is justified on a number of grounds, not the least being development time, accuracy, and stability. In the industrial environment, it also affords a single unit that can be replaced or repaired quickly without the need of specialized personnel, saving downtime, and, ultimately, costs. The chapter discusses primarily the methods of measurement but also summarizes the structure of the network analyzer with particular emphasis on vector network analyzers.

Appendices: Four appendices document some aspects of the work that did not lend themselves to inclusion in the various chapters and add additional material. These include the issue of radiation safety due to the open nature of the resonator used (Appendix A), a short discussion on material properties (Appendix B), and some elements of the method of simulation employed in the design and optimization of the sensors (Appendix C). Appendix D is a short exposé of some elements of electromagnetic theory. These include elements of wave propagation in lossless, low loss, and lossy dielectrics; the Poynting theorem and the Poynting vector; and elements of reflection, transmission, and refraction at interfaces between various materials. These concepts are used throughout the book either implicitly or explicitly, and in this sense, the purpose of the appendix is to complete the theory.

A note on references

The work reported in this text is based on my own work, on work of some of my students, and, of course, many of the theoretical discussions and practical aspects of measurements have been published by others. I have tried to properly refer and acknowledge all sources I used. In the interest of readability, I opted to not include references in the text but, rather, to include an annotated bibliography section where necessary. I felt as well that annotated references are more useful in that they guide the reader to either verify or expand on the material given. Another aspect is the preferred references to books rather than scientific publication. The reason for that is readability and the desire to base the exposition on mature work rather than current research.

Acknowledgments

I wish to acknowledge the contributions to this work by four students. Mr. Scott DuFore did some of the earlier work on testing and verification of the stripline resonator sensor for fabric coating analysis including in-plant testing as part of his Masters' thesis. Mr. Omar Bhuiya worked extensively on the simulation of the rubber thickness sensor. Some of his results are reported here as part of the design and optimization, whereas a full account of the simulations is reported in his Masters' thesis. Dr. Nader Farhat worked extensively on the optimization of the sensors and on design and execution of measurements. Their contribution is gratefully acknowledged. Dr. Julian Gomes contributed to understanding of the coupling to the transmission line resonator and to some of the simulations used in the present work.

Other people worked on various aspects of the design and implementation of the two sensors described in this work. I wish to acknowledge the work of Dr. Louis Cabrera and Mr. Ken Kot who did much of the mechanical work and testing that led to the final design and performed many of the measurements reported. Finally, Dr. Michael Madaras will always be remembered as the one who initiated the project that led to the design of the sensors described in this work and who was a constant force behind their execution over many long months of work. May he rest in peace.

Chapter 1
Introduction to microwaves

1.1 General

As mentioned in the preface, the subject of this work is microwaves and their use for the specific purpose of gauging. In that capacity, we will need to work with a number of electromagnetic structures whose understanding is crucial to proper understanding of the principles, possibilities, and limitations of the gauging methods to follow. These include the general topics of propagation of electromagnetic waves on transmission lines and both transmission line resonators and cavity resonators. We will also touch on waveguides, especially as these relate to measurement of material properties. Another aspect of electromagnetics that finds its way into the present work is the concept of probes or antennas. It is assumed that the reader is familiar with the general concepts of electromagnetic waves and wave propagation as well as elements of propagation on transmission lines and on waveguides. Although it is not possible to include here the theory associated with electromagnetics, some of the more important issues are discussed, especially in Chapters 2 and 3 where we introduce the elements most intimately tied to the present work.

The purpose of this short introduction is to discuss in general terms and with a minimum of theory the concepts involved in microwave gauging and to identify issues that will be addressed later, including frequency ranges, measurements and properties of electromagnetic fields in the microwave region.

1.2 The microwave domain

Microwaves occupy a portion of the electromagnetic spectrum bordering infrared radiation on the high end of its range at 300 GHz. The lower range is somewhat less well defined, but it is usually set at 300 MHz or in terms of wavelengths, between 1 m and 1 mm. The electromagnetic spectrum in Table 1.1 shows the accepted designations of various ranges within the spectrum of electromagnetic waves. Only frequencies above 3 kHz are shown, and the upper range shows the infrared, visible light, ultraviolet, and X-ray ranges. The spectrum of course continues into higher frequencies including alpha, beta, and gamma radiation and cosmic rays. The microwave domain is a small range of the spectrum between 300 MHz and 300 GHz. This range is somewhat arbitrary but has been accepted in standards and

Table 1.1 The electromagnetic spectrum

Frequency	Designation	Usage
0.003–0.03 (MHz)	VLF	
0.03–0.3 (MHz)	LF	LF radio
0.3–3 (MHz)	MF	AM radio
3–30 (MHz)	HF	SW radio
30–300 (MHz)	VHF	TV, FM radio
300–3,000 (MHz)	UHF	TV, radar
3–30 (GHz)	SHF	Satellites, radar
30–300 (GHz)	EHF	Radar
300–3,000 (GHz)	mm and sub-mm waves	
3–420 (THz)	Infrared	
420–790 (THz)	Visible light	
790 THz–30 PHz	Ultraviolet	
30 PHz–3,000 EHz	X-rays	

MHz = 10^6 Hz, GHz = 10^9 Hz, THz = 10^{12} Hz, PHz = 10^{15} Hz, EHz = 10^{18} Hz.

Table 1.2 IEEE radar band designations

Band	Frequency (GHz)	Name
HF	0.003–0.03	High frequency
VHF	0.03–3	Very high frequency
UHF	0.3–1	Ultrahigh frequency
L	1–2	Long wave
S	2–4	Short wave
C	4–8	Compromise (between C and X)
X	8–12	Crosshair
Ku	12–18	Kurz-under
K	18–27	Kurz
Ka	27–40	Kurz-above
V	40–75	
W	75–110	
mm or G	110–300	Millimeter

is used to identify microwaves and microwave applications. At the lower end of the band, microwaves overlap the Ultra-High Frequency (UHF) band and at the higher end, they border the far infrared region. In addition to the general spectrum, microwaves are often identified with subdomains or bands. These are designated by letters to allow simple identification of equipment and operating frequencies. The standard IEEE radar band designations are shown in Table 1.2. The names in the third column are customary but are seldom used. There is a separate military band designation shown in Table 1.3. Additional designations exist either by organizations such as the International Telecommunication Union (ITU), Nato, broadcasting organization or by usage such as the bands used in waveguides, shown in

Table 1.3 New military microwave band designation

Band	Frequency (GHz)	Wavelength (cm) (in vacuum)
A	0.1–0.25	300.0–120.0
B	0.25–0.5	120.0–60.0
C	0.5–1.0	60.0–30.0
D	1.0–2.0	30.0–15.0
E	2.0–3.0	15.0–10.0
F	3.0–4.0	10.0–7.5
G	4.0–6.0	7.5–5.0
H	6.0–8.0	5.0–3.75
I	8.0–10.0	3.75–3.0
J	10.0–20.0	3.0–1.5
K	20.0–40.0	1.5–0.75
L	40.0–60.0	0.75–0.5
M	60.0–100.0	0.5–0.3

Table 1.4 Waveguide bands

Band	Frequency (GHz)
R	1.7–2.6
D	2.2–3.3
S	2.6–3.95
E	3.3–4.9
G	3.95–5.85
F	4.9–7.05
C	5.85–8.2
H	7.05–10.1
X	8.2–12.4
Ku	12.4–18
K	18–26.5
Ka	26.5–40
Q	33–50
U	40–60
V	40–75
E	60–90
W	75–110
F	90–140
D	110–170
G	140–120
Y	325–500

Table 1.4. The latter is intended to follow the modes and bandwidths of standard rectangular cross-section waveguides. Note also that bands can overlap, again to cover standard waveguides. The relative location of the microwave domain in the electromagnetic spectrum is shown schematically in Figure 1.1.

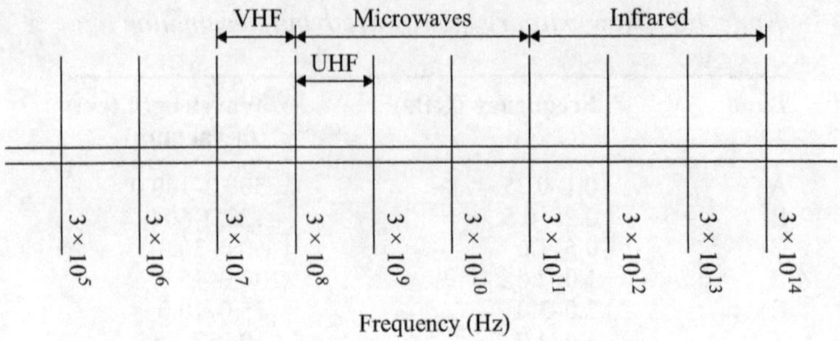

Figure 1.1 *Relative location of the microwave domain in the electromagnetic spectrum*

As examples of various uses, many microwave ovens operate in the L and S bands of Table 1.2 (a typical frequency is 2.450 GHz). Radar detectors used by police operate in the X and K bands (10 or 24 GHz). Communication with satellites is normally done in the C and K bands (4–30 GHz).

In the present work, we use microwaves at their very lowest end—around 350–500 MHz. This choice will be further explained in Chapter 4, but at this point, it is sufficient to indicate that the choice of operating frequency is a compromise between the need for high sensitivity, which requires high frequencies, and the need for a relatively large spatial coverage, which could be best done at lower frequencies.

1.3 History

It is of some interest to note that the first electromagnetic experiment to show the existence of waves as predicted by Maxwell is, in a way, a microwave experiment. When Heinrich Hertz used a spark transmitter to transmit electromagnetic waves, the wavelengths he used were smaller than 1 m (about 60 cm or a frequency of about 500 MHz), well within the microwave range. Propagation of waves in waveguides was shown theoretically as early as 1897 by Lord Rayleigh, even though they could not be realized experimentally at the time for lack of appropriate generators. However, 3 years earlier, Sir Oliver Lodge, observed wave guiding when he surrounded a spark generator of the type used by Hertz by a conducting tube. Waveguides were not actually used until about 1936. In the meantime, the theory of transmission lines was developed by Oliver Heaviside based on the work by Maxwell. The first reliable microwave sources were developed in the 1930s for radar applications. The first device was the magnetron, a microwave source that, in one form or another, still finds applications in high power microwave systems and, in vast quantities, in microwave ovens. The development of the klystron in the 1930s was another important step, because it could be used as a microwave amplifier. The importance of these two microwave sources is amply proven by their continuous use ever since their invention. Solid-state microwave devices are coming into wide use, especially for low-power applications, but their earliest use can be traced to the use of the crystal diode in the 1930s. Active devices became available in the 1960s with the development of negative resistance devices and field effect transistors. The use of gallium arsenide (GaAs) and indium phosphide (InP) has also benefited this aspect of solid state device development. Currently, there are many relatively simple, reliable, and inexpensive microwave devices including radio-frequency-integrated circuits (RFICs), microwave integrated circuits (MICs), and monolithic integrated circuits MICs that the designer can use. At high power levels, one finds vacuum tubes such as klystrons, magnetrons, and traveling wave tubes, devices that cannot be replaced with semiconductor devices.

Passive devices were also developed in parallel with the development of microwave sources. These include a variety of antennas, couplers, filters, attenuators, waveguides, cavity resonators, absorbers, terminators, rotators, and others.

Microwave circuit development followed steps similar to other circuits at lower frequencies. As examples, the idea of heterodyne receivers was used starting with the earliest radar equipment, while MICs find applications in many communication systems.

Measurement equipment was, for a long time, one of the stumbling blocks of microwave work. This however has been largely solved by the availability of computer-controlled network and spectrum analyzers. Modern microwave equipment is as reliable and as accurate as any other equipment although, in most cases, it tends to be more expensive.

1.4 Advantages and disadvantages of microwaves for testing, measurements, and gauging

Testing with microwaves is dominated by the basic properties of microwaves. Since their penetration in good conducting materials is minimal, they are mainly used to test nonconducting materials. This includes dielectric and lossy dielectric materials. Testing and measurements on conducting materials are limited to dimensional testing such as thickness gauging and surface measurements such as testing for surface conditions and flaws.

On the other hand, microwaves are affected by a large number of material properties. In lossless or lossy dielectrics, such diverse properties, as density, porosity, material composition, uniformity of the material, delamination of layers, moisture, and contamination content, are only some of the properties that can be measured and, more importantly, that can affect a measurement. The range of nonmetallic materials in which this is possible is extensive and growing steadily. This includes ceramics, plastics, polymers, and composites, as well as organic materials such as wood products, foods, or biological materials. Measurements and testing in all dielectrics and lossy dielectrics are possible.

The spatial resolution that can be expected of microwave tests depends on the wavelength of the wave and the type of measurement undertaken. For microwaves and millimeter waves, this is of the order of at most, 1 mm. This resolution indicates the ability of the test to discern closely spaced discontinuities in the materials. However, more sensitive measurements are possible by correlating them with changes in amplitude or phase. This is the case with dimensional measurements where variations in thickness of a few microns are measurable. High-resolution testing can be achieved by scanning of the microwave field and by moving the source. This is similar to synthetic aperture radar (SAR) methods and can be used for imaging. If this is done in the near field of antennas, it may even be called "micrometry."

Another particular property of microwave testing is the means by which energy is coupled into the testing environment. This can be as simple as a horn antenna or an open microwave guide. In some other cases, an aperture is used for this purpose or, as with microwave cavities, a simple probe or loop serves the purpose. In all cases, the coupling can be done through air, free space, or a convenient dielectric. Impedance matching can also be employed if necessary. While most testing is done in what may be termed the "near-field" environment (close to the antenna or source), measurements in the far field are also possible where the waves propagate through a medium such as air and interact with the test sample. This is particularly applicable to scattering methods, including radar and radar-like testing applications.

Because of the influence of so many effects and properties on the wave, the testing environment can be, and often is, noisy. Reflections from near and far surfaces, edges, and other artifacts in the material are often encountered, resulting in noise and loss of resolution. One of the most sensitive method of testing, and the one pursued in this work, is resonance. In classical resonant methods, the microwave fields are confined within a space by conducting walls in what are called cavity resonators. Power is coupled into the cavity to compensate for losses and the fields in the cavity then interact with any material or structure introduced into the cavity. Because the fields can be very high and the resonant frequency highly dependent on the physical dimensions and on material properties, the measurements can be extremely sensitive. In general, what is measured is the resonant frequency and that in turn correlated with properties of interest such as permittivity or dimensions. In the present work, use is made of open or partially open resonators based on transmission lines operating in the low microwave frequency range. The open structure offers some unique advantages such as testing of continuous materials, for example, fabrics or dielectric sheets but also has distinct disadvantages such as reduced sensitivity and interaction with structures outside the test region. Much of the work reported in the following chapters deals with means of mitigating these disadvantages and increasing the sensitivity of the system.

Microwave radiation is highly directive, and, because of the short wavelengths involved, the devices used are often very compact. While many of the applications are in high-power communication and radar system, low-power applications are just as common. In this range, the choice of microwave sources is relatively wide and includes low-power solid-state devices.

Of primary interest in this work is the interaction of microwaves with materials. This takes the form of absorption in materials, scattering, attenuation, and transmission. These effects are exploited in various testing arrangements to allow for quantitative measurements in materials. In the context of resonant structures, the most important aspect is absorption. The absorption of microwaves in water is well known and widely used in microwave ovens. These ovens depend on this effect because it is typical of the whole microwave range. Thus, while most microwave ovens operate at specific frequency bands because of regulation, they can also operate at other frequencies. The absorption in water can be used either directly or indirectly for testing of moisture and related effects (e.g., curing, drying). Other materials absorb radiation at specific frequencies. Often, these are narrow bands that allow chemical analysis of materials. These resonant frequencies are extremely useful for material characterization and identification. Sometimes, even traces of materials can alter the resonant frequency of a microwave cavity, leading to detection of materials in trace amounts. This again may be used for testing or detection. Typical applications of this type are contamination tests and detection of explosives.

Because microwaves border on the one hand the high-frequency radio range and on the other the low infrared range, they have properties of both. More than any other frequency range, the microwave range is sometimes analyzed using circuit theory and sometimes using wave theory. Transmission lines are almost always analyzed as distributed parameter circuits, while the aspects of refraction, transmission, and propagation in waveguides are analyzed using wave theory.

Another aspect of microwaves is the special nature of the components used. The familiar conductors are now replaced with guiding structures such as transmission lines or waveguides. In transmission lines, we still use circuit concepts including voltage and current. Hence, power propagates along the lines in the familiar environment of circuits although, unlike circuits, there are additional concepts of speed of propagation and delays on lines. In other microwave structures, power is transmitted not through the flow of current but through propagation of fields. The fields are guided in the required direction by guiding structures. These can take the form of hollow conductors, parallel plates, or dielectric slabs. The impression one gets of a certain "plumbing" character involved in microwaves is not without basis. A waveguide or even a coaxial transmission line cannot be bent in sharp corners as one would do with a wire. The guides and lines must be properly terminated and matched to loads and sources, and modification of propagation properties is often accomplished by physical structures in the waveguides.

Microwave radiation has other properties that are less important in testing than, for example, in communication. One of these is the large bandwidth possible, because of the high frequency of the waves. Microwaves also penetrate easily through the ionosphere, with obvious applications to satellite communication. In addition, microwaves penetrate into the body, causing a variety of effects, including heating of tissue. The safety aspects of microwave measurements and gauging cannot be neglected and, in fact are regulated by standards and regulating agencies such as the ITU or the Federal Communication Commission (FCC).

1.5 Energy associated with microwaves

The radiation energy associated with microwaves can be estimated considering the quantum equivalent photon. The energy of a photon is equal to hf where h is the Planck constant ($h = 4.14 \times 10^{-15}$ eV). Thus, the maximum energy of a photon in the microwave range is roughly 1.2×10^{-3} eV at the top of the frequency range (minimum is about 1.0×10^{-6} eV at the lower frequency range). This energy is relatively low and is much lower than the energy needed for ionization. The energy is much lower than the energy in molecular links. Thus, because it cannot break these links, it is considered a nonionizing form of radiation.

The danger from microwave radiation is considered to be primarily due to absorption rather than due to its intrinsic energy. Exposure to microwaves is not unlike absorption in a microwave oven, except for the levels encountered. For this reason, the safety levels of radiation are often defined on the surface, in terms of power per unit area (W/m^2). There is also some evidence that nonthermal effects of microwave radiation play a role, but this issue is still controversial and not very well researched.

The exposure levels allowed should serve as guidelines only. While there may be no harmful effects due to allowable levels of microwave radiation, the levels in the USA are much higher than in other countries. As a rule, one should avoid all exposure to microwave radiation if only because of its absorption effects. Exposure to microwave radiation has other consequences, some that are not immediately obvious. For example, pacemakers wearers can experience interference from Radio Frequency (RF) sources including microwave sources. Other devices may also be interfered with although this is by no means unique to microwaves.

The allowable energy density exposure in industry in the USA is 1 mW/cm^2 or 10 W/m^2. As a means of understanding the thermal effects of this radiation level (nonthermal effects are not as well defined and are still being debated), it is useful to compare this radiation level with radiation from the sun. The maximum sun radiation is about 1,400 W/m^2 or 140 mW/cm^2. To compare the fields associated with the two types of radiation, these two energy densities are viewed as the result of a Poynting vector and the equivalent electric and magnetic field intensities are calculated. For the allowable microwave exposure in the USA of about 1 mW/cm^2, the electric and magnetic field intensities are calculated from the time averaged power density:

$$P_{av} = \frac{E^2}{2\eta_0} = 10 \quad (\mathrm{W/m^2}) \tag{1.1}$$

where η_0 is the intrinsic impedance of free space and is equal to 377 Ω. From these, the electric field intensity is

$$E = \sqrt{2 \times 377 \times 10} = 88.83 \quad (\mathrm{V/m}) \tag{1.2}$$

The magnetic field intensity is

$$H = \frac{E}{\eta_0} = \frac{88.83}{377} = 0.23 \quad (\mathrm{A/m}) \tag{1.3}$$

For comparison, the electric and magnetic field intensities associated with radiation from the sun are

$$P_{av} = \frac{E^2}{2\eta_0} = 1{,}400 \quad (\mathrm{W/m^2}) \tag{1.4}$$

$$E = \sqrt{2 \times 377 \times 1{,}400} = 1{,}027 \quad (\mathrm{V/m}) \tag{1.5}$$

$$H = \frac{E}{\eta_0} = \frac{1{,}027}{377} = 2.72 \quad (\mathrm{A/m}) \tag{1.6}$$

A direct comparison of these values would indicate that microwaves, even in the highest range, produce fields that are much lower than those produced by the sun. This, however, does not necessarily mean that microwaves are less "dangerous" than the sun's radiation at the levels given. There are two aspects of microwave energy that make them different than energy from the sun: one is the high absorption in water, and therefore the heating effects of microwave power. The second is its deeper penetration in tissue as well as penetration through clothing. Whereas normal clothing protects against the sun, it does not protect against microwave radiation.

Because of the concern for radiation, the designs developed in this work were evaluated for microwave exposure. The fields produced by the systems were measured and compared with standards and regulations to ensure compliance. These measurements are given in Appendix A together with requirements published by the FCC and by the American Conference of Governmental Industrial Hygienists.

1.6 Properties of fields at high frequencies

The properties of electromagnetic fields in any frequency range can be deduced from Maxwell's equations and material properties. However, the high-frequency range has some common properties that will be emphasized throughout this work. These properties form the cornerstone of measurements at high frequencies. For this reason, they will be discussed qualitatively and briefly here.

The first and foremost point is that the fields are always part of a wave. The electric field intensity and magnetic field intensity are used to describe a wave. Their relations with each other and with materials define a variety of properties such as mode of propagation, reflection, refraction, transmission, and impedance. The wave nature of the fields also forces us to think in terms of speeds of propagation in different materials and delays in signals (or phase retardation) because of these speeds. The notion that an electromagnetic disturbance propagating instantaneously is convenient in the static or quasistatic domains but not in the dynamic case where propagation times are extremely important and, in fact, often serve for measurements and testing. Simple aspects of low-frequency fields such as impedance matching take a whole new meaning at high frequencies. At low frequency, mismatched impedances mean mainly inefficiency, but, at high frequencies, this also means reflection of waves back into the transmitting medium and standing waves. Standing waves can only exist in the dynamic environment and are only important at higher frequencies.

Flow of currents is also different than at low frequencies. First, we must deal with displacement currents. Thus, the whole space can support currents, not only conductors. In many cases, displacement currents are dominant or the only currents in existence. Without these displacement currents, many phenomena cannot be explained, including the existence of the wave itself. Just consider the simple dipole antenna to see that the low frequency, or quasistatic model is inadequate (in the quasistatic model, currents must close through conductors, but the dipole antenna is open circuited at both ends). Conduction currents are dominated by the skin effect and flow only close to the surface of conductors. At microwave frequencies, the skin depth is of the order of a few microns. A regular solid conductor is, therefore, useless at these frequencies except for its role as a guide for waves (waves are guided along the conductor) and to enclose and contain fields within cavities. If large currents exist, the surface of the conductor must be large. Reduction of impedance now means reduction of the "surface impedance" of the conductor. Thus, a thin layer of highly conductive material on a substrate (which can be nonconducting) is more efficient than a thick, heavy conducting wire. The idea of conductivity itself, while the same as low-frequency conductivity, takes a different meaning. Because of the skin depth, many of the conductors we use are often considered to be perfect conductors. Low-conductivity materials, like sea water, in the microwave range are considered to be highly conducting (or, more accurately—high loss materials, since conductivity is most often associated with losses). Materials that we normally view as lossless such as insulating dielectrics or water must be considered as lossy dielectrics since the losses, even when relatively small are important.

With waves as the primary concept at high frequencies, we also need to deal with wave behavior of voltages and currents in circuits and on transmission lines leading to a new and different way of treating the common circuit parameters, especially as they occur on transmission lines. The behavior of circuits now must be treated using distributed parameter concepts leading to finite speed of propagation and delays in circuits.

Material properties in the microwave range are considerably more complex and, to a certain extent, more prominent than at low frequencies. Many of the properties which are normally considered frequency independent must now be altered to include frequency dependence. The best example of this is the permittivity of dielectrics. Both the permittivity and permeability are complex numbers where the imaginary part is associated with losses. In addition to conduction losses, we can talk of dielectric losses and of attenuation of waves due to material properties.

Generation of electromagnetic waves is accomplished with antennas, so one must also delve into properties of antennas and the propagation of these electromagnetic waves in space and in materials. Coupling of power from antennas or probes into structures such as waveguides or transmission lines must be properly defined and controlled to ensure proper operation of transmission line and waveguide circuits including resonators.

This rich and complex system of properties complicates matters considerably, but it also allows a very wide range of measurements and tests that are not possible at low frequencies. These measurements are only limited by our instruments and our imagination. By direct or indirect measurement of material properties and associated effects, virtually any effect can be measured. From the measurement point of view, this is fortunate since a basically simple method allows measurements as diverse as simple dimensional measurements of a sheet of metal or the curing condition of a polymer. Techniques for measurements are also diverse, ranging from simple time-of-flight measurements to spectroscopy to holography.

1.7 Microwaves and mechanics

A unique feature of work with microwaves is the role mechanical quantities have in testing. A big part of this has to do with the fact that the analysis of transmission lines, waveguides, and resonators is based on a distributed parameter model, and therefore, dimensions are often critical to performance of the system and to measurements within it. In some cases, it is hard to escape the feeling of dealing with a "plumbing" system especially when waveguides must be used. Connectors and transmission lines become critical components in the system, and none can ever be considered lossless. Waveguides must be accurately machined and connectors must match perfectly. Transmission lines must be low loss at the frequencies of interest. The second part is the interaction of waves with structures. That means that any object, material or structure, either intentionally or unintentionally in the path of the waves will interact with the waves and, in the context of measurements will affect the results. Conducting structures can act as antennas and can reradiate power. By doing so they alter the distribution of fields. In some cases, when measurements are done with microwaves in open air, special considerations such as calibration procedures must be followed to mitigate the effects of the test fixtures and objects in the vicinity of the fixture. In some applications, vibrations in structures can have considerable effects on measurements. In general one must pay close attention to the mechanical and structural aspects of a microwave system. A third aspect that must be taken into consideration is the effect of environmental conditions on measurements. Material properties within the space of interest define the properties of the waves including attenuation, changes in phase, reflection, transmission and resonance, if any. Material properties, especially permittivity, are often temperature dependent and are almost always frequency dependent. Humidity plays an important role as well. It is often necessary to evaluate the permittivity of materials as part of the measurement regime as variations in permittivity are often the source of discrepancies between measurements and expected results.

1.8 Instrumentation and instruments

One factor that often discourages research and development of microwave systems is the instrumentation of these systems. Instrumentation at microwave frequencies tends to be complex and expensive as one would expect of any instrument that must operate at very high frequencies and large bandwidths. Often too, the equipment is specialized and produced in small quantities adding to cost. In many cases, instruments must be calibrated often if meaningful measurements are to be obtained. The use of fixtures is common, and often fixtures are specialized pieces of equipment designed for specific measurements. All this means that personnel must have proper training in microwave measurements, sometimes specialized to a class of measurements. The dominant test instrument is the network analyzer, which itself may be geared toward a frequency range or a class of applications. Scalar-network analyzers are used to measure quantities based on amplitudes, whereas vector-network analyzers also analyze phases and the quantities associated with them. Some instruments are geared toward the needs in communication, whereas others are general purpose. In general, network analyzers are superb, highly accurate instruments with a cost to match. In addition to network analyzers, one encounters spectrum analyzers, again with the bandwidth and accuracy needed. To this, one must add counters, sources, and an array of auxiliary components including test fixtures, connectors, attenuators, filters, and the like.

On the circuit scale, microwave components are also specialized not only because of the frequency range and bandwidth they need to operate but also because of issues of impedance matching and control of losses. Nevertheless, an array of active and passive components is available. These include discrete and integrated components of every type needed for design of circuits. Integrated circuits also exist, often implementing specialized circuits such as splitters, mixers, or oscillators. One can expect these to be more expensive than the equivalent low-frequency components and have different operational requirements as they are based on different technologies.

The gauging methods and instrumentation used in this work are good examples of adaptation of classical methods to the unique environments of measurement of material properties and of thickness gauging. The methods, their limitations, and their advantages will be introduced gradually, culminating in systems in the industrial environment. The measurements are undertaken using a network analyzer, which is incorporated as an integral part of the gauging system. This approach, while certainly costly, has the distinct advantage of producing a very accurate system and of reducing the development time of the systems.

Chapter 2
Transmission lines and transmission line resonators

2.1 Introduction

Transmission lines are arrangements of conductors whose purpose is the transfer of power or information from a source to a load, both of which should be viewed as generic. That is, the source may well be a generator but it can equally be the output of a device such as a transmitter, a receiving antenna, or an amplifier, whereas the load can be any device or system that receives the power or information such as a resistor, a transmitting antenna, or an actuator. Transmission lines differ from the common circuit theory approach to transfer of power. In circuits, we usually assume a lumped parameter model, that is, a line, connecting two points in a circuit has some resistance, capacitance, inductance, and conductance that depend on the type of line, materials, and dimensions, but these are the total values for the whole line. That is, one can model the line as in Figure 2.1 where R, L, G, and C are, respectively, the total resistance, total inductance, total capacitance, and total conductance of the line of length l. Often too, these properties are neglected in favor of an "ideal" line assuming that the properties of the line do not influence the operation of the circuits to which it connects. In other words, we often neglect the resistance, capacitance, inductance, and conductance of the line. For example, the wiring in a house is certainly necessary, and it has physical properties that depend on type of conductors, lengths, thicknesses, etc., but we rarely even think of these. They are there to allow operation of devices but their properties are not supposed to influence the operation of the devices. As a consequence of this model, we must also assume that power transfer along a line is instantaneous, that is, as if the load and the generator were connected without the intermediary of a line. This approach is essentially a DC approach, extended to low frequency systems and is appropriate under these conditions. It relies on the fact that the lines are relatively short and frequencies are low. In more exact terms, we use this lumped circuit approach when the physical dimensions of the circuit are small compared to the shortest wavelength at which the circuit operates. The limitations of this approach are also clear: when frequencies are high (short wavelengths) or when lines are long, the lumped parameters approach breaks down and we must consider the effects of waves propagating along the line.

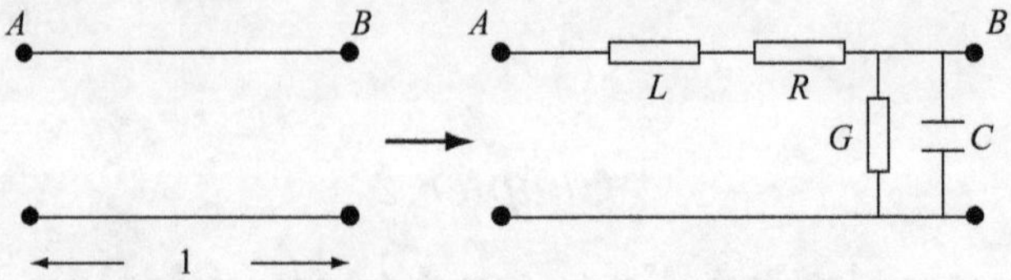

Figure 2.1 A section of transmission line of length l and its equivalent low frequency model

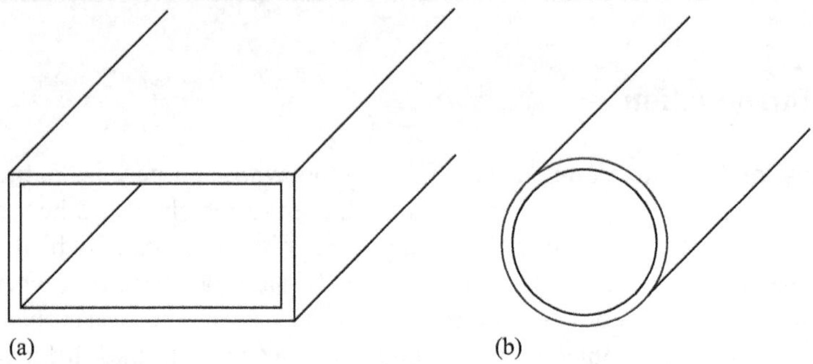

Figure 2.2 Examples of waveguides: (a) rectangular and (b) cylindrical cross-section waveguides

Transmission line theory, on the other hand, assumes a distributed parameter approach, that is, the transmission line still has the same basic parameters as in Figure 2.1, but these are distributed along the line. The model used to develop the relations on transmission lines is shown in Figure 2.4. It consists of a sequence of segments of very short lengths Δz, each segment characterized by the four parameters, but now, these are defined per unit length of the line. In effect, the line is now a circuit with passive components distributed along the line. This model will allow us to characterize the behavior of the line at any frequency and for any length.

Thus, the theory of transmission lines may be viewed as a more general approach to treatment of transfer of energy on lines. Many of the methods used here will be familiar from circuit theory, and in most cases, the results will be in terms of voltages and currents on the line. At times, it will become convenient to shift the discussion to electric and magnetic fields and hence waves on the line. The relation between circuit and field values will also be discussed to facilitate this approach.

One can also generalize the idea of transmission line to any structure that can transfer power along a path. In this category, one can then include waveguides. Just like a transmission line that guides voltage and current waves (and hence power), one can guide electromagnetic waves. In fact, in the previous paragraph, we already mentioned the fact that classical transmission lines can be viewed as guiding propagation of electric and magnetic fields along the line. Waveguides then are a special type of transmission line in which conductors, if any, take a different role. Figure 2.2 shows two waveguides, one rectangular and one circular in cross-section, made as a hollow tube. At its most fundamental level, one can understand the guiding property simply by the fact that waves are contained within the conducting shells and hence are constrained to move along the waveguide just like talking into a tube only allows propagation of sound waves along the tube.

Of course, there are significant differences between two-conductor transmission lines and waveguides (only one conductor and in some cases, no conductors at all). The most obvious is that now we cannot talk of voltages and currents except as abstractions, in the sense that if power propagates, then one can talk of equivalent current and voltages. Then, there is the fact that in transmission lines propagation is transverse electromagnetic (TEM)—voltage and currents, or more accurately, the electric and magnetic fields are orthogonal to each other and to the direction of propagation. In waveguides, the effect is one of waves "bouncing" off the walls and this produces non-TEM waves. Either TE waves (in which, the magnetic field has a component in the direction of propagation) or TM waves (the electric field has a component in the direction of propagation) may exist. These have other effects in terms of impedance, speed of propagation, and losses. Also, unlike transmission lines, in which propagation occurs at all frequencies, propagation in waveguides can only occur above a minimum cutoff frequency related to the physical dimensions of the waveguide's cross-section. For this reason, one cannot postulate a low frequency model for waveguides. In general, waveguides are only practical at very high frequencies. As is the case with transmission lines, a discontinuous section of a waveguide, shorted at its ends, forms a resonator. These resonators are called cavity resonators for obvious reasons. They will resonate at frequencies that depend on the dimensions of the cavity. We will describe waveguides here for the simple reason that the resonators we use both resemble cavity resonators, and the method of analysis we employ is based on cavity resonators. However, we will keep that discussion short since much of it parallels that of transmission lines once one makes the connection between propagating electric and magnetic fields and propagating currents and voltages.

2.2 The transmission line

A transmission line is a physical connection between two locations through two or more conductors. We must indicate at the outset that any transmission of energy through conducting or nonconducting media may be considered a transmission line. Also, any guiding of energy by physical structures such as waveguides may be included in this general definition. However, we will restrict our discussion here to conducting lines with the following properties:

1. The transmission line is made of two conductors in any configuration.
2. The electric and magnetic field intensities in the line are perpendicular to each other and perpendicular to the direction of propagation of power. This type of propagation is called TEM propagation and has all the properties of plane waves.

Examples of lines that we may consider are parallel conducting wires such as the two-wire power cable used to power an appliance or the overhead power transmission line made of thick cables and suspended from towers. Another common type of transmission line is the coaxial transmission line shown in Figure 2.3(d). It is made of two coaxial conductors: an inner, thin, solid conductor and an outer hollow cylindrical conductor. The latter is usually stranded to allow flexibility, and the two conductors are insulated with some dielectric material. Dimensions of coaxial cables and their properties vary, but a good example of an often-used coaxial cable is the antenna cable in televisions, input cables for oscilloscopes, or input leads in audio equipment. In this work, coaxial transmission lines will be used as testing lines and to feed antennas used as probes in resonators, which in turn excite the resonator and measure the fields within it. Coaxial lines are often the choice in many applications because their fields are contained within the dielectric separating the two conductors and as such they cannot interfere and cannot be interfered with. It is this property that makes the coaxial cable so important in testing and measurements.

A third type of transmission line that we will concern ourselves with is the parallel strip line shown in Figure 2.3(e). This line may be made of two strips, very close to each other, such as strips on printed circuit boards or of two parallel plates in a number of possible configurations. We will make considerable use of this type of line and of modifications to the basic structure. This type of line is one of a class of lines often referred to as planar structures that has gained popularity because of the ease with which they can be made by lithographic means and hence integrated with electronics. A separate discussion of planar transmission lines and their connection to this work is given in Chapter 3 where we also discuss transmission line resonators and properties of waveguides and microwave cavity resonators.

Although each line has its own properties and parameters, the discussion at this point will be general and will encompass all lines that satisfy the above requirements. In doing so, we first discuss infinite lines, followed by finite, load terminated lines. The lossless (ideal) line is discussed first since it is the simplest, followed by lossy or attenuating lines. Special attention is given to shorted and open transmission lines since these then lead to the ideas of resonance.

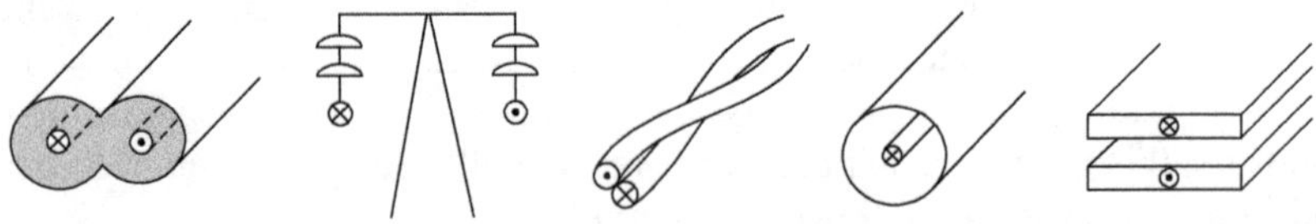

(a) (b) (c) (d) (e)

Figure 2.3 Examples of transmission lines: (a) simple two-lead, insulated cable, (b) overhead power line, (c) twisted pair, (d) the coaxial transmission line, and (e) parallel plate transmission line (strip line)

2.3 Transmission line parameters

A transmission line has three types of parameters:

1. Dimensional parameters: These include length, dimensions of each conductor (thickness, width, diameter, etc.), spacing between lines, thickness of dielectrics, and the like. These parameters define the physical configuration of the line but also play a role in defining its electrical properties.
2. Material parameters: The line is made of conductors and dielectrics. The electrical properties of these materials are their conductivities, permittivities, and permeabilities. These obviously affect the way a line performs its task.
3. Electric parameters: These are the resistance, capacitance, inductance, and conductance per unit length of the line. The four line parameters are as follows:
   R: Series resistance of the line in Ohms per unit length (Ω/m).
   L: Series inductance of the line in Henrys per unit length (H/m).
   C: Shunt capacitance of the line in Farads per unit length (F/m).
   G: Shunt conductance of the line in Siemens per unit length (S/m).

Before we discuss the properties of transmission lines, it is important to be able to define the various line parameters. These can be evaluated in different ways and depend on the type of transmission line. One simple way is to measure the parameters for any given transmission line configuration. This has the distinct advantage of allowing parameter evaluation for arbitrary configurations. But it comes at the cost of considerable effort in experiment setup, expensive equipment, and time considerations. Another approach that is valid under certain conditions is theoretical calculations. In principle, the parameters can be evaluated from known electromagnetic relations. In simple configurations, one can obtain closed-form formulae, which are very useful. Of course, their accuracy depends heavily on material properties of the conductors and any dielectric included in the model as well as on dimensions. Table 2.1 summarizes the properties of some of the common, classical transmission lines based on simplified assumptions on the transmission lines. A third method often used is to forego the evaluation of the parameters and evaluate, measure, or simulate the overall performance of the transmission line or the transmission line circuit as a whole. This is particularly attractive in simulation of transmission line components using methods such as the finite element method or the finite difference method. We will have recourse to this approach in Chapter 4 where we simulate transmission line resonators. In this work, we will have recourse to a class of special transmission lines; for the sake of simplicity, we delay their discussion until later in the following chapter. These lines, often called planar structures, are modifications of the parallel plate transmission line. Although these are unique structures and often more difficult to analyze, the theory discussed here applies to them and indeed to any transmission line.

Table 2.1 Transmission line parameters for some common, classical transmission lines

Two-wire line [Figure 2.3(a)]	**Coaxial line [Figure 2.3(d)]**	**Parallel plate line [Figure 2.3(e)]**
a = radius of conductor, d = distance between centers of conductors	a = radius of inner conductor, b = inner radius of outer conductor	w = width of plates, d = distance between plates
$R = 1/\pi a\delta\sigma_c$	$R = 1/2\pi\delta\sigma_c[1/a + 1/b]$	$R = 2/w\delta\sigma_c$ (Ω/m)
$L = (\mu/\pi)\cosh^{-1}(d/2a)$	$L = (\mu/2\pi)\ln(b/a)$	$L = \mu d/w$ (H/m)
$G = \pi\sigma/(\cosh^{-1}(d/2a))$	$G = 2\pi\sigma/(\ln(b/a))$	$G = \sigma w/d$ (S/m)
$C = \pi\varepsilon/(\cosh^{-1}(d/2a))$	$C = 2\pi\varepsilon/(\ln(b/a))$	$C = w\varepsilon/d$ (F/m)

Notes:

1. If $(d/2a)^2 \gg 1$, $\cosh^{-1}(d/2a) \approx \ln(d/a)$. For widely separated, two-wire, thin lines, this approximation can be used to simplify the expressions. σ_c and μ_c are the conductivity and permeability of the conductor, respectively. σ, μ, and ε are the properties of the dielectric between the conductors.
2. The parameters in the table were obtained under simplifying assumptions.

2.3.1 Calculation of line parameters

Given accurate geometrical data and material properties, it is usually possible to calculate the line parameters either from circuit principles or from field principles. However, the calculation of resistance, capacitance, inductance, and conductance per unit length of a pair of conductors depends on our ability to evaluate these quantities in relatively complex environments. Although it is not the purpose of this section to show how these are calculated, it is worth noting that even seemingly simple configurations can lead to rather complex calculations. For example, consider the calculation of the capacitance per unit length of the parallel plate transmission line. We can start with a circuit approach by assuming the plates form a parallel plate capacitor. The capacitor has width w, length $l = 1$ m, separation d, and a dielectric of permittivity ε between the plates. The capacitance per unit length is then that shown in Table 2.1. This calculation, while simple and useful in the sense that it indicates dependencies of the capacitance per unit length on dimensions and properties, assumes that $w \gg d$. In practical transmission lines, such as traces on a printed circuit boards, this condition is unlikely to be satisfied leading to significant errors in the calculation of the parameters and therefore in the properties of the transmission lines. More accurate parameters can be obtained using methods such as conformal mapping, especially for relatively simple geometries such as the parallel plate transmission line. In more complex geometries, one must use numerical methods and these are usually based on field principles. In the example of the parallel plate transmission line, one can use methods like the finite element method or the method of moments to evaluate accurate parameters for and practical transmission line. For example, the capacitance per unit length of a parallel plate transmission line of length 1 m, width 50 mm, and separation of 2 mm with air between the plates is 88.45 pF when calculated using the formula in Table 2.1. A method of moments calculation of the same geometry produces a capacitance of 111.76 pF. This is an error of over 25%. The same considerations apply to other parameters although some calculations may be more accurate than others. The parameters for the coaxial transmission line are much more accurate and, as a matter of fact, the capacitance per unit length is exact. The other parameters for the coaxial line are not exact but are rather accurate.

Most of the parameters in Table 2.1 seem to be independent of frequency. The resistance per unit length is frequency dependent because the AC resistance is related to the skin depth in the conductor and that in turn depends on frequency. But other parameters are also, indirectly, dependent on frequency. Permittivity of dielectrics is frequency dependent; hence, capacitance per unit length is also frequency dependent. In most practical materials, conductivity and permeability are not frequency dependent; hence, inductance per unit length is independent of frequency. For these reasons, the parameters should be evaluated very carefully to ensure the line properties are accurate.

A final word on commercially available transmission lines: manufacturers typically provide data on line parameters in the range of frequencies the line is intended to work in. In many cases however, this data is limited to geometrical dimensions, losses, and characteristic impedance. In some cases, the per unit length parameters are available as well. These are usually sufficient for design since the line parameters are only needed to evaluate the losses per unit length and the characteristic impedance.

2.4 The transmission line equations

As discussed above, the lumped parameter approach to transmission lines is not feasible. Instead, we define the transmission line equations using a distributed parameter approach. The transmission line is viewed as being made of a large number of short segments, each of length Δl as shown in Figure 2.4 which also shows the parameters of one segment. In this notation, $R\Delta l$ is the resistance of the

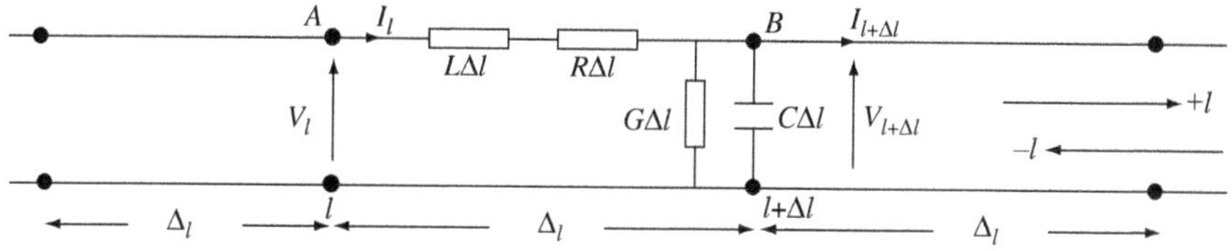

Figure 2.4 A transmission line viewed as a distributed parameter circuit built of segments of arbitrary but small length Δl. One segment is shown in detail. Note the general direction l. Later, we will replace this with a specific coordinate

line of length Δl, $L\Delta l$ is the inductance, $C\Delta l$ is the capacitance, and $G\Delta l$ is the conductance, where R, L, C, and G are given per unit length. The total series impedance of the line segment is therefore

$$Z = R\Delta l + j\omega L\Delta l \ (\Omega) \tag{2.1}$$

and the parallel line admittance is

$$Y = G\Delta l + j\omega C\Delta l \ (1/\Omega) \tag{2.2}$$

These parameters can now be used to build a transmission line of any length, as shown in Figure 2.4. The Δ notation was used to indicate that the segment of line used is arbitrary but must be small compared to wavelength. The circuit equations are written using Kirchhoff's laws for one of the segments to obtain the transmission line equations, assuming for the moment that both current and voltage are phasors. The voltage across the line segment of length Δl can be written in terms of the voltages at points A and B and the current in the segment. With the notation in Figure 2.4, we have,

$$V(l+\Delta l) - V(l) = -I(l)[R\Delta l + j\omega L\Delta l) \ (\mathrm{V}) \tag{2.3}$$

Dividing both sides by Δl

$$\frac{V(l+\Delta l) - V(l)}{\Delta l} = -I(l)[R + j\omega L] \tag{2.4}$$

The term on the left-hand side becomes the derivative of V with respect to l if we let Δl tend to zero. Thus, since Δl is arbitrarily small, we may write

$$\frac{dV(l)}{dl} = -I(l)[R + j\omega L] \tag{2.5}$$

This relation holds at any point on the line. Similarly, the current in the segment can be written in terms of the current at points A and B and the voltage at point B as

$$I(l+\Delta l) - I(l) = -V(l+\Delta l)[G\Delta l + j\omega C\Delta l] \ (\mathrm{A}) \tag{2.6}$$

Following steps identical to (2.4) and (2.5), we get

$$\frac{dI(l)}{dl} = -V(l+\Delta l)[G + j\omega C] \tag{2.7}$$

To obtain an equation of the same form as for the voltage in (2.5), we expand the term $V(l+\Delta l)$ in a Taylor series about l as $V(l+\Delta l) = V(l) + (dV(l)/dl)\Delta l/1! + (d^2V(l)/dl^2)(\Delta l)^2/2! + \ldots$. Neglecting all terms that contain Δl gives an approximation $V(l+\Delta l) \approx V(l)$. Substitution of this in (2.7) gives

$$\frac{dI(l)}{dl} = -V(l)[G + j\omega C] \tag{2.8}$$

The transmission line equations are the current and voltage relations in (2.5) and (2.8), respectively. These are two coupled first-order differential equations. Before attempting to solve for current and voltage, we can eliminate one of the variables and obtain separate equations for $V(l)$ and $I(l)$. To do so, we substitute $I(l)$ from (2.5) into (2.8) and $V(l)$ from (2.8) into (2.5). From (2.5),

$$I(l) = -\frac{dV(l)}{dl}\frac{1}{[R+j\omega L]}\ (\text{A}) \tag{2.9}$$

Substitution of this into (2.8) gives

$$\frac{d^2V(l)}{dl^2} - V(l)[G+j\omega C][R+j\omega L] = 0 \tag{2.10}$$

Similarly, substituting $V(l)$ from (2.8) into (2.5), we get

$$\frac{d^2I(l)}{dl^2} - I(l)[G+j\omega C][R+j\omega L] = 0 \tag{2.11}$$

These two equations are wave equations. These can be written as

$$\frac{d^2V}{dl^2} - \gamma^2 V = 0 \tag{2.12}$$

and

$$\frac{d^2I}{dl^2} - \gamma^2 I = 0 \tag{2.13}$$

where

$$\gamma = \alpha + j\beta = \sqrt{[G+j\omega C][R+j\omega L]} \tag{2.14}$$

The first of these is the wave equation for the voltage on the line, and the second is the wave equation for current in the line. Therefore, γ is the propagation constant in analogy with the definition of the propagation constant for a wave propagating in space. α is called the attenuation constant, β is called the phase constant. The solution is identical to that of propagation of plane waves. For the general transmission line described here, the solution for voltage and current can be written as

$$V(l) = V^+e^{-\gamma l} + V^-e^{\gamma l}\ (\text{V}) \tag{2.15}$$

$$I(l) = I^+e^{-\gamma l} + I^-e^{\gamma l}\ (\text{A}) \tag{2.16}$$

Direct substitution of these solutions into (2.12) and (2.13) shows they are correct. The solution to these equations has two parts: one propagating in the positive l direction, the other in the negative l direction, along the line, exactly as for plane waves. V^+ and V^- are the amplitudes of the voltage waves propagating in the positive and negative l directions, respectively. For the current solution, I^+ and I^- are the respective amplitudes of the current waves. The amplitudes of the forward and backward propagating waves, V^+ and V^-, can be calculated from the terminal voltages on the transmission line as we shall see later.

So far, we have defined one characteristic quantity of the line: the propagation constant in (2.14). Now that we obtained the voltages and currents on the line, we can define the second characteristic quantity of any transmission line: the characteristic line impedance.

The *characteristic line impedance* Z_0 of a transmission line is defined as the ratio between the forward-propagating voltage amplitude and the forward-propagating current amplitude:

$$Z_0 = \frac{V^+}{I^+} \ (\Omega) \tag{2.17}$$

To evaluate the characteristic impedance in terms of the line parameters (since these are known and independent of line current), we substitute the general solution from (2.15) and (2.16) into the transmission line relations in (2.5) and (2.8). Starting with (2.5), we get

$$\frac{d\left(V^+e^{-\gamma l} + V^-e^{\gamma l}\right)}{dl} = -\left(I^+e^{-\gamma l} + I^-e^{\gamma l}\right)[R + j\omega L] \tag{2.18}$$

or, after evaluating the derivatives,

$$-\gamma V^+e^{-\gamma l} + \gamma V^-e^{\gamma l} = -\left(I^+e^{-\gamma l} + I^-e^{\gamma l}\right)[R + j\omega L] \tag{2.19}$$

Similarly, using (2.8), we get

$$-\gamma I^+e^{-\gamma l} + \gamma I^-e^{\gamma l} = -\left(V^+e^{-\gamma l} + V^-e^{\gamma l}\right)[G + j\omega C] \tag{2.20}$$

Now, suppose, first, that only a forward-propagating wave exists by setting $V^- = 0$, $I^- = 0$ in (2.19) and (2.20). We get

$$-\gamma V^+e^{-\gamma l} = -I^+e^{-\gamma l}[R + j\omega L] \ \text{and} \ -\gamma I^+e^{-\gamma l} = -V^+e^{-\gamma l}[G + j\omega C] \tag{2.21}$$

Thus, the characteristic impedance can be written as

$$Z_0 = \frac{V^+}{I^+} = \frac{R + j\omega L}{\gamma} = \frac{\gamma}{G + j\omega C} \ (\Omega) \tag{2.22}$$

The first form is obtained from the first expression in (2.21) and the second from the second expression. Also, by substituting for γ from (2.14), we obtain

$$Z_0 = \sqrt{\frac{R + j\omega L}{G + j\omega C}} \ (\Omega) \tag{2.23}$$

Now suppose that, only a backward-propagating wave exists. By setting $V^+ = 0$, $I^+ = 0$ in (2.19) and (2.20), we get

$$\gamma V^-e^{\gamma l} = -I^-e^{\gamma l}[R + j\omega L] \ \text{and} \ \gamma I^-e^{\gamma l} = -V^-e^{\gamma l}[G + j\omega C] \tag{2.24}$$

Dividing each of these two equations by I^-, we can write

$$\frac{V^-}{I^-} = -\frac{R + j\omega L}{\gamma} = -\frac{\gamma}{G + j\omega C} = -Z_0\ (\Omega) \tag{2.25}$$

We can summarize these results as follows:

$$Z_0 = \frac{V^+}{I^+} = -\frac{V^-}{I^-} = \frac{R + j\omega L}{\gamma} = \frac{\gamma}{G + j\omega C} = \sqrt{\frac{R + j\omega L}{G + j\omega C}}\ (\Omega) \tag{2.26}$$

The characteristic impedance Z_0 is independent of location on the line and only depends on line parameters—thus, the name characteristic impedance. The characteristic impedance is, in general, a complex value. However, although all other line parameters are given in per meter units, the characteristic impedance is a line property, independent of length. In other words, for any given line, if we were to measure the characteristic impedance, the above value would be obtained for any length of line and at any location on the line.

Using (2.26), the line current given in (2.16) can be written as

$$I(l) = \frac{V^+}{Z_0}e^{-\gamma l} - \frac{V^-}{Z_0}e^{\gamma l}\ (\mathrm{A}) \tag{2.27}$$

Finally, we also mention that the wavelength and phase velocity for any propagating wave are, respectively, given as

$$\lambda = \frac{2\pi}{\beta}\ (\mathrm{m}),\ v_p = \frac{\omega}{\beta}\ (\mathrm{m/s}) \tag{2.28}$$

The quantity βl has units of radians. It is called the *electrical length* of the line and may be considered an additional line parameter.

The discussion in this section assumed time-harmonic quantities. This was done on purpose, since phasor calculations are usually simpler to perform and the final result is also simpler. More important, this choice allowed us to use known results for TEM wave propagation. In turn, this choice shows that propagation along transmission lines is similar to transmission in free space and other materials, as long as the basic assumptions of TEM waves are satisfied. Both plane waves in materials and waves in transmission lines satisfy these conditions. Thus, we can expect that other parameters such as reflection and transmission of energy as well as the reflection and transmission coefficients should be similar. We will discuss these topics separately.

Instead of using the time-harmonic forms for voltage and current, we could start with the time-dependent voltage and current to obtain the time-dependent transmission line equations following essentially identical steps as above. One reason for this is that many of the properties we require, including phase and attenuation constants, wavelength, wave number, and the like, can only be properly defined for time-harmonic fields. If quantities are not time harmonic, the time dependent form must be used. Although we will use the time-harmonic forms exclusively, the following section discusses, briefly, the transmission line in the time domain for the sake of completeness.

2.4.1 Time-domain transmission line equations

Although we will use the transmission line equations almost exclusively in the frequency domain, it is nevertheless useful to derive here the time-domain transmission line equations. One can envision the use of the time-domain equations in instances when a single frequency cannot describe the behavior of the line. However, it should also be remembered that the line parameters themselves are frequency dependent (see, e.g., the expression for R in Table 2.1) and a complete, exact analysis in the time domain is rather difficult. In most cases, it is easier to transform the time-domain signal into the frequency domain and analyze the transmission line at the individual harmonics, recalculating the line parameters at each harmonic if necessary. However, if we assume the line parameters to be constants, analysis in the time domain is possible. On the other hand, simulation in the time domain is common because it can capture effects that are not possible or difficult to obtain in the frequency domain. In particular, this is the case with transmission lines that incorporate nonlinear elements or transmission lines with discontinuities or multifrequency operation on transmission lines. We will use these ideas in simulation of behavior of transmission line resonators through use of the finite-difference time-domain method in Chapter 4 (see also Appendix C).

To obtain the time-domain transmission line equations, we start with Figure 2.4, but with time dependent voltages $V(l, t)$, $V(l+\Delta l, t)$ $I(l, t)$, and $I(l+\Delta l, t)$. With these and with the fact that the potential across an inductor is $LdI(l,t)/dt$ and the current in a capacitor is $CdV(l,t)/dt$, we write by applying Kirchhoff's laws:

$$V(l+\Delta l,t)-V(l,t)=-I(l,t)R\Delta l-L\Delta l\frac{dI(l,t)}{dt}\ \text{(V)} \tag{2.29}$$

$$I(l+\Delta l,t)-I(l,t)=-V(l,t)G\Delta l-C\Delta l\frac{dV(l,t)}{dt}\ \text{(A)} \tag{2.30}$$

Dividing each equation by Δl and allowing Δl to tend to zero, we obtain

$$\frac{dV(l,t)}{dl}=-I(l,t)R-L\frac{dI(l,t)}{dt} \tag{2.31}$$

$$\frac{dI(l,t)}{dl}=-V(l,t)G-C\frac{dV(l,t)}{dt} \tag{2.32}$$

We can now rewrite these equations so that each is a function of a single variable as follows:

First, we take the derivative with respect to l on both sides of (2.31) and (2.32)

$$\frac{d^2V(l,t)}{dl^2}=-R\frac{dI(l,t)}{dl}-L\frac{d}{dl}\left(\frac{dI(l,t)}{dt}\right)=-R\frac{dI(l,t)}{dl}-L\frac{d}{dt}\left(\frac{dI(l,t)}{dl}\right) \tag{2.33}$$

$$\frac{d^2I(l,t)}{dl^2}=-G\frac{dV(l,t)}{dl}-C\frac{d}{dl}\left(\frac{dV(l,t)}{dt}\right)=-G\frac{dV(l,t)}{dl}-C\frac{d}{dt}\left(\frac{dV(l,t)}{dl}\right) \tag{2.34}$$

Substituting $dI(l,t)/dl$ from (2.32) into (2.33) and $dV(l,t)/dl$ from (2.31) into (2.34), we get

$$\frac{d^2V(l,t)}{dl^2} - LC\frac{d^2V(l,t)}{dt^2} - (LG+RC)\frac{dV(l,t)}{dt} - RGV(l,t) = 0 \qquad (2.35)$$

$$\frac{d^2I(l,t)}{dl^2} - LC\frac{d^2I(l,t)}{dt^2} - (LG+RC)\frac{dI(l,t)}{dt} - RGI(l,t) = 0 \qquad (2.36)$$

Equations (2.31) and (2.32) are equivalent to (2.5) and (2.8), whereas (2.35) and (2.36) are equivalent to (2.10) and (2.11). In fact, one can obtain (2.5), (2.8), (2.10), and (2.11) from (2.31), (2.32), (2.35), and (2.36), respectively, by simply replacing d/dt by $j\omega$ and d^2/dt^2 by $(j\omega)^2 = -\omega^2$.

2.5 Types of transmission lines

The transmission line equations in Section 2.4 were obtained for a completely general transmission line. As can be seen, the equations are rather involved. The propagation constant as well as the line impedance are complex and are not always easy to evaluate. Both a phase constant and an attenuation constant exist; therefore, we can expect the waves along the line to decay due to attenuation as well as change their phases. The fact that both a forward- and backward-propagating wave exists indicates that the line may be finite in length whereby the backward-propagating wave is due to a reflection from the load, a connection on the line, or any other discontinuity that may exist.

For practical applications, we distinguish between a number of special types of transmission lines in addition to the above general lossy line. These include the *lossless transmission line* and the *infinitely long transmission line* as well as the so-called distortionless transmission line. The wave characteristics on these lines are simplified because of the assumptions associated with the three types of lines, but, more importantly, they represent useful, practical lines. These are described briefly next since many of their properties are useful either as exact representation or, often as practical approximations to lines. Following these, we discuss the shorted and open transmission lines as introduction to transmission line resonators. In addition, the concepts associated with planar transmission line structures are discussed since these will be used later in this work as resonators.

2.5.1 The lossless transmission line

A lossless transmission line is a line for which both the series resistance and the shunt conductance are zero ($R = 0$, $G = 0$). In practice, this implies that the line is made of perfect conducting materials and perfect dielectrics. Although no practical line satisfies these conditions exactly, many lines satisfy them approximately. The implications of these conditions are that the attenuation constant is zero, the propagation constant is purely imaginary, and the characteristic impedance of the line is real.

If we substitute $R = 0$ and $G = 0$ in the propagation constant in (2.14), we get

$$\gamma = j\beta = j\omega\sqrt{LC} \tag{2.37}$$

Similarly, the characteristic impedance of the line [from (2.23)] is real and equal to

$$Z_0 = \sqrt{\frac{L}{C}}\ (\Omega) \tag{2.38}$$

A number of propagation parameters can now be easily evaluated. The phase and attenuation constants are found from the propagation constant:

$$\beta = \omega\sqrt{LC}\ \text{(rad/m)},\ \alpha = 0 \tag{2.39}$$

The wavelength is defined as

$$\lambda = \frac{2\pi}{\beta} = \frac{2\pi}{\omega\sqrt{LC}}\ \text{(m)} \tag{2.40}$$

and the speed of propagation of the wave along the line (phase velocity) is

$$v_p = \frac{\omega}{\beta} = \frac{1}{\sqrt{LC}}\ \text{(m/s)} \tag{2.41}$$

Because the dielectric is lossless, the phase velocity may also be written as

$$v_p = \frac{1}{\sqrt{\mu\varepsilon}}\ \text{(m/s)} \tag{2.42}$$

From this, the following relation is obtained:

$$\mu\varepsilon = LC \tag{2.43}$$

In particular, the phase constant and the phase velocity only depend on the inductance and capacitance per unit length. The voltage or current waves propagate along the line without attenuation at a speed dictated by the inductance and capacitance per unit length of the line.

Lossless transmission lines are often invoked in analysis first because they offer the simplest forms for the various quantities and second, because transmission lines designed for use at high frequencies, such as connections to antennas or to instruments often have very low losses. When relatively short lines are used, the low loss approximation is justified and reflects actual performance very well.

2.5.2 *The long transmission line*

A long transmission line is a line that for practical purposes may be considered to be infinite. The infinite transmission line is characterized by transmission without backward-propagating waves since a backward-propagating wave can only exist if the incident wave is reflected from a discontinuity in the wave's path. The long line may be lossy or lossless. For a lossy line, the voltage and current waves are found from (2.15) and (2.16) by removing the backward-propagating wave:

$$V(l) = V^+ e^{-\gamma l} \text{ (V)} \text{ and } I(l) = I^+ e^{-\gamma l} = \frac{V^+}{Z_0} e^{-\gamma l} \text{ (A)} \tag{2.44}$$

The propagation constant γ is given in (2.14), and the characteristic impedance of the line is given in (2.26).

If the long line is lossless, the voltage and current waves are

$$V(l) = V^+ e^{-j\beta l} \text{ and } I(l) = \frac{V^+}{Z_0} e^{-j\beta l} \tag{2.45}$$

The phase constant is given in (2.39) and the characteristic impedance in (2.38).

The infinite transmission line cannot be realized physically, but it will prove to be a convenient approximation for very long lines or for short lines before the forward wave has reached the load or, indeed, for shorter lines with high loss in which the reflected waves die out before they can reach the test point.

2.5.3 *The distortionless transmission line*

The propagation constant and characteristic impedance for general lossy lines were obtained in (2.14) and (2.23), respectively. These are rather complicated expressions and are frequency dependent. Whenever transmission lines are used for propagation of a single frequency wave (monochromatic wave), the fact that the line impedance and propagation constant are frequency dependent is less important, but when a wave has a range of frequencies, such as in the communication of information, each frequency component will be attenuated differently, the phase of each component will propagate at different speeds, and each component will see a different line impedance. This inevitably leads to distortion of the signals on the line.

The question is: how can we design a general lossy line so that the attenuation constant, phase velocity, and characteristic impedance of the line are independent of frequency? If we can do that, we would obtain a distortionless transmission line. To do so, we note that if $R/L = G/C$, the propagation constant in (2.14) becomes

$$\gamma = j\omega\sqrt{LC}\sqrt{1+\frac{R}{j\omega L}}\sqrt{1+\frac{R}{j\omega L}} = j\omega\sqrt{LC}\left[1+\frac{R}{j\omega L}\right] = j\omega\sqrt{LC} + R\sqrt{\frac{C}{L}} \tag{2.46}$$

From this, the attenuation and phase constants are

$$\alpha = R\sqrt{\frac{C}{L}} \text{ (Np/m)}, \ \beta = \omega\sqrt{LC} \text{ (rad/m)} \tag{2.47}$$

and, therefore, the phase velocity is

$$v_p = \frac{\omega}{\beta} = \frac{1}{\sqrt{LC}} \text{ (m/s)} \tag{2.48}$$

Thus, the first two conditions (i.e., that the attenuation constant and phase velocity are independent of frequency) are satisfied. What about the characteristic impedance? If we substitute the condition $R/L = G/C$ in (2.23), we get

$$Z_0 = \sqrt{\frac{R + j\omega L}{RC/L + j\omega C}} = \sqrt{\frac{L}{C}}\ (\Omega) \tag{2.49}$$

The characteristic impedance is also constant, and the above requirements are satisfied. Thus, for a line to be distortionless, the line parameters must be designed so that

$$\frac{R}{L} = \frac{G}{C} \tag{2.50}$$

With this condition, the distortionless transmission line has the same phase constant and characteristic impedance as the lossless line but a nonzero, constant attenuation. Again, as was noted in Section 2.5.1 about lossless lines, good quality transmission lines (mostly coaxial lines) are often designed as distortionless lines. Clearly as well, lossless lines are necessarily distortionless.

2.5.4 The low-resistance transmission line

It was mentioned before that a transmission line is made of two conductors in a given configuration. In a line of this type, it is often possible to assume that the conductivity of the conductor is so high as to have negligible resistance. In other words, the propagation on the transmission line is not affected by the conductor itself. The conductors are required only to guide the waves, but all propagation parameters are affected by the properties of the dielectric alone. Substituting $R = 0$ in (2.14) and (2.23), we get

$$\gamma = j\omega\sqrt{LC}\sqrt{1 + \frac{G}{j\omega C}} \tag{2.51}$$

$$Z_0 = \sqrt{\frac{j\omega L}{G + j\omega C}} \tag{2.52}$$

Since the conductor's effect can be neglected, we can view this as a TEM wave propagating in a lossy dielectric material with properties ε_r, μ_r, and σ as if the conductors were not there.

For a general lossy dielectric, the propagation constant is

$$\gamma = j\omega\sqrt{\mu\varepsilon}\sqrt{\left[1 + \frac{\sigma}{j\omega\varepsilon}\right]} \tag{2.53}$$

where σ is the conductivity of the dielectric and ε its dielectric constant. Direct comparison between (2.53) and (2.51) gives the following two relations:

$$LC = \mu\varepsilon \text{ and } \frac{\sigma}{\varepsilon} = \frac{G}{C} \tag{2.54}$$

where σ is the conductivity of the dielectric between the conductors. These two relations are important for two reasons:

1. They hold for lossless and lossy transmission lines even if the series resistance is not zero. This can be easily verified for the three transmission lines listed in Table 2.1.
2. The relations provide one of the simplest methods of evaluating some of the parameters of the line. If, for example, C is known, L and G can be evaluated directly. This is useful because in many cases, one of the line parameters is easier to evaluate than the other two. In such cases, these two relations provide a simple means of finding the line parameters.

Note also that if, in addition, $G = 0$, the line becomes lossless.

In most practical cases, the transmission lines are relatively short and, with some exceptions, are viewed as lossless. This is not to say that they are lossless but rather that over short distances, the losses are minimal. Better quality transmission lines, especially when they need to operate at higher frequencies, are specified with their losses, usually given in dB/m or dB/km. When the line cannot be assumed to be lossless, especially if the line is long, the distortionless line is often assumed. Both the lossless and distortionless lines have real characteristic impedance, given in (2.49).

2.6 The field approach to transmission lines

The discussion in the previous sections was in terms of general line parameters and therefore applies to any transmission line of the type considered here—two conductor transmission lines. The primary variables were the voltage and current of the line. This choice is natural if we view the line as a distributed parameter circuit. It is, however, possible to arrive at exactly the same results from a field point of view. In this case, the primary variables are the electric and magnetic field intensities and the discussion is much the same as that for plane waves in space (see Appendix D). One advantage of using field variables is that these are vectors and, therefore, the direction of propagation at any point is always available and indicates the direction in which net power is transferred. This is particularly useful in analysis of waveguides where voltages and currents can only be postulated as equivalent rather than physical quantities. To demonstrate this approach, we look now at the wave characteristics on a parallel plate transmission line.

Suppose the transmission line shown in Figure 2.5 is given. The line is very long and $w \gg d$. The material between the plates is a general dielectric. At a given instant in time, the potential between the two plates and the currents in the plates are as shown. For the given condition, the electric field intensity points from the upper plate to the lower plate (x direction) and the magnetic field intensity is parallel to the plates, pointing in the y direction. Because of our assumption that $w \gg d$, we may assume that the electric field is everywhere perpendicular to the plates (no fringing at the edges) and the magnetic field intensity is everywhere parallel to the plates.

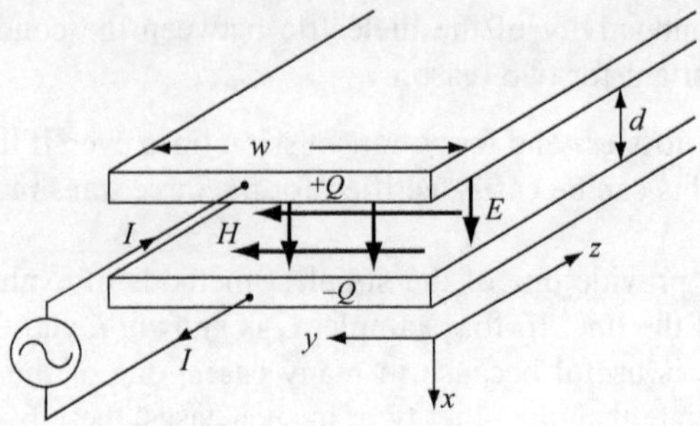

Figure 2.5 Relations between current and charge on the conductor and the electric and magnetic field intensities in the dielectric of a parallel plate transmission line

We know that the two fields are a solution to the source-free wave equation since there are no sources in this domain and propagation takes place; that is, the fields obey the general Maxwell equations. Also, because the transmission line is infinite in extent, in the z direction, there can only be a forward-propagating wave. Without knowing what the electric field intensity amplitude is, we can write, in general terms:

$$\mathbf{E} = \hat{\mathbf{x}} E_0 e^{-\gamma z} \ (\text{V/m}) \tag{2.55}$$

where we have replaced the generic coordinate l with z. The magnetic field intensity is perpendicular to the electric field intensity and, using the intrinsic impedance of the dielectric, we can write

$$\mathbf{H} = \hat{\mathbf{y}} \frac{E_0}{\eta} e^{-\gamma z} \ (\text{A/m}) \tag{2.56}$$

where η is the intrinsic impedance (also called the wave impedance) of the dielectric between the plates. The wave is a TEM wave (**E** and **H** are perpendicular to each other and to the direction of propagation). The direction of propagation of the wave is in the positive z direction, as shown by the Poynting vector:

$$\mathcal{P}(z) = \mathbf{E} \times \mathbf{H} = \hat{\mathbf{x}} E_0 e^{-\gamma z} \times \hat{\mathbf{y}} \frac{E_0}{\eta} e^{-\gamma z} = \hat{\mathbf{z}} \frac{E_0^2}{\eta} e^{-2\gamma z} \ (\text{W/m}^2) \tag{2.57}$$

The electric field intensity E_0 was arbitrarily chosen, but, in practice, its sources are the charge distribution on the conducting surfaces and the current density in the conducting plates. The voltage between the two plates can be written as

$$V = \int_{l_1} \mathbf{E} \cdot d\mathbf{l}_1 \ (\text{V}) \tag{2.58}$$

and the current in one of the plates (upper) as

$$I = \int_{l_2} \mathbf{H} \cdot d\mathbf{l}_2 \ (\text{A}) \tag{2.59}$$

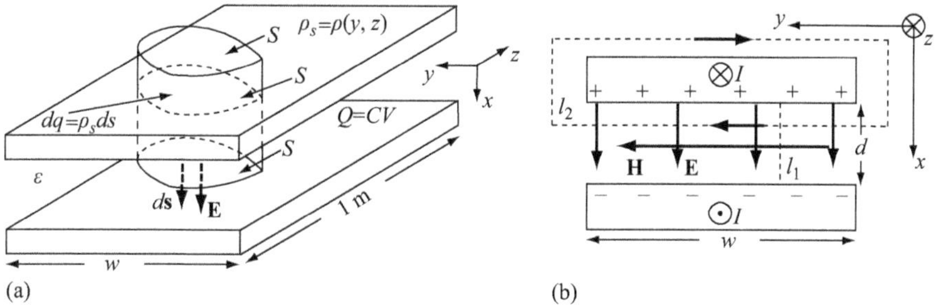

Figure 2.6 (a) Calculation of charge density using Gauss's law and (b) calculation of current density using Ampere's law

where the contours l_1 and l_2 are shown in Figure 2.6(b). These are the line voltage and current and may be substituted in (2.15) and (2.16) to obtain the transmission line voltage and current in terms of the electric and magnetic field intensities.

We can now calculate the charge density and the current density in the conductors that will produce the required electric and magnetic fields from (2.58) and (2.59). This calculation is not absolutely necessary for the discussion here, but it emphasizes two important points:

1. The sources of the fields produced by the transmission line are the charges and currents in the line.
2. The charge and current distributions must be of a form that produces these fields; not all charge and current distributions will produce a propagating wave in the transmission line.

Suppose that a charge distribution exists on the upper and lower plates as shown in Figure 2.6(a). To calculate the electric field intensity, we use Gauss's law. A small volume, with two surfaces parallel to the upper plate is defined as shown in Figure 2.6(a). The electric field intensity outside the plates is zero (as for parallel plate capacitors) and the electric field intensity between the plates is given by (2.55). Taking a surface S as shown, we get from Gauss's law

$$\int_s \mathbf{E} \cdot d\mathbf{s} = \int_s (\hat{\mathbf{x}} E_0 e^{-\gamma z}) \cdot (\hat{\mathbf{x}} ds) = \frac{1}{\varepsilon} \int_s \rho \, ds \tag{2.60}$$

or

$$\rho(y, z) = \varepsilon E_0 e^{-\gamma z} \ (\mathrm{C/m^2}) \tag{2.61}$$

Thus, the charge density is uniform in the y direction (independent of y) but varies along the line. This variation is better seen if the charge density is written in the time domain as

$$\rho(y, z, t) = \mathrm{Re}\left\{ \varepsilon E_0 e^{-(\alpha + j\beta)z} e^{j\omega t} \right\} = \varepsilon E_0 e^{-z} \cos(\omega t - \beta z) \ (\mathrm{C/m^2}) \tag{2.62}$$

In other words, the charge distribution must be cosinusoidal in the z direction. The attenuation constant produces a decaying charge density magnitude with distance. If propagation is without attenuation, then $\alpha = 0$ and there is no decay in amplitude of the electric field intensity. The charge density distribution on the lower plate is the same as on the upper plate but opposite in sign.

The current density in the line is calculated from Ampere's law. Using the upper plate again and assuming some current density in the plate, we can enclose this current density with an arbitrary contour as shown in Figure 2.6(b). The magnetic field intensity outside the plates is zero and between the plates is given by (2.56). In our case, **H** is in the positive y direction, as is $d\mathbf{l}$. Thus, the current density is in the positive z direction (**H** and **J** are always perpendicular to each other). Since the current is uniform in the y direction in this case, we can write $I(y, z) = wJ(y, z)$ and, performing the integration in (2.59) with the field in (2.56), we get

$$\frac{E_0}{\eta} e^{-\gamma z} = J(y, z) \text{ (A/m)} \tag{2.63}$$

This gives the magnitude of the current density in the upper plate. This current must be in the positive z direction to produce a magnetic field intensity in the positive y direction (based on our notation in Figure 2.5); therefore,

$$\mathbf{J}(x, y) = \hat{\mathbf{z}} \frac{E_0}{\eta} e^{-\gamma z} \text{ (A/m)} \tag{2.64}$$

The current density in the lower plate is the same in magnitude but in the negative z direction. The variation of current density along the line is also cosinusoidal, as for the charge density.

2.7 Finite transmission lines

By a finite transmission line is meant a line of finite length with a generator of some sort at one end and a load at the other. Both the generator and the load should be viewed in generic terms: the load may actually be a short circuit, an open circuit, or another transmission line. The generator may be an actual source, the output of another transmission line or, perhaps, a receiving antenna. The configuration we discuss here is shown in Figure 2.7.

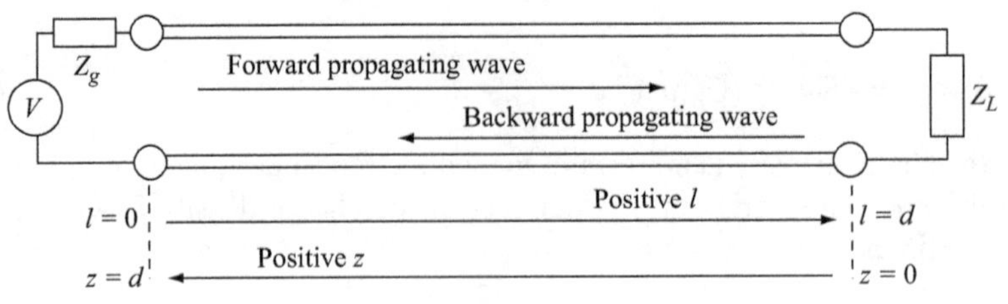

Figure 2.7 A finite transmission line with the reference shifted to the load

Until now, we discussed only infinite lines or made no specific reference to the length of the line. Now, we have to discuss the distance on the line with respect to the fixed points of the line: These are the locations of the load and the generator. Thus, we seek a reference point to which to relate all our calculations. We could choose either the generator or the load for this purpose, but it is common to use the load as a reference point. This choice is partly arbitrary, partly based on convenience, and mostly on convention. At any rate, the only important point here is to be consistent and not flip between points of reference.

We must be careful now: inspecting Figure 2.7, the positive z direction is toward the generator. On the other hand, the positive direction of propagation of power must be from the generator toward the load since energy is naturally transferred from generator to load. We recall that the wave solutions on a general transmission line are

$$V(l) = V^+e^{-\gamma l} + V^-e^{\gamma l} \text{ (V)} \tag{2.65}$$

$$I(l) = I^+e^{-\gamma l} + I^-e^{\gamma l} \text{ (A)} \tag{2.66}$$

In these equations, l is positive toward the load ($l = 0$ at the generator). The first term is the forward-propagating wave (from generator to load) and the second is the backward-propagating wave (from load to generator). This convention was used for infinitely long transmission lines.

Our new convention, which corresponds to the common usage for finite transmission lines, requires that the forward propagating wave propagates in the negative z direction and the backward propagating wave propagates in the positive z direction. Thus, to create our reference system, we replace $+l$ by $-z$ and $-l$ by $+z$:

$$V(z) = V^+e^{\gamma z} + V^-e^{-\gamma z} \text{ (V)} \quad \text{and} \quad I(z) = I^+e^{\gamma z} + I^-e^{-\gamma z} \text{ (A)} \tag{2.67}$$

The first term in each relation is still the forward-propagating wave and the second is the backward-propagating wave. The relation between the z and l notations and the forward- and backward-propagating waves is shown in Figure 2.7 for a transmission line of length d, connected to a generator and a load. The two sets of equations also indicate what is involved in choosing a particular point of reference.

2.7.1 The load reflection coefficient

First, we recall the definition of the characteristic impedance Z_0. This was defined for an infinite transmission line as the ratio between the forward-propagating voltage wave and the forward-propagating current wave. Thus, for any line, the characteristic impedance is

$$Z_0 = \frac{V^+e^{\gamma z}}{I^+e^{\gamma z}} = \frac{V^+}{I^+} = -\frac{V^-}{I^-} \text{ } (\Omega) \tag{2.68}$$

as was shown in (2.26). This impedance is characteristic of the line and has nothing to do with generator or load. Similarly, the propagation constant γ is independent of load or generator, as are the parameters R, L, C, and G.

Since the load is very important for our analysis and since it is one of the few variables an engineer has any control over, it is only natural that we should wish to analyze the transmission line behavior in terms of the load impedance and the line's variables. Thus, we first calculate the load impedance:

$$Z_L = \frac{V_L}{I_L} \ (\Omega) \tag{2.69}$$

where V_L and I_L are the total load voltage and total load current, respectively. By total voltage and current, is meant that it is the sum of forward and backward voltages and currents, respectively.

The load is located at $z = 0$. In terms of the current and voltage of the line, this becomes

$$Z_L = \frac{V(0)}{I(0)} = \frac{V^+ + V^-}{I^+ + I^-} = \frac{V^+ + V^-}{V^+/Z_0 - V^-/Z_0} = Z_0 \frac{V^+ + V^-}{V^+ - V^-} \ (\Omega) \tag{2.70}$$

Note that if only forward-propagating waves exist ($V^- = 0$), the load impedance must be equal to the characteristic impedance of the line. This condition defines matching between load and line. Matching in transmission lines only requires that the load and line impedances be equal, unlike circuits where matching also means maximum transfer of power (conjugate matching). Under matched conditions ($Z_L = Z_0$), there are no backward propagating waves.

On the other hand, if $Z_L \neq Z_0$, there will be both forward-propagating and backward-propagating waves. At the load ($z = 0$), we can calculate the backward propagating wave amplitude V^- from (2.70) as

$$V^- = V^+ \frac{Z_L - Z_0}{Z_L + Z_0} \ (\text{V}) \tag{2.71}$$

The backward-propagating wave is due to the reflection of the forward-propagating wave at the load. Thus, we define the *load reflection coefficient* as

$$\Gamma_L = \frac{V^-}{V^+} = \frac{Z_L - Z_0}{Z_L + Z_0} \tag{2.72}$$

It is important to remember that this is the reflection coefficient at the load only. At other locations on the line, the reflection coefficient is, in general, different, and we should never confuse the load reflection coefficient with any other reflection coefficient that may be convenient to define. The load reflection coefficient will always be denoted with a subscript L as in (2.72). Note also that, in general, the load reflection coefficient is a complex number since it is the ratio of the complex amplitudes V^- and V^+. Thus, we can also write the reflection coefficient as

$$\Gamma_L = |\Gamma_L| e^{j\theta_\Gamma} \tag{2.73}$$

where θ_Γ is the phase angle of the reflection coefficient. This form will become handy later in our study.

2.7.2 Line impedance and the generalized reflection coefficient

After calculating the characteristic impedance and the reflection coefficient at the load, we can now tackle the question of the impedance at any other point on the line. This is an important question because it will allow us to connect the line to, say, a generator, ensuring that the line is matched to the generator, or to connect one line to another. These are questions of practical engineering importance. The simple example in Figure 2.8 shows the concepts involved. A loudspeaker is to be connected to a power amplifier through a transmission line. We know that for optimal operation, the output of the amplifier must be matched to the load. At the amplifier, the load consists of the speaker and the line and the amplifier must be matched to the line. We defer the question of matching for now, but for any attempt at matching, we must be able to calculate the input impedance of the line.

This input impedance, which, in general, is different than the characteristic impedance of the line, must in some way depend on the load impedance. That this must be so should be obvious from our experience: suppose the above amplifier is matched to the line for the given load. If we now change the load, say by shorting the speaker, the system is not matched any more. In fact, by shorting the load, we may well have damaged the amplifier. It is, therefore, important to be able to calculate the line impedance for any load condition. Before continuing, we distinguish between two terms associated with impedance of the line. These are as follows:

Input line impedance is the impedance at the input or generator side of the line. In the above example, this impedance is the impedance of the line at the end, which is connected to the source (amplifier in this case). This impedance will always be denoted as Z_{in}.

Line impedance is the impedance at any point on the line. The distinction between the two terms is shown in Figure 2.8. The line impedance will be denoted as $Z(z)$. The distinction is not terribly important since if we were to cut the line at the points A–A', the line impedance would then become the input impedance. We will, however, distinguish between the two terms wherever appropriate.

To calculate the line impedance, we need to calculate the total voltage and total current at any point on the line and divide the voltage by current. Using Figure 2.9(a) as a guide, the voltage and current at point z on the line are

$$V(z) = V^+e^{\gamma z} + V^-e^{-\gamma z}\ (\text{V}),\ I(z) = \frac{V^+}{Z_0}e^{\gamma z} - \frac{V^-}{Z_0}e^{-\gamma z}\ (\text{A}) \tag{2.74}$$

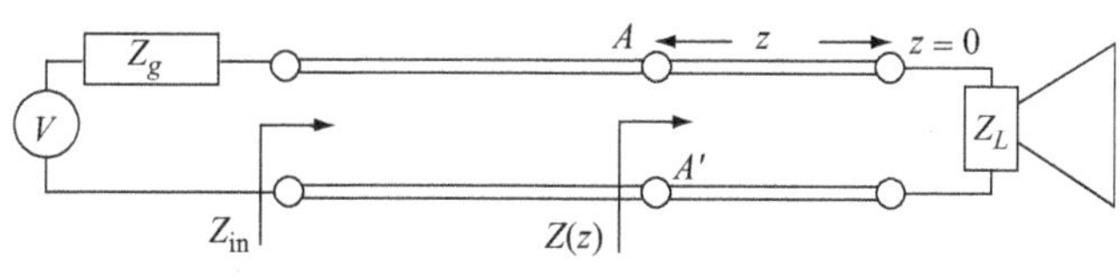

Figure 2.8 Distinction between load, input, and line impedances

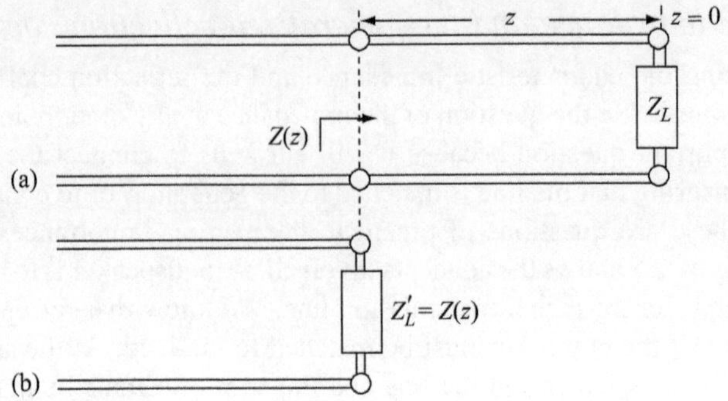

Figure 2.9 Method of calculation of the generalized reflection coefficient: (a) the impedance on the line at a general point, viewed as a new load to the line to the left of the point and (b) the new line and load

where we made use of (2.27) to rewrite the current in terms of voltage and characteristic impedance. We can divide $V(z)$ by $I(z)$ to obtain $Z(z)$, but this would not be very helpful now because the result would be in terms of both the forward and backward waves. Instead, we use the load reflection coefficient in (2.72) to write

$$V(z) = V^+(e^{\gamma z} + \Gamma_L e^{-\gamma z}) \text{ (V)} \quad \text{and} \quad I(z) = \frac{V^+}{Z_0}(e^{\gamma z} - \Gamma_L e^{-\gamma z}) \text{ (A)} \tag{2.75}$$

The line impedance at point z is

$$Z(z) = \frac{V(z)}{I(z)} = Z_0 \frac{(e^{\gamma z} + \Gamma_L e^{-\gamma z})}{(e^{\gamma z} - \Gamma_L e^{-\gamma z})} \text{ } (\Omega) \tag{2.76}$$

This expression is quite useful because it requires only knowledge of the reflection coefficient at the load, the characteristic impedance of the line, and the value of z (distance from the load). We will make considerable use of this expression here and in the following chapter.

Another way to look at the expression in (2.76) is to use the definition of the reflection coefficient in (2.72) and substitute it in (2.76). Doing so and rearranging terms gives

$$\begin{aligned} Z(z) &= Z_0 \frac{((Z_L + Z_0)e^{\gamma z} + (Z_L - Z_0)e^{-\gamma z})}{((Z_L + Z_0)e^{\gamma z} - (Z_L - Z_0)e^{-\gamma z})} \\ &= Z_0 \frac{Z_L(e^{\gamma z} + e^{-\gamma z}) + Z_0(e^{\gamma z} - e^{-\gamma z})}{Z_0(e^{\gamma z} + e^{-\gamma z}) + Z_L(e^{\gamma z} - e^{-\gamma z})} \text{ } (\Omega) \end{aligned} \tag{2.77}$$

Now, we can use the identities $(e^{\gamma z} + e^{-\gamma z})/2 = \cosh \gamma z$ and $(e^{\gamma z} - e^{-\gamma z})/2 = \sinh \gamma z$ and write

$$Z(z) = Z_0 \frac{Z_L \cosh \gamma_z + Z_0 \sinh \gamma_z}{Z_0 \cosh \gamma_z + Z_L \sinh \gamma_z} = Z_0 \frac{Z_L + Z_0 \tanh \gamma z}{Z_0 + Z_L \tanh \gamma z} \text{ } (\Omega) \tag{2.78}$$

where the relation $\tanh \gamma z = \sinh \gamma z / \cosh \gamma z$ was used. With these relations, we can now calculate the line impedance at any location, including at the input of the line.

Now, we can argue as follows: if the line impedance at a point on the line is equal to $Z(z)$, then cutting the line at this point and replacing the cut section by an equivalent load equal to $Z(z)$ should not change the conditions on the line to the left of the cut. This is shown in Figure 2.9(a). The equivalent line in Figure 2.9(b) can be viewed as a new line with load impedance $Z(z)$. There is no reason we cannot calculate the reflection coefficient at this point on the line using (2.72) with $Z(z)$ instead of Z_L. Using (2.72) and (2.75), we get for the reflection coefficient at point z on the line

$$\Gamma(z) = \frac{V^-(z)}{V^+(z)} = \frac{V^+\Gamma_L e^{-\gamma z}}{V^+ e^{\gamma z}} = \frac{\Gamma_L e^{-\gamma z}}{e^{\gamma z}} = \Gamma_L e^{-2\gamma z} \tag{2.79}$$

or using the form in (2.73) and also the relation $\gamma = \alpha + j\beta$,

$$\Gamma(z) = \Gamma_L e^{-2\gamma z} = \Gamma_L e^{-2\alpha z - j2\beta z} = |\Gamma_L| e^{-2\alpha z} e^{j\theta_\Gamma} e^{-j2\beta z} \tag{2.80}$$

The reflection coefficient $\Gamma(z)$ is called the *generalized reflection coefficient* to distinguish it from the load reflection coefficient. The generalized reflection coefficient on a general, lossy line can be viewed as having an amplitude $|\Gamma_L|$ at the load, which decays exponentially (for a lossy line) as we move toward the generator, and a phase which varies linearly with z and is equal to

$$\phi_{\Gamma(z)} = \theta_\Gamma - 2\beta z \text{ (rad)} \tag{2.81}$$

Although these relations are rather general, we will, for the most part, use lossless transmission lines. This simply means that $\alpha = 0$ and $\gamma = j\beta$, but doing so will simplify analysis considerably.

2.7.3 The lossless, terminated transmission line

In all of the above relations, we assumed a general lossy transmission line in which the propagation constant is a general complex number. There was no reason to do otherwise since we could always replace γ by $\alpha + j\beta$ to obtain the expressions in terms of the attenuation and phase constants α and β for any condition. However, the expression in (2.78) requires the use of hyperbolic sine, cosine, and tangent functions. If the line is lossless, then $\alpha = 0$ and $\gamma = j\beta$. Under these conditions, the voltage and current on the line [setting $\gamma = j\beta$ in (2.75)] are

$$V(z) = V^+\left(e^{j\beta z} + \Gamma_L e^{-j\beta z}\right) \text{ (V)} \quad \text{and} \quad I(z) = \frac{V^+}{Z_0}\left(e^{j\beta z} - \Gamma_L e^{-j\beta z}\right) \text{ (A)} \tag{2.82}$$

Similarly, the line impedance of a lossless transmission line is found by setting $\gamma = j\beta$ in (2.78):

$$Z(z) = Z_0 \frac{Z_L + jZ_0 \tan\beta z}{Z_0 + jZ_L \tan\beta z} = Z_0 \frac{Z_L \cos\beta z + jZ_0 \sin\beta z}{Z_0 \cos\beta z + jZ_L \sin\beta z} \ (\Omega) \tag{2.83}$$

where $\tanh(j\beta z) = j\tan(\beta z)$ was used. In general, the line impedance is a complex value, as we should expect. The latter expression is also useful in that it indicates explicitly the periodic nature of the line impedance and that the period is directly related to the term βz, which in Section 2.4 we called the *electrical length* of the transmission line. Not surprisingly, the electrical length of the line plays an important role in line behavior.

The generalized reflection coefficient for lossless lines was obtained in (2.80). Like the line impedance, the reflection coefficient is periodic along the line. This is best seen if the exponential function is written as $e^{-j2\beta z} = \cos 2\beta z - j\sin 2\beta z$. The generalized reflection coefficient now is

$$\Gamma(z) = \Gamma_L e^{-j2\beta z} = |\Gamma_L| e^{j\theta_\Gamma} e^{-j2\beta z} = |\Gamma_L|(\cos(\theta_\Gamma - 2\beta z) - j\sin(\theta_\Gamma - 2\beta z)) \quad (2.84)$$

Thus, the generalized reflection coefficient for lossless transmission lines can be viewed as having constant amplitude equal to that of $|\Gamma_L|$ but varying in phase along the line as

$$\phi_{\Gamma(z)} = \theta_\Gamma - 2\beta z \quad (2.85)$$

Because of this phase angle, the generalized reflection coefficient has maxima and minima along the line. However, it is more convenient to talk of maxima and minima in voltage or current, or both. Consider (2.82). Rearranging the terms, we get the voltage on the line as

$$V(z) = V^+\left(e^{j\beta z} + \Gamma_L e^{-j\beta z}\right) = V^+ e^{j\beta z}\left(1 + \Gamma_L e^{-j2\beta z}\right) = V^+ e^{j\beta z}(1 + \Gamma(z))\ (\mathrm{V}) \quad (2.86)$$

Similarly, the current on the line is

$$I(z) = \frac{V^+}{Z_0}\left(e^{j\beta z} - \Gamma_L e^{-j\beta z}\right) = \frac{V^+}{Z_0} e^{j\beta z}\left(1 - \Gamma_L e^{-j2\beta z}\right) = \frac{V^+}{Z_0} e^{j\beta z}(1 - \Gamma(z))\ (\mathrm{A}) \quad (2.87)$$

Now, we can discuss the maximum and minimum magnitudes of the voltage. First, we note that the term $e^{j\beta z}$ varies between -1 and $+1$. Thus, its magnitude is 1. Similarly, the generalized reflection coefficient $\Gamma(z)$ varies between $-\Gamma(z)$ and $+\Gamma(z)$ because the term $e^{-j2\beta z}$ varies between -1 and $+1$. Thus, we can write the maximum and minimum magnitudes of voltage as

$$V_{\max} = |V^+|(1 + |\Gamma(z)|)\ (\mathrm{V}) \quad (2.88)$$

$$V_{\min} = |V^+|(1 - |\Gamma(z)|)\ (\mathrm{V}) \quad (2.89)$$

The same can be done for the current. Following identical steps but starting with (2.87), we get

$$I_{\max} = \frac{V_{\max}}{|Z_0|} = \left|\frac{V^+}{Z_0}\right|(1 + |\Gamma(z)|)\ (\mathrm{A}) \quad (2.90)$$

$$I_{\min} = \frac{V_{\min}}{|Z_0|} = \left|\frac{V^+}{Z_0}\right|(1 - |\Gamma(z)|)\ (\mathrm{A}) \quad (2.91)$$

The ratio between the maximum and minimum voltage (or current) is called the *standing wave ratio* (SWR) and is defined as

$$\mathrm{SWR} = \frac{V_{\max}}{V_{\min}} = \frac{I_{\max}}{I_{\min}} = \frac{1+|\Gamma(z)|}{1-|\Gamma(z)|} \quad \text{(dimensionless)} \tag{2.92}$$

The SWR varies between 1 and ∞. If the reflection coefficient is zero (no reflected waves), the SWR is 1. If the magnitude of the reflection coefficient is 1 the SWR is ∞. Thus, a matched load produces no reflected waves and the line should have an SWR of 1.

Sometimes, the SWR is known or may be measured. In such cases, the magnitude of the generalized reflection coefficient can be calculated from the SWR as

$$|\Gamma(z)| = \frac{\mathrm{SWR}-1}{\mathrm{SWR}+1} \tag{2.93}$$

This expression can be substituted in (2.88) and (2.89) to obtain the minimum and maximum voltage on the line in terms of the SWR:

$$V_{\max} = |V^+|(1+|\Gamma(z)|) = |V^+|\left(1+\frac{\mathrm{SWR}-1}{\mathrm{SWR}+1}\right) = |V^+|\left(\frac{2\mathrm{SWR}}{\mathrm{SWR}+1}\right) \ (\mathrm{V}) \tag{2.94}$$

$$V_{\min} = |V^+|(1-|\Gamma(z)|) = |V^+|\left(1-\frac{\mathrm{SWR}-1}{\mathrm{SWR}+1}\right) = |V^+|\left(\frac{2}{\mathrm{SWR}+1}\right) \ (\mathrm{V}) \tag{2.95}$$

From the last three equations, it is apparent that the effect of the SWR is as follows:

1. The larger the SWR, the larger the maximum voltage and the lower the minimum voltage on the line.
2. If SWR = 1, the reflection coefficient is zero. In this case, $V_{\max} = V_{\min} = |V^+|$. The magnitude of the voltage on the line does not vary. The phase of course varies. This corresponds to a matched load.
3. If SWR = ∞, the magnitude of the reflection coefficient equals 1 ($\Gamma(z) = -1$ or $\Gamma(z) = +1$). In this case, $V_{\max} = 2|V^+|$ and $V_{\min} = 0$. We will see shortly that this corresponds to either a short circuit ($\Gamma(z) = -1$) or an open circuit ($\Gamma(z) = +1$). This condition in plane waves was called a complete standing wave.

Now that we have all the tools to calculate the reflection coefficient anywhere on the line as well as the SWR, we can return to the equations for current and voltage and see how these behave along the line. The basis of calculation is (2.86) and (2.87). Voltage and current anywhere on the line (including at the load) are

$$V(z) = V^+e^{j\beta z}\left(1+\Gamma_L e^{-j2\beta z}\right) = V^+e^{j\beta z}\left(1+|\Gamma_L|e^{j\theta_\Gamma}e^{-j2\beta z}\right) \ (\mathrm{V}) \tag{2.96}$$

$$I(z) = \frac{V^+}{Z_0}e^{j\beta z}\left(1-\Gamma_L e^{-j2\beta z}\right) = \frac{V^+}{Z_0}e^{j\beta z}\left(1-|\Gamma_L|e^{j\theta_\Gamma}e^{-j2\beta z}\right) \ (\mathrm{A}) \tag{2.97}$$

We can also calculate the voltage and current at the load. These are obtained by setting $z = 0$:

$$V_L = V^+\left(1+|\Gamma_L|e^{j\theta_\Gamma}\right) \ (\mathrm{V}) \quad \text{and} \quad I_L = \frac{V^+}{Z_0}\left(1-|\Gamma_L|e^{j\theta_\Gamma}\right) \ (\mathrm{A}) \tag{2.98}$$

To completely characterize the voltage and current waves, we must find the locations of the minima and maxima on the line. Suppose we plot the voltage and current starting at the load and going toward the generator. For any given load, the load reflection coefficient is known and we can calculate the voltage and current at the load [Eq. (2.98)] and the maximum and minimum voltage and current [from (2.88)–(2.91)]. We could, in fact, use (2.96) and (2.97) to plot the voltage and current directly. The only other bit of information needed is the location of minima and maxima in the voltage and current waves. These are found as follows.

From inspection of (2.96) and (2.97), the minimum in voltage must occur at locations on the line at which the phase $\theta_\Gamma - 2\beta z$ equals $-\pi$, -3π, -5π, etc. The general condition to be satisfied (taking z to be positive to the left and away from the load) is

$$\theta_\Gamma - 2\beta z = -(2n+1)\pi, \; n = 0, 1, 2, \ldots \tag{2.99}$$

This condition can be verified by direct substitution in (2.89) or (2.96). On the other hand, the current is maximum at this point because of the negative sign in front of Γ_L in (2.91) or (2.97). The location of the first minimum in voltage (maximum in current) occurs at

$$\theta_\Gamma - 2\beta z_{\min} = -\pi \;\rightarrow\; z_{\min} = \frac{\theta_\Gamma + \pi}{2\beta} \; (\mathrm{m}) \tag{2.100}$$

The next minimum occurs at

$$\theta_\Gamma - 2\beta z_{\min} = -3\pi \;\rightarrow\; z_{\min} = \frac{\theta_\Gamma + 3\pi}{2\beta} = \frac{\theta_\Gamma + \pi}{2\beta} + \frac{\pi}{\beta} \; (\mathrm{m}) \tag{2.101}$$

From the definition of wavelength, we can also write these relations in terms of the wavelength by using the relation

$$\lambda = \frac{2\pi}{\beta} \;\rightarrow\; \frac{\pi}{\beta} = \frac{\lambda}{2} \;\rightarrow\; \frac{1}{2\beta} = \frac{\lambda}{4\pi} \tag{2.102}$$

Thus, the conditions for minima are

For the first minimum:

$$z_{\min} = \frac{\lambda}{4\pi}(\theta_\Gamma + \pi) \; (\lambda) \tag{2.103}$$

The unit (λ) shows that the distance is indicated in wavelengths. For any minimum:

$$z_{\min} = \frac{\lambda}{4\pi}(\theta_\Gamma + (2n+1)\pi) \; (\lambda), \; n = 0, 1, 2, \ldots \tag{2.104}$$

This has the advantage of being described in terms of increments of $\lambda/2$.

The maxima occur at a distance of $\lambda/4$ on each side of a minimum. We know this must be so since the conditions on the line repeat at increments of $\lambda/2$. Between every two minima there is a maximum. Thus, we can calculate the location of the

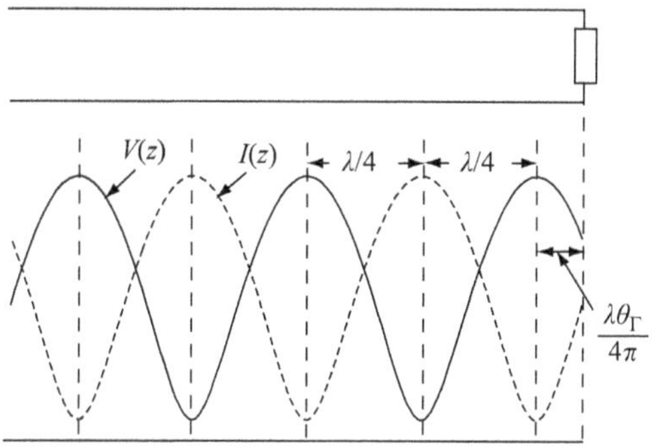

Figure 2.10 Locations of voltage maxima and minima on a transmission line and the relation between voltage and current minima and maxima on the line

first voltage maximum by adding $\lambda/4$ in (2.104). Voltage maxima (current minima) occur at

$$z_{\max} = \frac{\lambda}{4\pi}(\theta_\Gamma + (2n+1)\pi) + \frac{\lambda}{4} \rightarrow z_{\max} = \frac{\lambda}{4\pi}(\theta_\Gamma + 2n\pi)\ (\lambda),\ n = 0, 1, 2, \ldots \tag{2.105}$$

The complete description of voltage and current on the line are now shown in Figure 2.10. Note in particular that the minima are sharper than the maxima. In other words, the voltage or current do not vary sinusoidally. Whenever measurements of SWR are required, the minima are usually easier to identify. Note also that Figure 2.10 assumes, arbitrarily, that $V_L > 0$ at the load. This does not have to be so: V_L can be negative or zero.

The maxima in line impedance occur at locations of voltage maxima (current minima) and minima in line impedance occur at location of voltage minima (current maxima).

From the foregoing discussion, it is clear that voltage and current are highly dependent on the load reflection coefficient and they vary from point to point. From (2.83), we can also tell that the line impedance varies from point to point. The above relations are general and apply to any load. The only restriction in the above discussion is that the line be lossless.

A number of particular solutions may be obtained for particular loads. These loads are useful because they lead to simple, practical solutions. These are as follows:

1. Matched load: $Z_L = Z_0$. The load reflection coefficient is zero ($\Gamma_L = 0$).
2. Short-circuited load: $Z_L = 0$. The load reflection coefficient is $\Gamma_L = -1$.
3. Open-circuit load: $Z_L = \infty$. The load reflection coefficient is $\Gamma_L = +1$.
4. Resistive load: $Z_L = R_L + j0$. The load reflection coefficient is real ($-1 < \Gamma_L < 1$).

These particular types of terminated transmission lines are discussed in the following sections.

2.7.4 The lossless, matched transmission line

A matched transmission line is a line on which the load is equal to the characteristic impedance of the line:

$$Z_L = Z_0 \tag{2.106}$$

Substitution of this condition in (2.72) results in a zero reflection coefficient at the load: $\Gamma_L = 0$. Thus, the line impedance anywhere on the line is

$$Z(z) = Z_0 \frac{Z_0 + jZ_0 \tan \beta z}{Z_0 + jZ_0 \tan \beta z} = Z_0 \ (\Omega) \tag{2.107}$$

Therefore, the impedance on the line for a matched load is constant and equal to Z_0.

The other relations on the line are also obtained by substituting $Z_L = Z_0$ and $\Gamma_L = 0$. Thus, the SWR on the line is SWR = 1 anywhere on the line. The voltage and current on the line are

$$V(z) = V^+ e^{j\beta z} \ (\mathrm{V}) \ \text{ and } \ I(z) = \frac{V^+}{Z_0} e^{j\beta z} \ (\mathrm{A}) \tag{2.108}$$

That is, the line voltage and current have only forward-propagating terms, as we expect with a zero reflection coefficient.

In summary, a matched load produces no reflected waves and, therefore, no standing waves on the line. All power on the line is transferred to the load.

2.7.5 The lossless, shorted transmission line

A shorted transmission line is characterized by $Z_L = 0$. From (2.72), the reflection coefficient is $\Gamma_L = -1$. For the same reason, SWR $= \infty$. The line impedance is now

$$Z(z) = Z_0 \frac{jZ_0 \tan \beta z}{Z_0} = jZ_0 \tan \beta z \ (\Omega) \tag{2.109}$$

The line impedance of a shorted transmission line is purely imaginary and varies between $-\infty$ and ∞. It has the following properties:

1. $\Gamma_L = -1$, SWR $= \infty$.
2. It is zero at the load and at any value $\beta z = n\pi$, $n = 1, 2, \ldots$. In terms of wavelength, the line impedance is zero at $z = n\lambda/2$, $n = 0, 1, 2, \ldots$ and is infinite at $z = n\lambda/2 + \lambda/4$, $n = 0, 1, 2, \ldots$.
3. The line impedance is purely imaginary and alternates between positive and negative values, as shown in Figure 2.11. The impedance is positive (inductive) for $n\lambda/2 < z < n\lambda/2 + \lambda/4$ and negative (capacitive) for $n\lambda/2 + \lambda/4 < z < n\lambda/2 + \lambda/2$, $n = 0, 1, 2, \ldots$. The line impedance changes from $+\infty$ to $-\infty$ at $z = n\lambda/2 + \lambda/4$.
4. A shorted transmission line behaves as an inductor or a capacitor, depending on the location on the line. A capacitance or an inductance may be designed by simply cutting a line of appropriate length as indicated in (3). In this sense, shorted transmission lines are viewed as circuit elements.

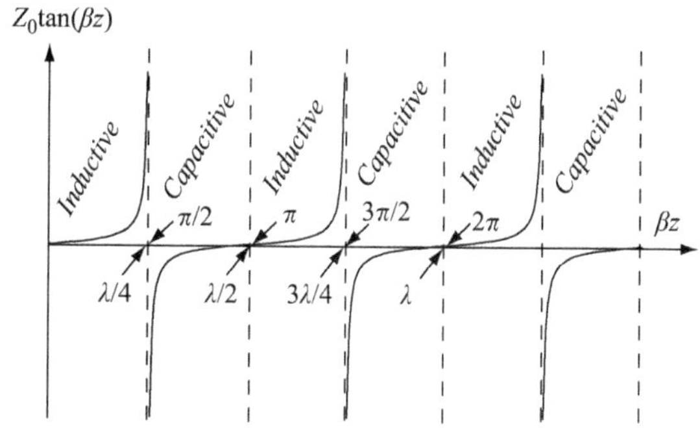

Figure 2.11 Line impedance on a shorted transmission line

5. The conditions on a shorted transmission line repeat at intervals of $\lambda/2$; that is, if we add or remove a section of length $\lambda/2$ (or any integer multiple of $\lambda/2$), the line impedance does not change.

The line voltage and line current are [setting $\Gamma_L = -1$ in (2.96) and (2.97)]

$$V(z) = V^+ e^{j\beta z}\left(1 - e^{-j2\beta z}\right) \text{ (V) and } I(z) = \frac{V^+}{Z_0} e^{j\beta z}\left(1 + e^{-j2\beta z}\right) \text{ (A)} \tag{2.110}$$

In particular, at the load ($z = 0$), we get

$$V_L = 0 \text{ (V) and } I_L = \frac{2V^+}{Z_0} \text{ (A)} \tag{2.111}$$

Thus, whereas the voltage at the load must be zero, the current must be twice the forward-propagating current. This, of course, is a consequence of the fact that there is no transfer of power into the load and the reflected current is equal in magnitude and phase to the forward current.

2.7.6 The lossless, open transmission line

An open transmission line may be assumed to have an infinite impedance as load. Since $Z_L \to \infty$, the reflection coefficient at the load is $\Gamma_L = +1$. For the same reason, SWR $= \infty$. Substitution of Z_L into the line impedance in (2.83) gives (since $Z_L \gg Z_0$):

$$Z(z) = Z_0 \frac{Z_L + jZ_0 \tan\beta z}{Z_0 + jZ_L \tan\beta z} = Z_0 \frac{Z_L}{jZ_L \tan\beta z} = -jZ_0 \cot\beta z \; (\Omega) \tag{2.112}$$

This result is very similar to the result for the shorted transmission line. The properties of this line are summarized as follows:

1. $\Gamma_L = +1$, SWR $= \infty$.
2. The line impedance is infinite at the load and at any value $\beta z = n\pi$, $n = 1, 2, \ldots$. In terms of wavelength, the line impedance is infinite at $z = n\lambda/2$, $n = 0, 1, 2, \ldots$. The line impedance is zero at $z = n\lambda/2 + \lambda/4$, $n = 0, 1, 2, \ldots$.

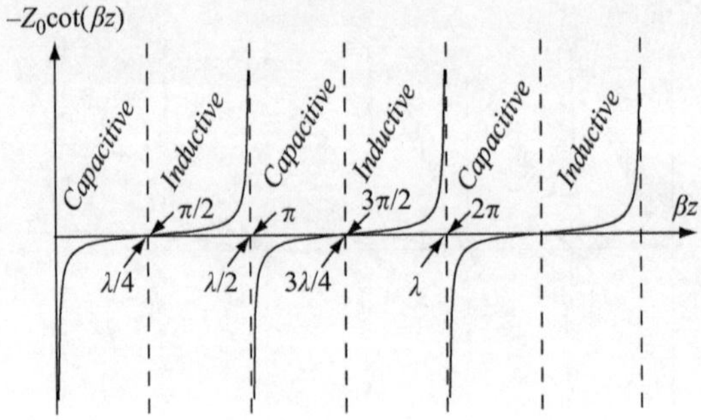

Figure 2.12 Line impedance on an open transmission line

3. The line impedance is purely imaginary and alternates between positive and negative values, as shown in Figure 2.12. The impedance is negative (capacitive) for $n\lambda/2 < z < n\lambda/2 + \lambda/4$ and positive (inductive) between $n\lambda/2 + \lambda/4 < z < n\lambda/2 + \lambda/2$, $n = 0, 1, 2, \ldots$. The line impedance changes from $+\infty$ to $-\infty$ at $z = n\lambda/2$.
4. An open transmission line behaves as an inductor or a capacitor, depending on the location on the line. A capacitance or an inductance may be designed by simply cutting a line of appropriate length as indicated in (3). Open transmission lines may also be viewed as circuit elements.
5. The conditions on an open transmission line repeat at intervals of $\lambda/2$; that is, if we add or remove a section of length $\lambda/2$ (or any integer multiple of $\lambda/2$), the line impedance is not affected.
6. The conditions on an open transmission line are identical to those of a shorted transmission line if their length differs by an odd multiple of $\lambda/4$. This can be seen by direct comparison of Figures 2.12 and 2.11.

The line voltage and line current on the open transmission line are

$$V(z) = V^+ e^{j\beta z}\left(1 + e^{-j2\beta z}\right)\ (\mathrm{V}) \ \text{ and } \ I(z) = \frac{V^+}{Z_0} e^{j\beta z}\left(1 - e^{-j2\beta z}\right)\ (\mathrm{A}) \tag{2.113}$$

In particular, at the load ($z = 0$), we get

$$V_L = 2V^+\ (\mathrm{V}) \ \text{ and } \ I_L = 0 \tag{2.114}$$

Thus, maximum voltage occurs at the load, whereas maximum current occurs at $\lambda/4$ from the load. Again, there is no transfer of power into the load and the reflected voltage is equal to the forward propagating voltage (and in the same phase).

Suppose now that we perform an experiment. First, we short a transmission line and obtain the line impedance at a point z. Then, we open the line and obtain the impedance at the same point. The shorted and open line impedances are those given in (2.109) and (2.112). If we take the product of these two impedances, we get

$$(jZ_0 \tan\beta z)\,(-jZ_0 \cot\beta z) = Z_0^2 \tag{2.115}$$

Perhaps a bit unexpected result but it gives us yet another way of calculating or measuring the characteristic impedance of a transmission line. The characteristic impedance of any lossless line is given as

$$Z_0 = \sqrt{Z_{\text{short}} Z_{\text{open}}} \ (\Omega) \tag{2.116}$$

where Z_{short} is the line impedance with shorted load and Z_{open} is the impedance with open load.

2.7.7 The lossless, resistively loaded transmission line

The discussion in Sections 2.7.1 and 2.7.2 was in terms of a general load, but it applies equally well for a resistive load: $Z_L = R_L + j0$. The reflection coefficient at the load is real: $-1 < \Gamma_L < 1$:

$$\Gamma_L = \frac{V^-}{V^+} = \frac{R_L - Z_0}{R_L + Z_0} \tag{2.117}$$

and since Z_0 is also real for lossless lines [see (2.38)], the reflection coefficient is real. It can be either positive or negative depending on the relative magnitudes of R_L and Z_0. The line impedance is now given as

$$Z(z) = Z_0 \frac{R_L + jZ_0 \tan\beta z}{Z_0 + jR_L \tan\beta z} = Z_0 \frac{R_L \cos\beta z + jZ_0 \sin\beta z}{Z_0 \cos\beta z + jR_L \sin\beta z} \ (\Omega) \tag{2.118}$$

This impedance is maximum at locations of maximum voltage and minimum at locations of minimum voltage, as described in Section 2.7.3. The main difference between a resistive load and a general load is that for a general load, the phase angle of the load reflection coefficient can have any value. On the other hand, for a resistive load, the phase angle can be either zero or $-\pi$. This can be seen from (2.117). There are two possible situations:

1. $R_L > Z_0$. In this case, Γ_L is real, positive and we can write

$$\Gamma_L = \frac{R_L - Z_0}{R_L + Z_0} \ \rightarrow \ \Gamma_L = |\Gamma_L| e^{j0} \tag{2.119}$$

Now, if we substitute $\theta_\Gamma = 0$ in (2.96) and (2.97), we obtain the general voltage and current waves on the line:

$$V(z) = V^+ e^{j\beta z}\left(1 + \Gamma_L e^{-j2\beta z}\right) \ (\text{V}) \tag{2.120}$$

$$I(z) = \frac{V^+}{Z_0} e^{j\beta z}\left(1 - \Gamma_L e^{-j2\beta z}\right) \ (\text{A}) \tag{2.121}$$

The voltage and current at the load are

$$V_L = V^+(1 + \Gamma_L) \ (\text{V}) \ \text{ and } \ I_L = \frac{V^+}{Z_0}(1 - \Gamma_L) \ (\text{A}) \tag{2.122}$$

The locations of voltage minima are now [see (2.103) and (2.104)]:

$$z_{\min} = \frac{\lambda}{4\pi}(2n+1)\pi \ (\lambda), \ n = 0, 1, 2, \ldots \tag{2.123}$$

Thus, the first minimum in voltage occurs at $n = 0$:

$$z_{\min} = \frac{\pi}{2\beta} = \frac{\lambda}{4} \ (\lambda) \tag{2.124}$$

Similarly, the locations of voltage maxima are (2.105)

$$z_{\max} = \frac{\lambda}{4\pi} 2n\pi \ (\lambda), \ n = 0, 1, 2, \ldots \tag{2.125}$$

The first voltage maximum is at the load ($z = 0$). The following voltage maxima (current minima) are at increments of $\lambda/2$ from the load. The voltage and current minima and maxima are shown in Figure 2.13(a).

2. $R_L < Z_0$. In this case, Γ_L is real and negative and we can write

$$\Gamma_L = \frac{R_L - Z_0}{R_L + Z_0} \rightarrow \Gamma_L = -|\Gamma_L| = |\Gamma_L| e^{-j\pi} \tag{2.126}$$

Now, if we substitute $\theta_\Gamma = -\pi$ in (2.96) and (2.97), we obtain the general voltage and current waves on the line:

$$V(z) = V^+ e^{j\beta z}\left(1 + |\Gamma_L| e^{-j\pi} e^{-j2\beta z}\right) \ (\mathrm{V}) \tag{2.127}$$

$$I(z) = \frac{V^+}{Z_0} e^{j\beta z}\left(1 - |\Gamma_L| e^{-j\pi} e^{-j2\beta z}\right) \ (\mathrm{A}) \tag{2.128}$$

The voltage and current at the load are

$$V_L = V^+(1 - |\Gamma_L|) \ (\mathrm{V}) \ \text{ and } \ I_L = \frac{V^+}{Z_0}(1 + |\Gamma_L|) \ (\mathrm{A}) \tag{2.129}$$

The locations of minima in voltage are

$$z_{\min} = \frac{\lambda}{4\pi}(2n\pi) \ (\lambda), \ n = 0, 1, 2, \ldots \tag{2.130}$$

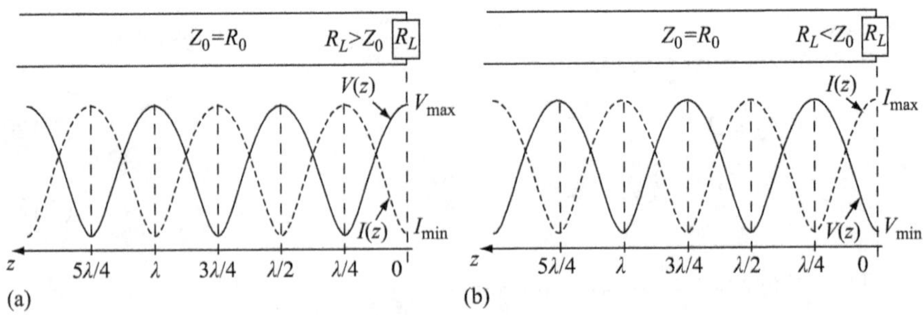

Figure 2.13 (a) Voltage and current maxima and minima for $R_L > Z_0$ and (b) voltage and current maxima and minima for $R_L < Z_0$

Thus, the first minimum in voltage occurs at $z = 0$. Subsequent minima occur at intervals of $\lambda/2$ from the load. The first voltage maximum occurs at $z = \lambda/4$ and the general relation for voltage maxima is

$$z_{max} = \frac{\lambda}{4\pi}(2n - 1)\pi \ (\lambda), \ n = 0, 1, 2, \ldots \tag{2.131}$$

The complete description of voltage and current on the line for $R_L < Z_0$ is shown in Figure 2.13(b). Note also that the maximum and minimum line impedance are given as

$$Z_{max} = Z_0 \frac{(1 + |\Gamma_L|)}{(1 - |\Gamma_L|)} = Z_0 \mathrm{SWR} \ (\Omega) \tag{2.132}$$

$$Z_{min} = Z_0 \frac{(1 - |\Gamma_L|)}{(1 + |\Gamma_L|)} = \frac{Z_0}{\mathrm{SWR}} \ (\Omega) \tag{2.133}$$

The properties of line impedance on a resistively loaded line are

1. $-1 < \Gamma_L < +1$, $1 < \mathrm{SWR} < \infty$.
2. The line impedance is maximum at locations of voltage maxima and minimum at locations of voltage minima. These locations are given in (2.123) and (2.125) for $R_L > Z_0$ and in (2.130) and (2.131) for $R_L < Z_0$.
3. The line impedance can be complex as can be seen from (2.118), but it is always real at locations of voltage maxima and voltage minima for any lossless line. The impedance at voltage maxima is $Z_{max} = Z_0$*SWR, whereas at voltage minima (current maxima), it is $Z_{min} = Z_0/\mathrm{SWR}$.
4. For $R_L > Z_0$, the first voltage maximum occurs at the load ($z = 0$) and the first voltage minimum at a distance $\lambda/4$ from the load. All conditions on the line repeat at intervals of $\lambda/2$.
5. For $R_L < Z_0$, the first voltage minimum occurs at the load and the first voltage maximum at a distance $\lambda/4$ from the load. All conditions on the line repeat at intervals of $\lambda/2$.

In effect, the main difference between a general load and a resistive load is the location of the minima and maxima. If the load is such that the magnitude of the reflection coefficient at the load is the same for resistive and arbitrary loads, the voltage and current on the line will be the same in both cases but displaced by the value of z_{min} in (2.123) or (2.103). In other words, if we take an arbitrary load and calculate all circuit parameters, we obtain the standing wave pattern for the line. The line can now be shortened by the magnitude of z_{min} or lengthened by $\lambda/2 - z_{min}$ to obtain an identical circuit but with a resistive loading which has the same reflection coefficient magnitude. We will use this property of transmission lines in the following chapter.

2.8 Power relations on a general transmission line

The power relation on a line can be written directly from the current and voltage on the line. The power at a distance z_0 from the load can be calculated by assuming that the load is at $z=0$ and the input is at $z=z_0$ as shown in Figure 2.14. For this condition, the line voltage and current for a general lossy line are given in (2.75). Setting $z=z_0$ gives the voltage and current as

$$V(z_0) = V^+(e^{\gamma z_0} + \Gamma_L e^{-\gamma z_0}) \text{ (V) and } I(z_0) = \frac{V^+}{Z_0}(e^{\gamma z_0} - \Gamma_L e^{-\gamma z_0}) \text{ (A)} \tag{2.134}$$

where Z_0 is the line characteristic impedance given in (2.23) and is, in general, a complex number. Now, the power entering this section of the transmission line is calculated from the current and voltage on the line at this point:

$$\begin{aligned}
P_i &= \frac{1}{2}\mathrm{Re}\left\{V_{z_0} I_{z_0}^*\right\} = \frac{1}{2}\mathrm{Re}\left\{[V^+(e^{\gamma z_0} + \Gamma_L e^{-\gamma z_0})]\left[\frac{V^+}{Z_0}(e^{\gamma z_0} - \Gamma_L e^{-\gamma z_0})\right]^*\right\} \\
&= \frac{|V^+|^2}{2}\mathrm{Re}\left\{\left(e^{(\alpha+j\beta)z_0} + |\Gamma_L|e^{j\theta_\Gamma}e^{-(\alpha+j\beta)z_0}\right)\frac{\left(e^{(\alpha+j\beta)z_0} - |\Gamma_L|e^{-j\theta_\Gamma}e^{-(\alpha+j\beta)z_0}\right)}{Z_0^*}\right\} \\
&= \frac{|V^+|^2}{2|Z_0|}\mathrm{Re}\left\{\left(e^{2\alpha z_0} + |\Gamma_L|e^{j(\theta_\Gamma - \beta z_0)} - |\Gamma_L|e^{-j(\theta_\Gamma - \beta z_0)} - |\Gamma_L|^2 e^{-2\alpha z_0}\right)e^{-j\theta z_0}\right\} \\
&= \frac{|V^+|^2}{2|Z_0|}\mathrm{Re}\left\{\left(e^{2\alpha z_0} + j2|\Gamma_L|\sin(\theta_\Gamma - \beta z_0) - |\Gamma_L|^2 e^{-2\alpha z_0}\right)e^{-j\theta z_0}\right\} \\
&= \frac{|V^+|^2}{2|Z_0|}\left(e^{2\alpha z_0} - |\Gamma_L|^2 e^{-2\alpha z_0}\right)\cos(\theta_{Z_0}) \text{ (W)}
\end{aligned} \tag{2.135}$$

where θ_{Z_0} is the phase angle of the characteristic impedance and θ_{z0} is the phase angle at $z=z_0$. To summarize

$$P_{z_0} = \frac{|V^+|^2}{2|Z_0|}\left(e^{2\alpha z_0} - |\Gamma_L|^2 e^{-2\alpha z_0}\right)\cos(\theta_{Z_0}) \text{ (W)} \tag{2.136}$$

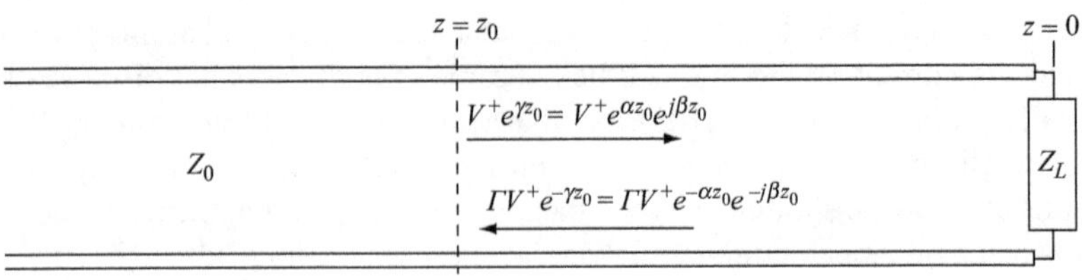

Figure 2.14 Notation used to calculate power relations on the transmission line

This relation has two components: the first is forward-propagating toward the load (in the negative z direction according to our convention which defines the load as reference) and the second in the positive z direction (from load to generator). Both are real powers and the total power is the sum of the two.

The power entering any section of the line may now be evaluated by setting the correct value for z_0. The power at the load may be found by setting $z_0 = 0$:

$$P_{\text{load}} = \frac{|V^+|^2}{2|Z_0|}\left(1 - |\Gamma_L|^2\right)\cos(\theta_{Z_0}) \text{ (W)} \tag{2.137}$$

The general power relation in (2.136) may be simplified under certain conditions. If there is only a forward-propagating wave, the forward-propagating voltage, current, and power are

$$V^+(z_0) = V^+e^{\gamma z_0} \text{ (V)},\ I^+(z_0) = \frac{V^+}{Z_0}e^{\gamma z_0} \text{ (A)},\ P^+(z_0) = \frac{|V^+|}{2|Z_0|}e^{2\alpha z_0}\cos(\theta_{Z_0}) \text{ (W)} \tag{2.138}$$

For the backward propagating wave alone

$$\begin{aligned} V^-(z_0) &= V^+\Gamma_L e^{\gamma z_0} \text{ (V)},\ I^-(z_0) = -\frac{V^+}{Z_0}\Gamma_L e^{\gamma z_0} \text{ (A)}, \\ P^-(z_0) &= \frac{|V^+|^2|\Gamma_L|^2}{2|Z_0|}e^{-2\alpha z_0}\cos(\theta_{Z_0}) \text{ (W)} \end{aligned} \tag{2.139}$$

For lossless lines, the attenuation constant is zero and the characteristic impedance is real. The power at any point on the line is therefore

$$P_{z_0} = \frac{|V^+|^2}{2Z_0}\left(1 - |\Gamma_L|^2\right) \text{ (W)} \tag{2.140}$$

It is worth mentioning again that this power is positive propagating from generator to load. Note that if the reflection coefficient is zero, all power on the line is transferred to the load, although this does not imply maximum power transfer from generator to load.

The instantaneous power on the line is calculated similarly by multiplying the instantaneous voltage by instantaneous current.

2.9 Passive transmission line circuits

A length of transmission line has a line impedance that depends on the length of the line and its properties. We saw as well that a real line has losses due to attenuation and shorted and open transmission lines behave either as inductors or capacitors. It is therefore obvious that by combining these elements one can build passive circuits—essentially *RLC* circuits. This has important implications since now, any low frequency *RLC* passive circuit has its equivalent in the high frequency range.

But transmission lines have a few extra "tricks" due to the finite speed of propagation on the line. These are associated with the phase on the line. It is therefore not surprising that a transmission line can behave as a resonator, an attenuator or that one can use transmission line segments to match a load to a line. The following sections discuss the use of transmission lines to design useful passive circuits that can operate successfully at high frequencies (and only at high frequencies). The first use we describe is matching of transmission lines using shorted and open transmission line stubs. Following these are power dividers and combiners, directional couplers and attenuators. These circuits are critical and described here because they have critical impact on measurements and in the context of gauging accurate circuits can make all the differences. These passive devices are integral to network and spectrum analyzers and some of them will be also discussed in Chapter 8 when we discuss these instruments. It should be understood that the purpose here is not the design of passive circuits but rather an understanding and a working knowledge of their operation. For this reason, only the salient features of the circuits are described, and there is no attempt to be exhaustive.

The design of transmission line resonators is left to Section 2.10 because resonators are at the heart of this work and there we need more than cursory descriptions. The theory, analysis, and application of transmission line resonators will be described in detail, leading to their application in the following chapters.

2.9.1 Impedance matching

We have described the shorted and open transmission lines in Sections 2.7.5 and 2.7.6. These lines have an impedance that is purely imaginary and can be either positive or negative. Thus, a segment of shorted transmission line of appropriate length will behave as an inductor or as a capacitor. A capacitance or inductance of almost any value may therefore be obtained by an open or shorted line. Similarly, an impedance of almost any value may be obtained by appropriate choice of lines and loads. It is, therefore, possible to use these line segments to build particular circuits with given properties. Examples of simple circuits are shorted or open stubs used for matching purposes as shown in Figure 2.15. Figure 2.15(a) shows a shorted stub of length l_1 connected at a distance d_1 from a mismatched load. The length of the stub and the location of the stub are used to adjust the real part of the line

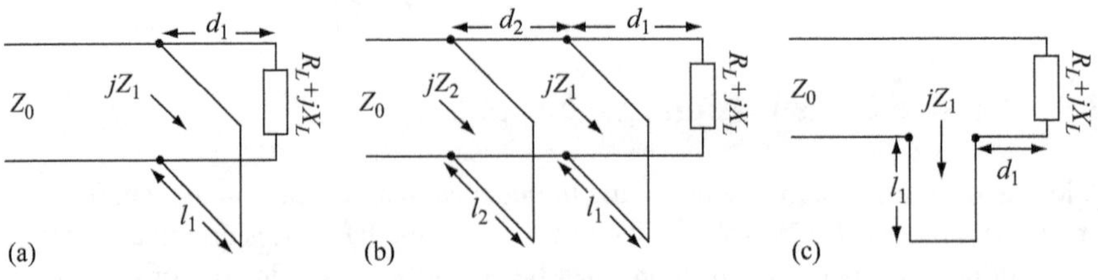

Figure 2.15 Matching circuits: (a) single short-circuited stub matching, (b) double short-circuited stub matching, and (c) series short-circuited stub matching

impedance and the imaginary part of the line impedance independently so that the impedance at the location of the stub equals Z_0 and hence match the load to the line. Although we do not deal here with all the details of matching, it suffices to say that d_1 is adjusted so that the real part of the input admittance to the segment of loaded line of length d_1 equals $1/Z_0$. Now the admittance of the stub is adjusted so that it equals the negative of the imaginary part of the segment. The connection of the shorted stub cancels the imaginary part of the segment admittance, and we are left with a matched line. Figure 2.15(b) is another method of matching with stubs whereby the two stubs are placed at fixed distances from the mismatched load (d_1 and d_2 are fixed) and matching is done by adjusting the lengths of the two stubs until the input impedance to the segment of length d_1+d_2 equals Z_0. Figure 2.15(c) is the simplest to understand but the least common in use because it requires cutting one conductor of the line and inserting a shorted line in series. This cannot always be done (e.g., in coaxial lines this is impossible). In essence, d_1 is adjusted until the real part of the input impedance of the segment of length d_1 equals Z_0, followed by adjustment of l_1 until its impedance equals the negative of the segment's imaginary part of the impedance. The net effect is that the segment of length d_1 exhibits an impedance Z_0, matching the load to the line.

There are other types of circuits that can be easily built and are useful in transmission line work. An example is the quarter wavelength transformer shown in Figure 2.16. This circuit is often used to match loads to transmission lines or to match two lines with different characteristic impedances. The operation of the circuit is deduced directly from Figure 2.16. The input impedance to a line segment of length z_0 and loaded with a load impedance Z_L is [from (2.83)]:

$$Z_{\text{in}} = Z_0 \frac{[Z_L \cos\beta z_0 + jZ_0 \sin\beta z_0]}{[Z_0 \cos\beta z_0 + jZ_L \sin\beta z_0]} \quad (\Omega) \tag{2.141}$$

where Z_L is the load impedance, Z_0 the characteristic impedance, z_0 the length, and β the phase constant on the line segment. Setting $z_0 = \lambda/4$ and $\beta z_0 = \beta\lambda/4 = (2\pi/\lambda)(\lambda/4) = \pi/2$ and replacing Z_L by Z_l and Z_0 by Z_t in (2.141), we get for the input impedance of the $\lambda/4$ section

$$Z_{\text{in}} = Z_t \frac{[Z_l \cos(\pi/2) + jZ_t \sin(\pi/2)]}{[Z_l \cos(\pi/2) + jZ_l \sin(\pi/2)]} = \frac{Z_t^2}{Z_l} \quad (\Omega) \tag{2.142}$$

Figure 2.16 A quarter-wavelength transformer located at distance d from load. The characteristic impedance of the transformer segment Z_t is selected to match the load impedance Z_l to the line of characteristic impedance Z_0

Referring now to Figure 2.16, where Z_l is the line impedance at a distance d from the load, we get the condition for matching using the quarter-wavelength transformer shown:

$$Z_t = \sqrt{Z_{in} Z_l} \ (\Omega) \tag{2.143}$$

Thus, two different transmission lines or any two impedances may be matched, provided a transformer of proper characteristic impedance Z_t can be found. The quarter-wavelength transformer is normally connected at a point of maximum or minimum voltage since the line impedance is real at that point but these are details of application.

Other examples of transmission line circuits are power dividers and combiners, attenuators couplers and many others, including resonant circuits. These are discussed next.

2.9.2 Power dividers

Power dividers and combiners are fundamental passive circuits in waveguide and transmission line networks. They allow one to perform simple tasks such as sampling the power on a line or connect two loads (such as antennas) to the same source with equal or unequal power in the two loads. Power dividers can be lossless (or nearly lossless) or can be lossy and either matched or mismatched. Power combiners are power dividers operating in "reverse" by combining power from two sources into a single load.

2.9.2.1 The lossless T-junction power divider

The name of this divider comes from its microwave implementation [see Figure 2.17(a)] even though in transmission lines it may not look like a T [see Figure 2.17(b)]. The principle can be easily understood from Figure 2.17(b). Assuming the three transmission lines are lossless, with $z_1 = z_0$, $z_2 = 1.5z_0$, $z_3 = 3z_0$ and with a voltage V at the junction, the input power into the junction is

$$P_i = \frac{V^2}{2Z_0} \tag{2.144}$$

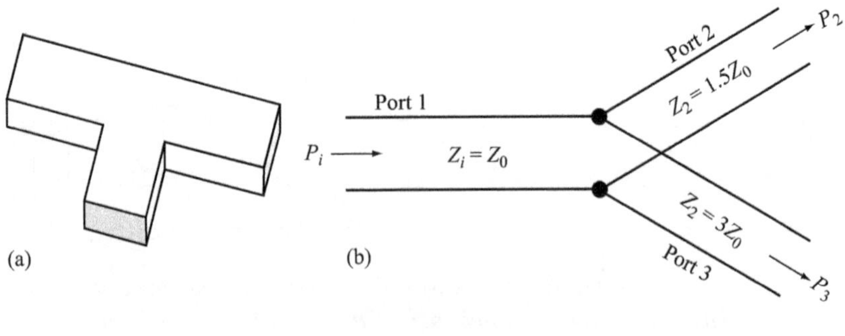

Figure 2.17 T-junction power divider: (a) waveguide T-junction and (b) transmission line T-junction

The powers in lines 2 and 3 are

$$P_2 = \frac{V^2}{2Z_2} = \frac{V^2}{2(1.5Z_0)} = \frac{V^2}{(3/2) \times 2Z_0} = \frac{2}{3}P_i \tag{2.145}$$

$$P_3 = \frac{V^2}{2Z_3} = \frac{V^2}{2(3Z_0)} = \frac{1}{3}P_i \tag{2.146}$$

That is, the power has been divided with 1/3 of the power into line 2 and 2/3 into line 1. The power can be divided equally provided $Z_2 = Z_3 = 2Z_0$. In principle, any division ratio can be achieved by selection of the impedances of lines 2 and 3.

Although very simple, there are a number of basic problems with this circuit. First, the junction may not be lossless. In addition, the junction offers a discontinuity in which case, the junction itself adds a susceptance to the circuit. This is shown in Figure 2.18 as *ja*. Now one is either obliged to match the junction by adding tuning elements that will cancel the susceptance or accept a reflection from the junction into line 1. Second, we note from Figure 2.17(b) that line 1 is matched since $Z_2 \| Z_3 = Z_i$. However, lines 2 and 3 are not matched. Looking into the junction from line 2, one sees the impedance of line 1 and line 3 in parallel. That is, the impedance seen by line 2 is $Z_0 \| 3Z_0 = 3Z_0/4$ and the reflection coefficient is

$$\Gamma_2 = \frac{3Z_0/4 - 3Z_0/2}{3Z_0/4 + 3Z_0/2} = -0.333 \tag{2.147}$$

Looking into the junction on line 3, the impedance is $Z_1 \| 1.5Z_2 = 3Z_0/5$

$$\Gamma_3 = \frac{3Z_0/5 - 3Z_0}{3Z_0/5 + 3Z_0} = -0.667 \tag{2.148}$$

This leads to the conclusion that the three lines cannot all be matched as long as the junction is lossless.

A third important problem is that the three lines are not isolated.

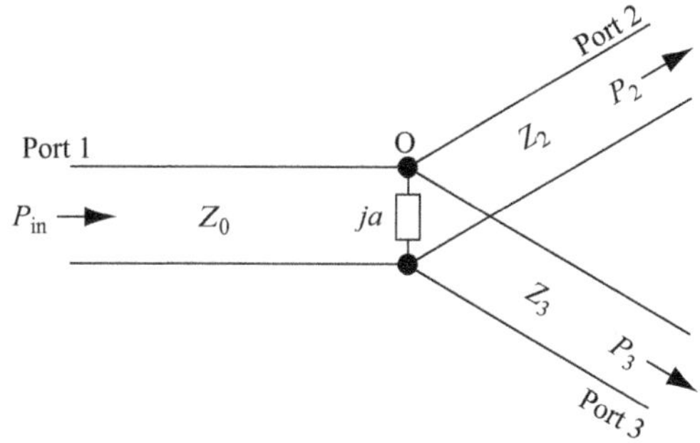

Figure 2.18 The effect of discontinuities at the lossless T-junction. The susceptance due to discontinuities can be compensated

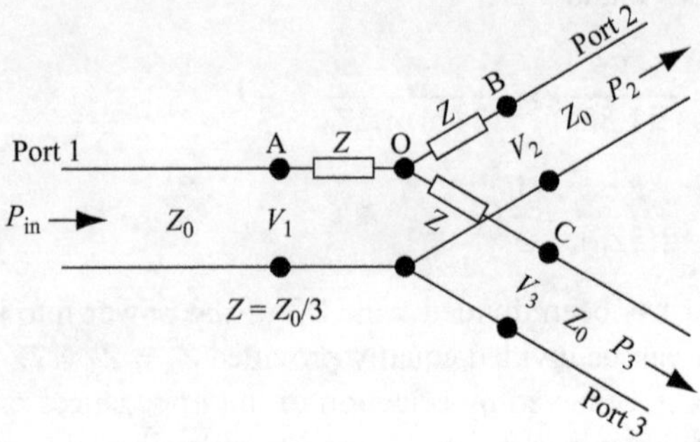

Figure 2.19 The lossy T-junction power divider. The power division is calculated at ports A, B, and C

2.9.2.2 The lossy T-junction power divider

A solution to the problem of matching the lines in a T-junction divider is to add lossy components (resistors) to the transmission line T-junction shown in Figure 2.19. The lines are all of impedance Z_0, and the resistance values are selected so that the input impedance to each lines equals Z_0 and power is divided by the required ratio. This is particularly simple to show for equal power division between lines 2 and 3. The junction now is defined between ports A, B, and C as shown in Figure 2.19 in which each of the resistor equals $Z_0/3$. The voltages across the junction are V_1, V_2, and V_3. Looking into the center point (junction O), each of lines 2 and 3 has an impedance of $Z_0 + Z_0/3 = 4Z_0/3$ and with the two lines in parallel, the impedance seen at O is $2Z_0/3$. The voltage at O is

$$V_O = V_1 \frac{2Z_0/3}{Z_0/3 + 2Z_0/3} = \frac{2}{3} V_1 \tag{2.149}$$

The voltages V_1 and V_2 are equal:

$$V_2 = V_3 = \frac{2}{3} V_1 \frac{Z_0}{Z_0 + Z_0/3} = \frac{2}{3} V_1 \times \frac{3}{4} = \frac{V_1}{2} \tag{2.150}$$

Now, the input power and the powers in line 2 and 3 are

$$P_i = \frac{V_1{}^2}{2Z_0}, \; P_2 = P_3 = \frac{\left(V_1{}^2/2\right)^2}{2Z_0} = \frac{V_1{}^2}{8Z_0} \tag{2.151}$$

Thus, given an input power P_i, the power in lines 2 and 3 are $P_i/4$ and half the power is lost in the resistors.

2.9.2.3 The Wilkinson power divider

A more complex divider that solves the issues with the lossless or lossy T-junction is the Wilkinson power divider shown in Figure 2.20. It consists of a $\lambda/4$ junction section and a resistive load across the output lines. The main requirements are that

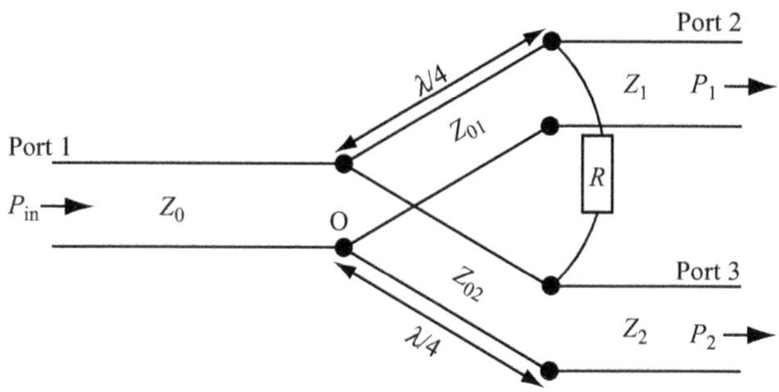

Figure 2.20 The Wilkinson power divider

the characteristic impedance of the $\lambda/4$ sections be $Z_0\sqrt{2}$ and the resistance be $2Z_0$. It is matched at all lines, there is isolation between the lines and, in addition, it is lossless when the lines are matched. The resistor only dissipates reflected power from lines 2 and 3 and these reflections only occur when the lines are mismatched. In addition, the Wilkinson power divider can be made with unequal power division $P = P_2/P_1$. The required impedances and resistance are

$$Z_{01} = Z_0\sqrt{\sqrt{P}(1+P)} \tag{2.152}$$

$$Z_{02} = Z_0\sqrt{\frac{1+P}{P\sqrt{P}}} \tag{2.153}$$

$$R = Z_0\left(\sqrt{P} + \frac{1}{\sqrt{P}}\right) \tag{2.154}$$

$$Z_1 = PZ_0,\ Z_2 = \frac{Z_0}{P} \tag{2.155}$$

The divider is matched to the input line of impedance Z_0 and to the two output lines with impedances PZ_0 and Z_0/P, respectively. For equal division, all three lines are of impedance Z_0 since then $P = 1$. In general, it is difficult to produce transmission lines with impedance that can be vastly different from each other because in classical transmission line both dimensions and material properties must be adjusted and the lines typically require connectors leading to discontinuities and poor control over the length of the $\lambda/4$ section. However, in some cases, such as in striplines (to be discussed in the following chapter), the impedance of lines can be controlled by controlling the width of the strips and/or distances from ground planes. In most cases, striplines are produced by lithographic techniques leading to accurate dimensions and hence accurate power division.

The Wilkinson power divider can be extended to N divisions as shown in Figure 2.21(a). In this implementation, the impedance of each $\lambda/4$ line is Z_0/N and each resistor equals Z_0. The problem with this method is the difficulty in designing the impedances of the $\lambda/4$ sections and that the resistances must go over lines,

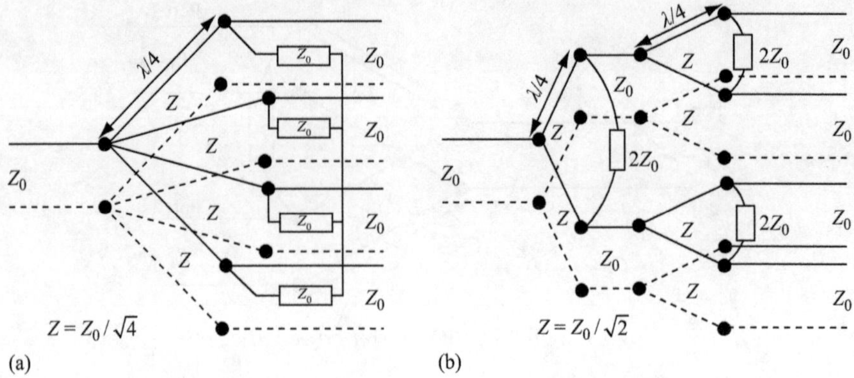

Figure 2.21 A 4-way equal split matched Wilkinson power divider: (a) basic method of implementation and (b) implementation by duplication of the 2-way divider

complicating its implementation by lithographic means. Another way is to simply repeat the divider as shown in Figure 2.21(b) for a 4-way divider. Obviously, this method is suited for division by integer values and in particular for 2^n, $n = 1, 2, 3, \ldots$ divisions. It is also possible to divide power unequally.

Because the power dividers are resistive, they are also linear and bidirectional. That is, they can also serve as power combiners. In all cases however, one must take into account the losses and the power dissipated in the divider itself. For example, in Figure 2.19, the power division was equal but each of the two output lines only propagates 1/4 of the input power whereas half the power was dissipated on the resistors.

2.9.3 Directional couplers

Given a transmission line (or waveguide) that transfers power from an input port to an output port, it is sometime necessary to couple some of that power into a second transmission line. The directional coupler is a little more than that; it couples power in a particular direction while still transferring power from the input port to the output port. The need for directional couplers arises from the fact that sometimes one needs to separate waves on a transmission line. An example is afforded by the need to measure the reflected wave on a line. This is shown in Figure 2.22(a). The incident wave propagates from port 1 (input port) to port 2, also called the through port. The reflected wave propagates in the opposite direction and is available on port 3 which is called the coupled port. The coupled port 3 must also be isolated from the input port 1. Thus, the coupler is a 4-port device, shown schematically in Figure 2.22(b). Port 4 is isolated. The coupler can be bidirectional in which case propagation is from port 2 to port 1 and the reflection is available on port 4, with port 3 isolated [dotted arrows in Figure 2.22(b)].

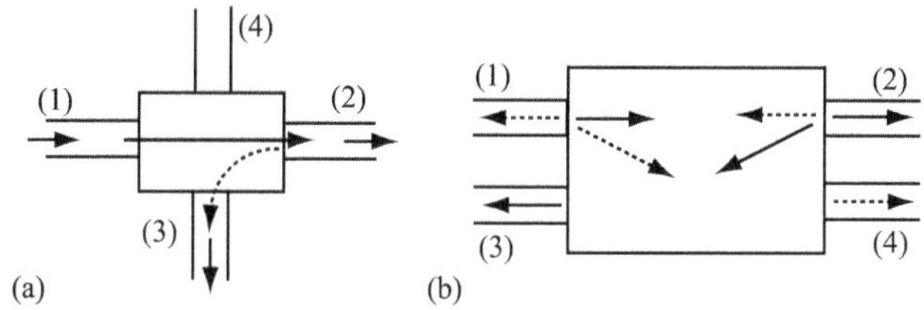

Figure 2.22 (a) Operating principle of the coupler and (b) schematic view of the coupler. Dotted arrows show the reverse operation for a bidirectional coupler

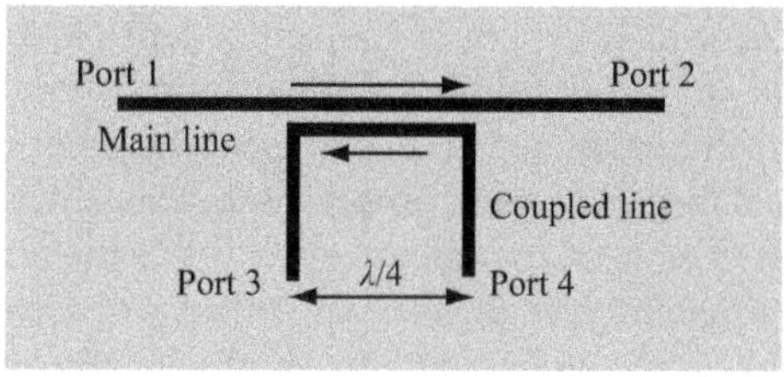

Figure 2.23 Schematic view of a transmission line coupler

In waveguides, coupling is done across a wall common to two waveguides through a hole or aperture. Depending on the location of the aperture or the relative angle between the two waveguides, it is possible to produce wave cancelation at the isolated port (two waves of equal amplitude but out of phase) and to enhance the waves at the coupled port (two waves in phase). In transmission lines, the coupling is done by two unshielded transmission lines in close proximity so that the fields produced by one line coupled into the second line. This is shown schematically in Figure 2.23. The two conductors are above a ground conductor so that the two lines have a common ground indicated by the dark plane in the figure. Under these conditions, it is possible to design the coupler with arbitrary coupling and with a properly isolated port. The $\lambda/4$ design shown guarantees that power in the coupled line propagates in the opposite direction to that in the main line hence only the reflected power will be present in the coupled line, propagating into port 3, leaving port 4 isolated as required. The schematic structure in Figure 2.23 is best implemented with striplines but it can be implemented with other types of lines.

There are many ways of implementing couplers and many types of couplers that are not discussed here. For the purpose of understanding their use in measurements, especially as they are used in network analyzers, this short description is sufficient.

2.9.4 Antennas and probes

Antennas are passive devices intended to generate and propagate power into a space such as into air but, of course, they can be embedded in dielectrics as well.

With few exceptions, antennas are designed for their far field properties, that is, the interest is in the fields generated by antennas at large distances. In this context that means many wavelengths away from the antenna. The near field, that is, the field in the immediate vicinity of the antenna is of secondary importance (again, with some exceptions). The antenna itself can be from a fraction of a wavelength in size to many wavelengths. It can take many physical shapes, including simple straight wires (electric dipoles or monopoles), loops (magnetic dipoles), conical, spiral, horn, patches, apertures, and almost any imaginable shape, each with its unique properties and performance characteristics. Many antennas are resonant, meaning that their frequency response is related to their electrical size, whereas other antennas are broadband operating over a wide frequency range. In general, the impedance of the antenna, the radiated power, and other properties are also related to their electrical size, which in turn is frequency dependent. As is well known, antennas serve an important function in communication with other applications in direction finding and in testing and measurements. It is not the intention in this section to discuss antennas and their properties beyond this general description simply because we have no need for them in this work in the true sense of antennas.

However, two particular types of antennas, called probes, are very important in microwaves and in this work. Probes are very short wire antennas or small loops as shown in Figure 2.24. Both of these are immediately recognized as an electric monopole antenna and a magnetic or loop dipole antenna. These names come from their field distribution, again as shown in Figure 2.24, where the electric field of the short wire antenna is similar to a point charge above a conductor (hence the name monopole), whereas the magnetic field of the loop is similar to a short magnet (a magnetic dipole). The dimensions are small compared to the wavelength meaning that they are not the best for applications such as communication except for very short distances. On the other hand, the electric probe is useful as a source of electric fields, whereas the loop generates magnetic fields, which can be used to couple power into waveguides and cavity resonators. The probes can also be used to sense electric and magnetic fields in various measurements. The fact that the probes are small means that they will have negligible effect on the structures they

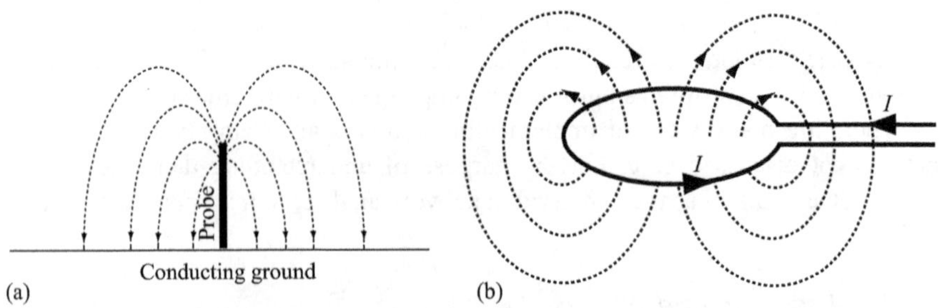

Figure 2.24 Electric and magnetic probes: (a) the electric probe with its electric field and (b) the magnetic probe and its magnetic field

couple power into or on the fields they are measuring. In some cases, the loop may only be a section of a circle and may in fact not be circular.

2.9.5 Attenuators

Attenuators are two-port passive devices whose function is to attenuate power. Thus, for example, given an input power P_i, the output power from the attenuator may be $P_i - 10$ dB, that is, the attenuator reduces the power by 10 dB. Attenuators are widely used in measurements to extend the dynamic range of the measuring equipment, in impedance matching, in amplifiers to control the output, and in many other applications. Attenuation is achieved by dissipating part of the input power so that the output power is at the level required by the device. One of the most obvious methods of constructing an attenuator is to use a lossy transmission line in which the attenuation depends on the length of the line. The problem with this is that the attenuator may be physically long, especially if high attenuation is needed and the attenuation is usually frequency dependent. Another method is to reflect part of the power, leading to reflection attenuators. There are other methods to achieve the same effect of which the use of resistors that are physically small compared to the wavelength is the simplest and one of the most common. As long as the resistors are small compared to the wavelength, they can be treated as lumped values, and the attenuation only depends on the lumped resistance values. In its simplest form, one can think of an attenuator as a simple resistive divider as shown in Figure 2.25. This particular attenuator is called an L-pad attenuator and may be used, for example, for impedance matching. Although not commonly used as power attenuators, it is very simple and forms the basis for other attenuators.

Given a voltage V_1 across port 1, the voltage across port 2 can be calculated from the voltage divider:

$$V_2 = V_1 \frac{R_2 \,||\, Z_{\text{out}}}{R_1 + R_2 \,||\, Z_{\text{out}}} \tag{2.156}$$

The input power is $V_1{}^2/Z_{\text{in}}$ and the output power is

$$P_{\text{out}} = \frac{V_2^2}{Z_{\text{out}}} = \frac{V_1^2}{Z_{\text{out}}}\left(\frac{R_2 \,||\, Z_{\text{out}}}{R_1 + R_2 \,||\, Z_{\text{out}}}\right)^2 \tag{2.157}$$

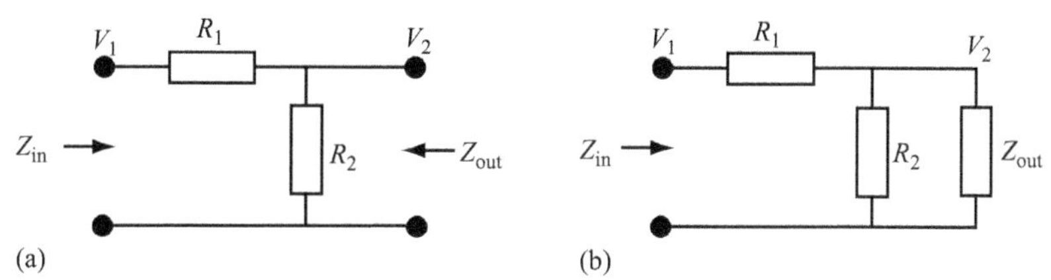

Figure 2.25 (a) L attenuator and (b) equivalent circuit used to calculate the voltage on port 2

The attenuation is

$$A = 10\log_{10}\frac{P_{out}}{P_{in}} = 10\log_{10}\left[\frac{Z_{in}}{Z_{out}}\left(\frac{R_2 \,||\, Z_{out}}{R_1 + R_2 \,||\, Z_{out}}\right)^2\right] \text{ (dB)} \tag{2.158}$$

Given the input and output impedances (i.e., the impedances of the lines connected to ports 1 and 2) and the required attenuation A, one can select the resistances R_1 and R_2 to accomplish the required attenuation. In most cases, it will also be necessary that the attenuator be matched to both the input and output lines. That means the following:

$$R_1 + R_2 \,||\, Z_{out} = Z_{in} \tag{2.159}$$

$$(R_1 + Z_{in}) \,||\, R_2 = Z_{out} \tag{2.160}$$

These then provide the necessary relations to design the attenuator in Figure 2.25. In most cases, $Z_{in} = Z_{out}$ (typically 50 Ω), simplifying somewhat the calculations.

In practical attenuators, it is often desired that the device be symmetric, that is, that attenuation in both direction is identical (the attenuator in Figure 2.25 is not symmetric). In addition, the attenuators may be balanced or unbalanced. In an unbalanced attenuator, the resistive elements are connected only to one conductor of the transmission line, whereas the ground conductor contains no resistive elements. A balanced attenuator will have resistors in the ground conductor identical to the resistors in the other conductor. The attenuator in Figure 2.25 is a nonsymmetric, unbalanced attenuator. Figure 2.26 shows the unbalanced classical T and π attenuators, whereas Figure 2.27 shows the equivalent balanced T and π

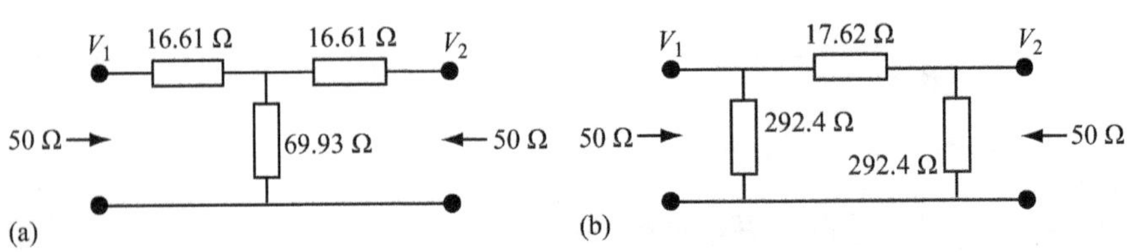

Figure 2.26 (a) 6-dB unbalanced T attenuator and (b) 3-dB unbalanced π attenuator

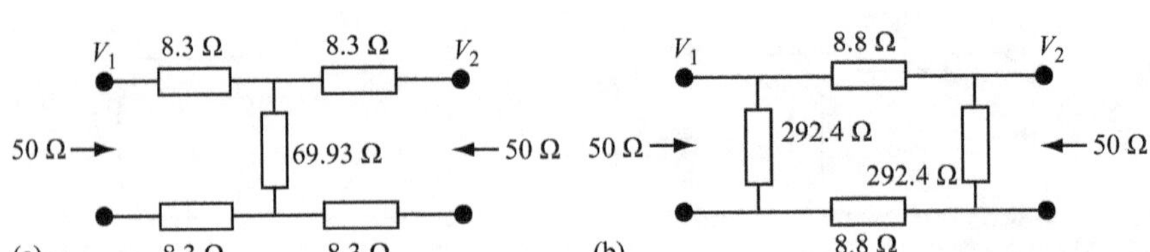

Figure 2.27 (a) 6-dB balanced T (or H) attenuator and (b) 3-dB balanced π (or O) attenuator

attenuators also called H and O attenuators (respectively) for obvious visual reasons. The T attenuators are shown with resistance values for 6 dB attenuation whereas the π attenuators are shown for 3-dB attenuation. Note that the attenuators in Figures 2.26 and 2.27 are clearly symmetric. Both the T and H attenuators can be designed for nonequal input and output impedances if necessary. Note also that the series resistance in the balanced T and π attenuators is half the series resistance of the respective unbalanced attenuator.

Attenuators come in many varieties. Fixed attenuators are designed with fixed resistances and are produced in commonly used values from about −3 to about −100 dB and for line impedances from 50 to 600 Ω. Variable attenuators use variable resistances to obtain attenuation within a given range. If the variable resistors are replaced with fixed resistors and switches, one obtains step attenuators. Attenuators are also rated by power. Because power is dissipated in the resistive elements, these must be able to dissipate the amount of power required, hence they can be physically large (a 3-dB attenuator will dissipate half the input power).

It should also be noted that the resistors are assumed to be ideal. Any deviation from that will introduce reactance in the circuits leading to degradation of performance including changes in phase. Resistances must be accurate or the attenuation will change representing an error from the nominal value of the attenuator and mismatch may occur at the input and output of the device leading to unwanted reflections.

2.9.6 Other circuits

There are of course many other passive devices, mirroring the devices one encounters in low frequency circuits. These include filters, phase shifters, isolators, circulators, terminators, and many more.

There are of course active devices as well, either specific to the microwave range or adapted from low frequency devices. These include tube devices such as klystrons and traveling wave tubes as well as semiconductor devices such as diodes, transistors, amplifiers, and integrated circuits. These will not be discussed here as they are out of the scope of the present work.

2.10 Transmission line resonators

Of particular importance in this work is the possibility of building resonant transmission lines. These resonant structures can then be used in a manner similar to other resonant structures such as closed resonant cavities based on shorted waveguide sections or circuit LC resonators. The equivalent circuit of resonators and hence of transmission line resonators are *RLC* circuits and are shown in Figure 2.28. Figure 2.28(a) and (b) shows generic series and parallel lossless resonators, whereas Figure 2.28(c) and (d) shows lossy resonators. These are easily recognized as circuits, but from the discussion of transmission line circuits in the previous section, it is easy to see how transmission lines may be analyzed in similar fashion. For example, using Figure 2.28(b), one can easily implement this as two stubs (shorted or open) in parallel. We will do so later in this chapter.

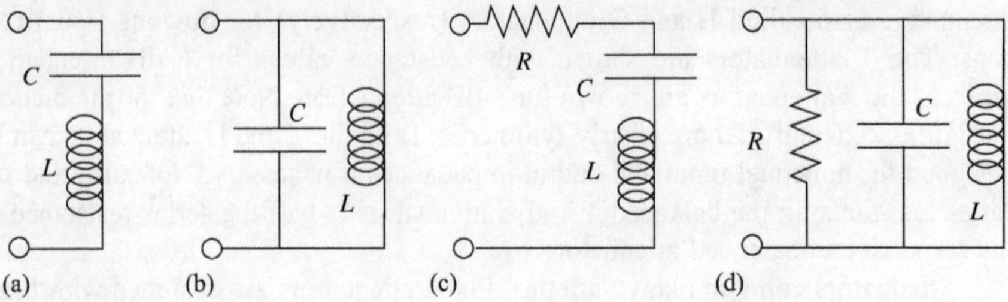

Figure 2.28 Resonating circuits: (a) series resonator, (b) parallel resonator, (c) lossy series resonator, and (d) lossy parallel resonator

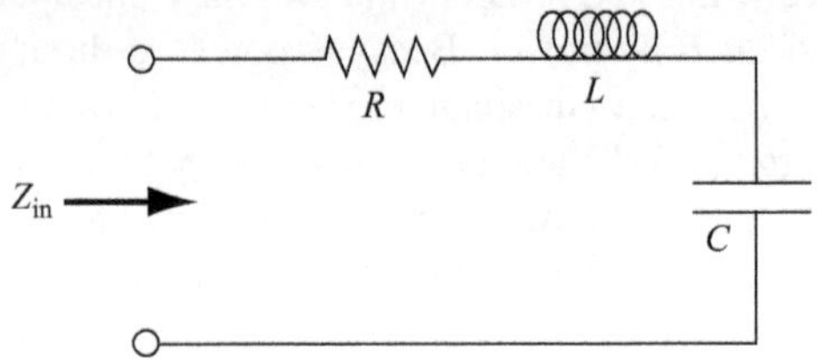

Figure 2.29 Series RLC circuit and its input impedance. The circuit is fed from a voltage source

2.10.1 The concept of resonance

There are two basic ways of defining resonance. One is in terms of circuit elements and the other in terms of energy. Both methods yield identical results, but one or the other is more useful depending on usage. In the case of transmission lines, the circuit approach is more convenient because circuit elements are easily defined, whereas the fields and the energy extend throughout the space around the lines and hence not easily accounted for (except, of course for coaxial lines). On the other hand, in the case of microwave cavity resonators, where the circuit elements are not obvious, the energy approach is preferred since the fields and energy are contained within the cavity between the conducting walls. We discuss both methods here starting with the circuit theory approach and then apply them to transmission line resonance. In particular, we show that the energy approach is useful for analysis especially when transmission line resonators are shielded.

We start with general circuits and only then relate these to transmission lines. This approach simplifies understanding and does not exclude generality.

2.10.2 The series RLC circuit

We analyze the series *RLC* circuit in terms of circuit elements and in terms of energy simply to show that both methods produce identical results. Consider the series *RLC* circuit in Figure 2.29. The input impedance is

$$Z_{in} = R + j\omega L + \frac{1}{j\omega C} = R + j\left(\omega L - \frac{1}{\omega C}\right) \tag{2.161}$$

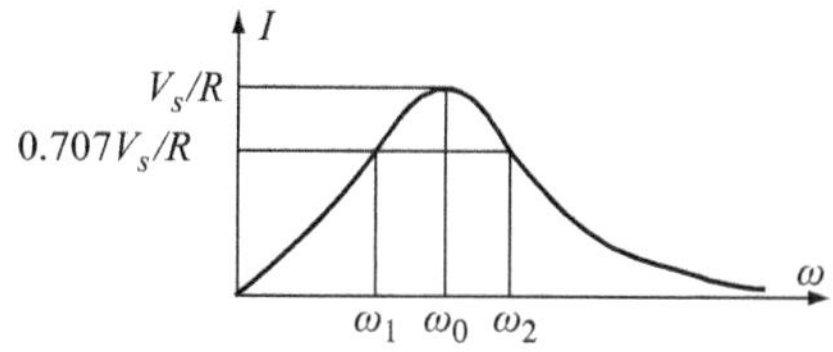

Figure 2.30 Response of the RLC circuit in Figure 2.18 for the current in the circuit. V_s is the voltage source amplitude that feeds the circuit

Resonance is defined as that frequency at which the imaginary part of the impedance (strictly speaking, of the transfer function of the circuit) is zero:

$$\omega_0 L - \frac{1}{\omega_0 C} = 0 \rightarrow \omega_0 = \frac{1}{\sqrt{LC}} \text{ (rad/s)} \tag{2.162}$$

or

$$f_0 = \frac{1}{2\pi\sqrt{LC}} \text{ (Hz)} \tag{2.163}$$

The resistance R does not affect the resonant frequency but it does dissipate power whereas an LC circuit does not.

The circuit exhibits a number of properties at resonance:

1. The impedance is purely resistive and its resistance equals R. In a lossless circuit, the input impedance is zero.
2. The magnitude of the input impedance is minimum at resonance
3. The source voltage V_s and the current I are in phase
4. The resistor voltage is equal to V_s but the voltages on the inductor and capacitor can be much larger than V_s.

The current amplitude in the circuit is shown in Figure 2.30. Consistent with the properties above, the current is maximum at resonance, whereas the voltage is minimum. Note also that an ideal resonant circuit behaves as a short circuit with infinite current and zero voltage. Although the height of the response at resonance is defined by the resistance, the shape of the response or its bandwidth is defined by the so-called half-power frequencies ω_1 and ω_2, the frequencies at which the power dissipated by the circuit is half the power at resonance. At resonance the power dissipated is

$$P(\omega_0) = \frac{I^2 R}{2} \tag{2.164}$$

At ω_1 and ω_2 (see Figure 2.30), the dissipated power is half of that

$$P(\omega_1) = P(\omega_2) = \frac{I^2 R}{2} = \frac{V^2}{2R} \tag{2.165}$$

where V is the amplitude of the source. By setting the impedance in (2.161) to $\sqrt{2}R$, we obtain

$$\sqrt{R^2 + \left(\omega L + \frac{1}{j\omega C}\right)^2} = \sqrt{2}R \tag{2.166}$$

Solving this for the half-power frequencies, we obtain

$$\omega_1 = -\frac{R}{2L} + \sqrt{\left(\frac{R}{2L}\right)^2 + \frac{1}{LC}} \tag{2.167}$$

$$\omega_2 = \frac{R}{2L} + \sqrt{\left(\frac{R}{2L}\right)^2 + \frac{1}{LC}} \tag{2.168}$$

By convention, the difference between the two half-power frequencies is called the bandwidth of the resonant circuit:

$$\text{BW} = \omega_2 - \omega_1 = \frac{R}{L} \tag{2.169}$$

The lower the resistance, the narrower the bandwidth. An ideal LC circuit will have zero bandwidth and infinite current amplitude. Since the bandwidth depends on R and L, we can also relate it to reactive (stored) energy and dissipated energy in the circuit. To do so, we define a quantity called quality factor (Q-factor) as the ratio of peak stored energy to energy dissipated in one period at resonance:

$$Q = 2\pi \frac{\text{peak stored energy}}{\text{energy dissipated/per period}} \tag{2.170}$$

Because at resonance the reactive energy oscillates between the inductor and the capacitor, we can calculate the peak energy stored and the energy dissipated in one cycle as

$$W_s = \frac{LI^2}{2}, \ W_d = \frac{I^2R}{2f} \tag{2.171}$$

where f is the frequency. Thus,

$$Q = 2\pi f_0 \frac{LI^2/2}{I^2R/2} = \frac{\omega_0 L}{R} \tag{2.172}$$

where f_0 is the resonant frequency. Because the peak energy stored in the capacitor equals the peak energy stored in the inductor, we also have

$$Q = \frac{\omega_0 L}{R} = \frac{1}{\omega_0 CR} \tag{2.173}$$

Now, the bandwidth can be written in terms of the resonant frequency and the quality factor:

$$\mathrm{BW} = \frac{\omega_0}{Q} \tag{2.174}$$

Again, consistent with the above, we can say that an ideal LC circuit has infinite quality factor and hence, the larger the losses in the circuit, the lower the quality factor.

The energy approach is very similar but perhaps more intuitive. Starting with the Poynting theorem, the input power into the circuit may be written as

$$P_{\text{in}} = P_{\text{loss}} + j2\omega(W_m - W_e) \tag{2.175}$$

where W_m and W_e are the time averaged magnetic and time averaged electric energy stored in the inductance and capacitance, respectively.

$$W_m = \frac{L|I|^2}{4}, \; W_e = \frac{C|V_c|^2}{4} = \frac{C|I/\omega C|^2}{4} = \frac{|I|^2}{4\omega^2 C} \tag{2.176}$$

where V_c is the voltage across the capacitor, I is the current in the circuit, and we have used the relation between current in and voltage across the capacitor ($I_C = j\omega C V_C$).

The input impedance is the input power divided by the RMS current squared:

$$Z_{\text{in}} = \frac{P_{\text{in}}}{(1/\sqrt{2})^2} = \frac{P_{\text{loss}} + j2\omega(W_m - W_e)}{I^2/2} \tag{2.177}$$

At resonance, the stored magnetic and stored electric energies are equal, hence,

$$W_m - W_e = \frac{L|I|^2}{4} - \frac{|I|^2}{4\omega_0^2 C} = 0 \rightarrow \omega_0 = \frac{1}{\sqrt{LC}} \tag{2.178}$$

and, at resonance

$$Z_{\text{in}} = \frac{P_{\text{loss}}}{I^2/2} = \frac{I^2 R/2}{I^2/2} = R \tag{2.179}$$

These are the same as those in (2.162) and (2.161), respectively. The Q-factor can be obtained in a similar manner using (2.170).

Since the resonant frequency as well as half power frequencies and bandwidth are related to the impedance of the circuit, it is often useful to analyze the impedance of the circuit in the vicinity of resonance. Rewriting the circuit impedance at a frequency $\omega = \omega_0 + \Delta\omega$ where $\Delta\omega$ is small:

$$\begin{aligned} Z_{\text{in}} &= R + j\left(\omega L - \frac{1}{\omega C}\right) = R + j\omega L\left(1 - \frac{1}{\omega^2 LC}\right) = R + j\omega L\left(1 - \frac{\omega_0^2}{\omega^2}\right) \\ &= R + j\omega L\left(\frac{\omega^2 - \omega_0^2}{\omega^2}\right) \end{aligned} \tag{2.180}$$

Expanding the term $\omega^2 - \omega_0^2$:

$$\omega^2 - \omega_0^2 = (\omega - \omega_0)(\omega + \omega_0) = \Delta\omega(2\omega - \Delta\omega) \simeq 2\omega\Delta\omega \tag{2.181}$$

where the fact that $\Delta\omega$ is small was invoked. Substituting this into (2.180),

$$Z_{\text{in}} \simeq R + j2L\Delta\omega \tag{2.182}$$

This form will be particularly useful in calculation of resonant frequencies of transmission lines, but it can also be used to simplify the calculation of half power frequencies. As mentioned above [see (2.166)], at half power the magnitude of the impedance equals $\sqrt{2}R$. Hence,

$$Z_{\text{in}} \simeq |R + j2L\Delta\omega| = \sqrt{2}R \tag{2.183}$$

or

$$R^2 + 4L^2(\Delta\omega)^2 = 2R^2 \rightarrow \Delta\omega = \frac{R}{2L} \tag{2.184}$$

That is, the half power frequencies are

$$\omega_1 = \omega_0 - \Delta\omega = \omega_0 - \frac{R}{2L} \tag{2.185}$$

$$\omega_2 = \omega_0 + \Delta\omega = \omega_0 + \frac{R}{2L} \tag{2.186}$$

These look somewhat different than the half power frequencies in (2.167) and (2.168) but, under the assumption that $\Delta\omega$ is small, (2.185) and (2.186) are good approximations to (2.167) and (2.168).

Note also that the bandwidth is $2\Delta\omega$ or

$$\text{BW} = 2\Delta\omega = \frac{R}{L} \tag{2.187}$$

Similarly, the Q-factor may be written as

$$Q = \frac{\omega_0}{\text{BW}} = \frac{\omega_0}{2\Delta\omega} = \frac{\omega_0 L}{R} \tag{2.188}$$

as previously obtained.

One can rewrite the half-power frequencies and the bandwidth in terms of the resonant frequency and the Q-factor as

$$\omega_1 = \omega_0 - \frac{\omega_0}{2Q}, \ \omega_2 = \omega_0 + \frac{\omega_0}{2Q}, \ \text{BW} = \frac{\omega_0}{Q} \tag{2.189}$$

Equations (2.167) and (2.168) may be similarly treated. Another modification that is useful, particularly in evaluation of the Q-factor in low loss circuits is to replace the resonant frequency ω_0 by a complex effective resonant frequency ω_c defined as

$$\omega_c = \omega_0\left(1 + \frac{j}{2Q}\right) \tag{2.190}$$

where ω_0 is the real resonant frequency (of the unperturbed cavity). The reason to do so is that one can then start with the solution for the lossless case (always easier to obtain) and then replace the resonant frequency with the complex effective resonant frequency to take into account the loss. For example, the impedance of the lossless circuit in is ($R = 0$):

$$Z_{\text{in}} = j2L(\omega - \omega_0) \tag{2.191}$$

Now replacing ω_0 with ω_c:

$$Z_{\text{in}} = j2L\left(\omega - \omega_0 - \frac{j\omega_0}{2Q}\right) = \frac{\omega_0 L}{Q} + j2L(\omega - \omega_0) = R + j2L\Delta\omega \tag{2.192}$$

The latter is exactly (2.182).

This is a perturbation method and is often used in analysis. We will discuss the perturbation method and its application to resonant cavities in the context of coupled transmission line resonators in Chapter 3.

2.10.3 Parallel resonant circuit

The basic parallel resonant circuit is shown in Figure 2.31. Although we could use the circuit parameters here as for the series resonant circuit, there is no need to repeat the process since the energy approach produces the same results. Starting with admittances of the components we write for the input impedance:

$$Z_{\text{in}} = \frac{1}{1/R + 1/j\omega L + j\omega C} \tag{2.193}$$

The input complex power into the circuit is

$$P_{\text{in}} = \frac{VI^*}{2} = \frac{|V|^2}{2Z_{\text{in}}^*} = \frac{|V|^2}{2}\left(\frac{1}{R} + \frac{j}{\omega L} - j\omega C\right) \tag{2.194}$$

The first term is the power loss in the resistance, the second the reactive magnetic power, and the third the reactive electric power. We can rewrite the latter two terms in terms of time averaged stored magnetic and electric energies by noting that the stored electric energy is

$$W_e = \frac{C|V|^2}{4} \tag{2.195}$$

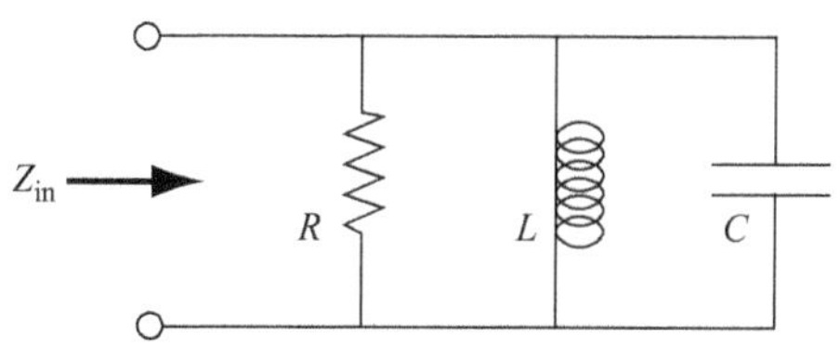

Figure 2.31 A parallel resonant circuit and its input impedance. The circuit is fed from a current source

and the time averaged stored magnetic energy is

$$W_m = \frac{|I_L|^2 L}{4} = \frac{|V|^2}{4\omega^2 L} \tag{2.196}$$

The power may now be written as

$$P_{\text{in}} = \frac{|V|^2}{2R} + j\frac{|V|^2}{2\omega L} - j\frac{|V|^2 \omega C}{2} = \frac{|V|^2}{2R} + j2\omega\left(\frac{|V|^2}{4\omega^2 L} - \frac{|V|^2 C}{4}\right)$$

$$= P_{\text{loss}} + j2\omega(W_m - W_e) \tag{2.197}$$

This relation is identical to (2.175). Since at resonance $W_m = W_e$, we get the resonant frequency as

$$\frac{|V|^2}{4\omega_0^2 L} - \frac{|V|^2 C}{4} \rightarrow \omega_0 = \frac{1}{\sqrt{LC}} \tag{2.198}$$

The impedance of the circuit may be written in terms of power as

$$Z_{\text{in}} = \frac{P_{\text{loss}} + j2\omega(W_m - W_e)}{I^2/2} \tag{2.199}$$

At resonance this is equal to

$$Z_{\text{in}} = \frac{P_{\text{loss}}}{I^2/2} = R \tag{2.200}$$

These relations are the same as for the series resonance, but we note that although a low loss series resonant circuit implies low series resistance, a low loss parallel resonator implies high parallel resistance (infinite in the ideal case).

The Q-factor is calculated from the definition in (2.170):

$$Q = \omega_0 \frac{2W_m}{P_{\text{loss}}} = \frac{R}{\omega_0 L} = \omega_0 RC \tag{2.201}$$

We note that in this case, the higher the resistance, the higher the Q-factor as one would expect from a parallel circuit. The bandwidth of the resonator is proportional to the reciprocal of Q:

$$\text{BW} = \frac{\omega_0}{Q} = \frac{L}{R} = \frac{1}{RC} \tag{2.202}$$

Assuming that the bandwidth is small, we can calculate the half power frequencies as

$$\omega_1 = \omega_0 - \frac{L}{2R} = \omega_0 - \frac{1}{2RC} \tag{2.203}$$

$$\omega_2 = \omega_0 + \frac{L}{2R} = \omega_0 + \frac{1}{2RC} \tag{2.204}$$

Again, as in (2.172), these can be rewritten in terms of the Q-factor:

$$\omega_1 = \omega_0 - \frac{\omega_0}{2Q}, \ \omega_2 = \omega_0 + \frac{\omega_0}{2Q}, \ \mathrm{BW} = \frac{\omega_0}{Q} \tag{2.205}$$

More accurate values that do not assume small bandwidth are obtained from the impedance relation by noting that the magnitude of the impedance at the half power points must equal $\sqrt{2}R$ as was done to obtain the half power frequencies in (2.167) and (2.168). Alternatively, by observing the duality between (2.161) and (2.193), we can simply replace R by $1/R$, L by $1/L$, and C by $1/C$ in (2.167) and (2.168) and obtain

$$\omega_1 = \sqrt{\left(\frac{1}{2RC}\right)^2 + \frac{1}{LC}} - \frac{1}{2RC} \tag{2.206}$$

$$\omega_2 = \sqrt{\left(\frac{1}{2RC}\right)^2 + \frac{1}{LC}} + \frac{1}{2RC} \tag{2.207}$$

These too can be rewritten in terms of the Q-factor.

Clearly, as R approaches infinity, these relations approach (2.203) and (2.204), respectively. The response of the parallel resonant circuit is shown in Figure 2.32. It is similar to the series circuit response except of course that here it is the voltage that changes with frequency and hence the impedance peaks at resonance.

The perturbation method can be applied here as well, but we will not pursue this since the results are the same and there is little value in "proving" that one can use the perturbation method.

Aside from the fact that a resonator resonates at a given frequency, the most important characteristic property is the Q-factor of the resonator. The Q-factor as calculated from the definition in (2.170) only depends on the properties of the resonator itself. When the resonator is connected to external circuits, as it must be, these external circuits "load" the resonator, that is, they add losses to the resonator by drawing currents. Necessarily, the effective Q-factor will be lower. To account for this, the Q-factor calculated in (2.170) is called the unloaded Q-factor if the cavity is unloaded; however, in practice, (2.170) is the loaded quality factor Q_L

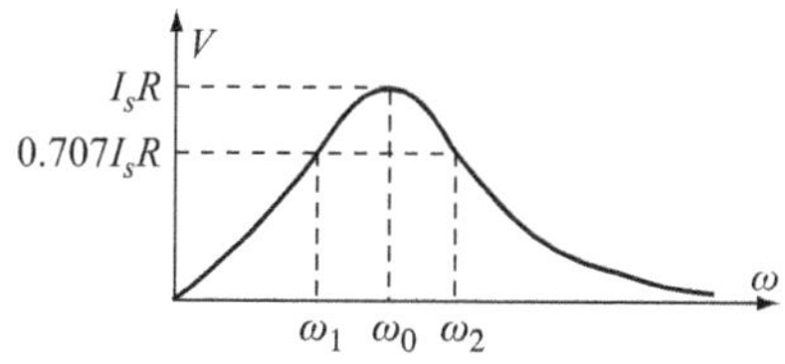

Figure 2.32 Response of the parallel RLC circuit in Figure 2.31 for voltage in the circuit. I_s is the current source amplitude that feeds the circuit

since the bandwidth measured depends on the loading. Thus, we use the loaded quality factor as

$$Q_L = \frac{\omega_0}{\text{BW}} = \frac{\omega_0}{\omega_2 - \omega_1} = \frac{f_0}{f_u - f_l} \tag{2.208}$$

where f_0 is the resonant frequency, f_u the upper 3 dB frequency, and f_l the lower 3 dB frequency so that $f_u - f_l$ is the bandwidth of the cavity resonator (or any resonant circuit). Again, it should be noted that this equals the unloaded quality factor if the cavity is not loaded.

There are two additional definitions for quality factor in addition to that in (2.170):

Loaded quality factor is defined as

$$Q_L = 2\pi \frac{\text{energy stored in the cavity}}{\text{energy lost in the cavity per cycle} + \text{energy lost in the external circuit per cycle}} \tag{2.209}$$

The external quality factor is defined as

$$Q_e = 2\pi \frac{\text{energy stored in the cavity}}{\text{energy lost in the external circuit per cycle}} \tag{2.210}$$

To calculate the external quality factor a load resistance R_L due to the external circuit is postulated which accounts for the loading and (2.172) or (2.201) is used to define an external Q-factor Q_e:

$$Q_e = \frac{\omega_0 L}{R_L} \text{ for series resonators} \tag{2.211}$$

$$Q_e = \frac{R_L}{\omega_0 L} \text{ for parallel resonators} \tag{2.212}$$

where R_L is the load resistance. The relation between the three quality factors is

$$\frac{1}{Q_L} = \frac{1}{Q_0} + \frac{1}{Q_e} \text{ or : } Q_L = \frac{Q_e Q_0}{Q_e + Q_0} \tag{2.213}$$

In this relation, the unloaded cavity was denoted as Q_0 to distinguish it from Q_e and Q_L.

In practical calculations or in measurements, Q_L is found from (2.208) and Q_e either from (2.211) or (2.212). Then, Q_0 is found from (2.213).

2.11 Series and parallel transmission line resonators

Now that the properties of resonant circuits have been established, we apply these concepts to transmission line resonators. Although these resonators may look very different in structure because they are essentially "geometric" structures formed by shorted and/or open transmission line sections, they operate exactly the same as any resonator. To do so, these sections represent series or parallel capacitors and inductors and, in lossy resonators, they also represent resistances. The conditions for resonance remain the same and the behavior at resonance as well.

There are two basic ways of building a transmission line resonator. The first method exploits the fact that at resonance, the input impedance of a series resonator is purely resistive and minimum [see (2.179) or (2.200)] or, if the resonator is lossless, zero. This means that any shorted transmission line of length $n\lambda/2$ ($n = 1, 2, 3, \ldots$) will resonate. Similarly, any open transmission line of length $n\lambda/4$ ($n = 1, 3, 5, \ldots$) must also resonate (see Sections 2.7.5 and 2.7.6 for properties of shorted and open transmission lines). Both of these produce minimum (zero) input impedance and hence are series resonators. Parallel resonators can be built using $n\lambda/2$ ($n = 1, 2, 3, \ldots$) open-circuit transmission lines or $n\lambda/4$ ($n = 1, 3, 5, \ldots$) short-circuited transmission lines since these lines will have maximum (infinite) input impedance. A second method of building a transmission line resonator is based on the fact that resonance is due to the capacitance and inductance [see (2.162) or (2.198)]. Based on this and the fact that shorted and open transmission line stubs have specific capacitances or inductances, it suffices to connect together two stubs of different lengths for the circuit to resonate. A parallel resonator can be built by connecting two line sections in parallel, with one line having a capacitive input impedance, the other with inductive input impedance and adjusting the lengths of the line sections to obtain the required resonant frequency (see Sections 2.7.5 and 2.7.6; Figures 2.11 and 2.12). The line sections can be, in principle, connected in series although parallel connection of transmission line sections is more practical. The resulting circuit is often called "tapped transmission line resonator." These two approaches to transmission line resonators are discussed next in some detail, starting with the impedance approach.

2.11.1 *Short-circuited λ/2 transmission line resonator*

Consider the half-wavelength shorted transmission line in Figure 2.33. Using (2.78) with $Z_L = 0$, the input impedance to the line is

$$Z_{\text{in}} = Z_0 \tanh(\gamma l) \tag{2.214}$$

where $\gamma = \alpha + j\beta$ is the propagation constant, α the attenuation constant (due to losses on the line), and $\beta = 2\pi/\lambda$ the phase constant. Or

$$Z_{\text{in}} = Z_0 \tanh(\alpha + j\beta)l = Z_0 \frac{\tanh \alpha l + j \tan \beta l}{1 + j \tanh \alpha l \tan \beta l} \tag{2.215}$$

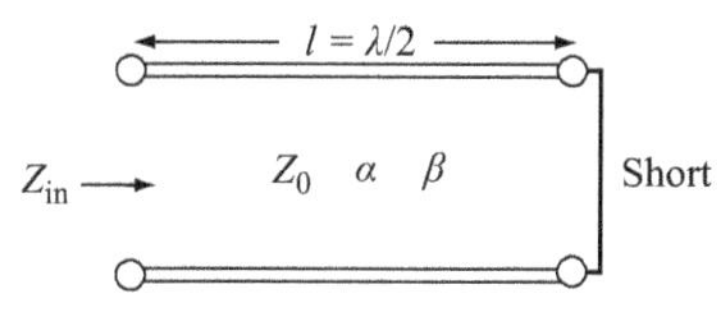

Figure 2.33 A half-wavelength shorted transmission line segment

We first note that at resonance $\beta l = (2\pi/\lambda)(\lambda/2) = \pi$, tan $\beta l = 0$, and hence,

$$Z_{\text{in}} = Z_0 \tanh \alpha l \tag{2.216}$$

The input impedance is real and clearly minimal since the attenuation constant in practical transmission line resonators is very small. In fact, we can approximate $\tanh \alpha l \approx \alpha l$ and write

$$Z_{\text{in}} \approx Z_0 \alpha l \tag{2.217}$$

This is then a series resonator with its equivalent resistance

$$R = Z_0 \alpha l = Z_0 \alpha \frac{\lambda}{2} \tag{2.218}$$

If we assume a frequency ω near resonance so that $\omega = \omega_0 + \Delta\omega$ where $\Delta\omega$ is small, then from the definition of the phase constant we have

$$\beta = \frac{\omega}{v_p} = \frac{\omega_0}{v_p} + \frac{\Delta\omega}{v_p} \tag{2.219}$$

Thus,

$$\beta l = \frac{\omega_0 l}{v_p} + \frac{\Delta\omega l}{v_p} = \pi + \frac{\Delta\omega\pi}{\omega_0} \tag{2.220}$$

where we used the fact that $\lambda = \lambda/2$ at resonance and hence $\beta l = (2\pi/\lambda)(\lambda/2) = \pi$. Thus,

$$\tan \beta l = \tan\left(\pi + \frac{\Delta\omega\pi}{\omega_0}\right) = \tan\left(\frac{\Delta\omega\pi}{\omega_0}\right) \approx \frac{\Delta\omega\pi}{\omega_0} \tag{2.221}$$

where the fact that $\Delta\omega\pi/\omega_0$ is a small angle was used. With these, the input impedance in (2.215) is

$$Z_{\text{in}} \approx Z_0 \frac{\alpha l + j(\Delta\omega\pi/\omega_0)}{1 + j\alpha l(\Delta\omega\pi/\omega_0)} \approx Z_0 \alpha l + jZ_0 \frac{\Delta\omega\pi}{\omega_0} \tag{2.222}$$

where we used the fact that $\alpha l \ll 1$ and $\Delta\omega\pi/\omega_0 \ll 1$. The first term on the right-hand side is the resistance of the circuit [see (2.217)]. Comparing this relation with (2.182), we conclude that the equivalent inductance is

$$L = \frac{Z_0 \pi}{2\omega_0} \tag{2.223}$$

From the relation for resonant frequency in (2.162), the equivalent capacitance becomes

$$C = \frac{1}{\omega_0^2 L} = \frac{2}{Z_0 \omega_0 \pi} \tag{2.224}$$

Again, from equivalence with the series resonant circuit, the quality factor

$$Q = \frac{\omega_0 L}{R} = \frac{\pi}{2\alpha l} = \frac{\beta}{2\alpha} \tag{2.225}$$

The latter relates the Q-factor to the attenuation constant on the line. Bandwidth and half power frequencies are immediately available from (2.185) to (2.187):

$$\mathrm{BW} = \frac{R}{L} = \frac{Z_0 \alpha l}{Z_0 \pi / 2\omega_0} = \frac{\omega_0}{Q} = \omega_0 \frac{2\alpha}{\beta} \tag{2.226}$$

Hence, the half power frequencies are

$$\omega_1 = \omega_0 - \omega_0 \frac{\alpha}{\beta} \tag{2.227}$$

$$\omega_2 = \omega_0 + \omega_0 \frac{\alpha}{\beta} \tag{2.228}$$

2.11.2 Open-circuited λ/2 transmission line resonator

Consider the half-wavelength open transmission line in Figure 2.34. Using (2.78) with $Z_L \to \infty$, the input impedance to the line is

$$Z_{\mathrm{in}} = Z_0 \frac{1}{\tanh(\gamma l)} \tag{2.229}$$

where $\gamma = \alpha + j\beta$ is the propagation constant, α the attenuation constant (due to losses on the line), and $\beta = 2\pi/\lambda$ the phase constant. Or

$$Z_{\mathrm{in}} = Z_0 \frac{1}{\tanh(\gamma l)} = Z_0 \frac{1 + j\tan\beta l \tanh\alpha l}{\tanh\alpha l + j\tan\beta l} \tag{2.230}$$

At resonance $\beta l = (2\pi/\lambda)(\lambda/2) = \pi$, tan $\beta l = 0$, and hence,

$$Z_{\mathrm{in}} = \frac{Z_0}{\tanh\alpha l} \tag{2.231}$$

Since the attenuation constant is small, the input impedance is real and clearly larger than Z_0. Because αl is small, we can approximate $\tanh\alpha l \approx \alpha l$ and write

$$Z_{\mathrm{in}} \approx \frac{Z_0}{\alpha l} \tag{2.232}$$

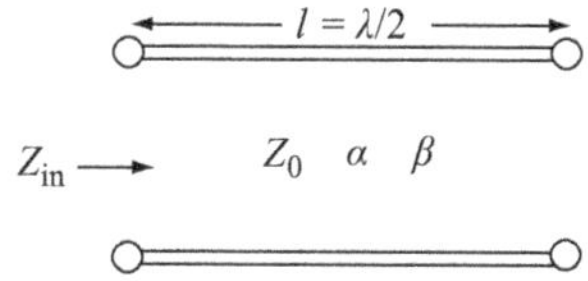

Figure 2.34 A half-wavelength open transmission line segment

This is therefore a parallel resonator with its equivalent resistance

$$R = \frac{Z_0}{\alpha l} = \frac{2Z_0}{\alpha\lambda} \tag{2.233}$$

At a frequency ω near resonance so that $\omega = \omega_0 + \Delta\omega$ where $\Delta\omega$ is small, then from the definition of the phase constant

$$\beta = \frac{\omega_0}{v_p} + \frac{\Delta\omega}{v_p} \to \beta l = \pi + \frac{\Delta\omega\pi}{\omega_0} \tag{2.234}$$

where again we used the fact that $l = \lambda/2$ at resonance, and hence, $\beta l = (2\pi/\lambda)(\lambda/2) = \pi$. Thus,

$$\tan\beta l = \tan\left(\pi + \frac{\Delta\omega\pi}{\omega_0}\right) = \tan\left(\frac{\Delta\omega\pi}{\omega_0}\right) \approx \frac{\Delta\omega\pi}{\omega_0} \tag{2.235}$$

where the fact that $\Delta\omega\pi/\omega_0$ is a small angle was used. With these, the input impedance in (2.230) is

$$Z_{\text{in}} = Z_0\frac{1}{\tanh(\gamma l)} = Z_0\frac{1}{\alpha l + j(\Delta\omega\pi/\omega_0)} = \frac{1}{(\alpha l/Z_0) + j(\Delta\omega\pi/Z_0\omega_0)} \tag{2.236}$$

Comparing this with the impedance of the parallel circuit resonator in (2.183), we write directly for the equivalent values:

$$R = \frac{Z_0}{\alpha l} \tag{2.237}$$

and

$$\frac{\Delta\omega\pi}{Z_0\omega_0} = \omega_0 C - \frac{1}{\omega_0 L} \tag{2.238}$$

or, since the impedance at resonance must be purely real,

$$C = \frac{\pi}{2\omega_0 Z_0}, \; L = \frac{1}{\omega_0^2 C} = \frac{2Z_0}{\omega_0\pi} \tag{2.239}$$

From equivalence with the parallel resonant circuit, the quality factor

$$Q = \frac{\omega_0 R}{L} = \frac{\omega_0(Z_0/\alpha l)}{2Z_0/\omega_0\pi} = \frac{\beta}{2\alpha} \tag{2.240}$$

Bandwidth and half power frequencies are immediately available from (2.237) to (2.239):

$$\text{BW} = \frac{L}{R} = \frac{2Z_0/\omega_0\pi}{Z_0/\alpha l} = \frac{2\alpha l}{\omega_0\pi} = \omega_0\frac{2\alpha}{\beta} \tag{2.241}$$

$$\omega_1 = \omega_0 - \omega_0\frac{\alpha}{\beta} \tag{2.242}$$

$$\omega_2 = \omega_0 + \omega_0\frac{\alpha}{\beta} \tag{2.243}$$

2.11.3 Additional properties of transmission line resonators

From the discussion in Sections 2.11.1 and 2.11.2, it is clear that the resonant frequency is defined by the length of the line, whereas the type of resonator is defined by the load (short or open). However, the situation is a bit more complex than that since we have assumed, at the outset, that the length of the line is $\lambda/2$ (or $\lambda/4$, depending of which type of resonator we start with). Given a length of transmission line, the resonant frequency in (2.239) or (2.224) is considered to be the fundamental mode of resonance. However, that is not the only resonant frequency possible. Suppose that we were to cut the resonator in Figure 2.33 or Figure 2.34 in two. The net effect is that the length l in (2.220) and (2.134) is reduced to $l/2$. The consequence of that is that the resonant frequency must be twice as high. That is, the line will resonate at $2\omega_0$. Similarly, extending the length of a line by a half wavelength will reduce the fundamental frequency to half its original value. In fact, a transmission line resonator of a given length will resonate in an infinite number of modes, both series and parallel. These properties arise simply from the fact that conditions on a transmission line repeat at intervals of half a wavelength.

From the properties of shorted and open lines, we recall that a $\lambda/4$ open transmission line has zero input impedance, whereas a $\lambda/4$ shorted transmission line has infinite input impedance. That means that a series transmission line resonator will resonate as a parallel transmission line at a frequency at which its length is $n\lambda/4$ ($n = 3, 5, 7, \ldots$) since under these conditions the input to the line behaves as that of an open $\lambda/2$ line. Similarly, a parallel transmission line resonator will resonate as a series transmission line resonator at a frequency at which the length of the line is $n\lambda/4$ ($n = 3, 5, 7, \ldots$). In both of these situations, the line behaves as an $n\lambda/2$ line, whereas the additional $\lambda/4$ length modifies the load.

To place these considerations in more concrete terms, one can view a transmission line resonator in terms of the natural frequencies of the transmission line section. Assuming a shorted transmission line section of length d, connected to an ideal voltage source (a generator with zero internal impedance), we can calculate the stored electric and magnetic energies as well as the dissipated time averaged power from which we can calculate the resonant frequencies and Q-factor in a process identical to the one discussed above, but now, we do not limit ourselves to the fundamental frequency. The voltage and current at any point on the line are found from (2.113):

$$V(z) = V^+\left(e^{j\beta z} - e^{-j\beta z}\right) = j2V^+\sin(\beta z) = j2V^+\sin\left(\frac{n\pi}{d}z\right) \qquad (2.244)$$

$$I(z) = \frac{V^+}{Z_0}\left(e^{j\beta z} + e^{-j\beta z}\right) = \frac{2V^+}{Z_0}\cos(\beta z) = \frac{2V^+}{Z_0}\cos\left(\frac{n\pi}{d}z\right) \qquad (2.245)$$

For a line of length d, the phase constant was written as $n\pi/d$ to ensure that the voltage is zero at the generator and at the load and the current is maximum as required for (series) resonance.

The time-average stored magnetic energy is due to the inductance per unit length, L:

$$W_m = \frac{1}{4}\int_{z=0}^{z=d} L|I(z)|^2 dz = \frac{1}{4}\int_{z=0}^{z=d} L\left(\frac{2V^+}{Z_0}\right)^2 \cos^2\left(\frac{n\pi}{d}z\right)dz = \left(\frac{2V^+}{Z_0}\right)^2 \frac{Ld}{8} \tag{2.246}$$

The time averaged stored electric energy is due to the capacitance per unit length, C:

$$W_e = \frac{1}{4}\int_{z=0}^{z=d} C|V(z)|^2 dz = \frac{1}{4}\int_{z=0}^{z=d} C(2V^+)^2 \sin^2\left(\frac{n\pi}{d}z\right)dz = (2V^+)^2 \frac{Cd}{8} \tag{2.247}$$

The dissipated power in the line is due to R and G or, more generally due to the attenuation constant:

$$\begin{aligned} P_{\mathrm{av}} &= \frac{1}{2}\int_{z=0}^{z=d} R|I(z)| + G|V(z)|^2 dz \\ &= \frac{1}{2}\int_{z=0}^{z=d} R|I(z)| + G|V(z)|^2 dz \\ &= \frac{1}{2}\int_{z=0}^{z=d} R\left|\frac{2V^+}{Z_0}\right|^2 + G|2V^+|^2 dz = (2V^+)^2 \frac{Rd/Z_0^2 + Gd}{4} \end{aligned} \tag{2.248}$$

Now, we can calculate the Q-factor and resonant frequency for the nth mode:

$$\begin{aligned} Q_n &= \frac{\omega_n(W_m + W_e)}{P_{\mathrm{av}}} = \frac{\omega_n\left((2V^+/Z_0)^2(Ld/8) + (2V^+)^2(Cd/8)\right)}{(2V^+)^2((Rd/Z_0{}^2 + Gd)/4)} \\ &= \frac{\omega_n/2((1/Z_0{}^2)L + C)}{(R/Z_0{}^2) + G} \end{aligned} \tag{2.249}$$

For a low loss transmission line, the characteristic impedance may be approximated as

$$Z_0 \approx \sqrt{\frac{L}{C}} \tag{2.250}$$

With this,

$$Q_n = \omega_n \frac{LC}{RC + GL} \tag{2.251}$$

and, since the input impedance (for an ideal transmission line) must vanish at resonances the resonant frequencies of the line are

$$\omega_n = \frac{1}{\sqrt{L_n C_n}} = \frac{n\pi}{d}\frac{1}{\sqrt{LC}}, \quad n = 0, 1, 2, 3, \ldots. \tag{2.252}$$

where L_n and C_n are the equivalent inductance and capacitance, respectively. From the magnetic energy, we have

$$\frac{1}{4}L_n|I|^2 = \frac{1}{8}Ld|I|^2 \;\rightarrow\; L_n = \frac{Ld}{2} \tag{2.253}$$

From (2.252) and (2.253),

$$C_n = \frac{2Cd}{n^2\pi^2} \tag{2.254}$$

and, for an *RLC* circuit [see (2.188)],

$$R_n = \frac{\omega_n L_n}{Q_n} = \frac{1}{2}\left(\frac{R}{L} + \frac{G}{C}\right)Ld \tag{2.255}$$

A proper model that accounts for the resonant frequencies and their corresponding Q-factors is shown in Figure 2.35. This is a lumped element model in which the resonators are in parallel, each representing a resonant frequency. The series resistance in each branch defines the Q-factor at that resonant frequency. Note that $n=0$ means $\omega_0 = 0$. This defines the DC response of the transmission line. Note that at DC the capacitance has been removed since based on (2.254) it becomes infinite and hence a short circuit.

From (2.252), it is clear that given a length, d, and a fundamental resonant frequency, ω_0, the higher order resonant frequencies occur at $2\omega_0$, $3\omega_0$, $\ldots$, $n\omega_0$.

The resonant frequencies and the model for an open transmission line may be obtained following similar arguments. In fact, all that is necessary is to recognize that for a resonant open line the current at the load and at the input to the line must be zero (for an ideal line), meaning that the impedance must be infinite. The current and voltage are [see (2.113)]:

$$V(z) = 2V^+\cos\left(\frac{n\pi}{d}z\right),\ I(z) = j\frac{2V^+}{Z_0}\sin\left(\frac{n\pi}{d}z\right) \tag{2.256}$$

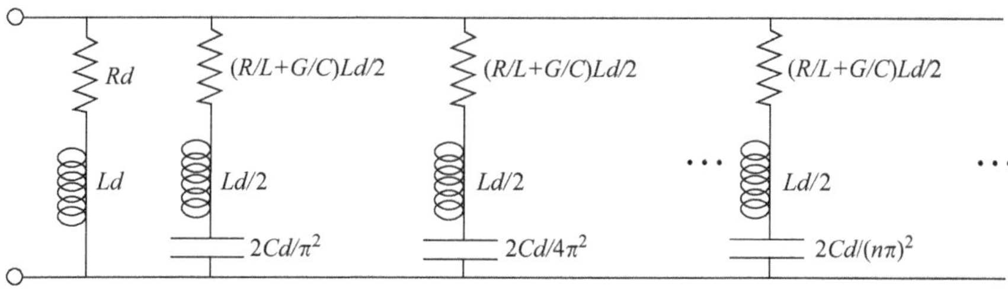

Figure 2.35 A lumped element model for the shorted transmission line that defines resonance and Q-factors for all modes of resonance

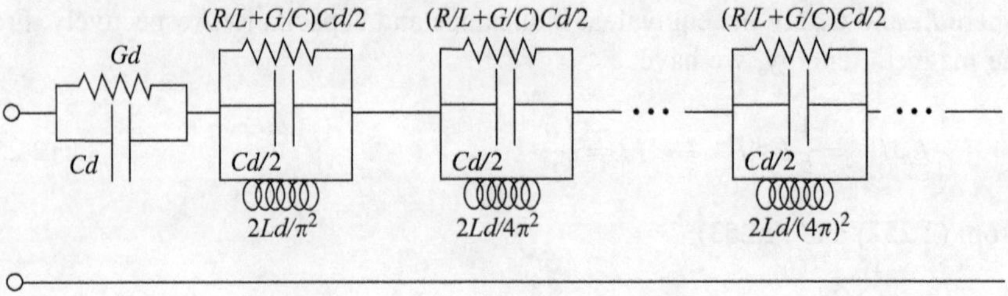

Figure 2.36 A lumped element model for the open transmission line that defines resonance and Q-factors for all modes of resonance

Comparing these to the current and voltage for the shorted line, it is easy to see that the time averaged electric and magnetic energy as well as the Q-factor are the same. However, equating now the electric energy to that of a capacitance, we have

$$\frac{1}{4}C_n|2V|^2 = \frac{1}{8}Cd|2V|^2 \rightarrow C_n = \frac{Cd}{2} \qquad (2.257)$$

and, as a consequence,

$$L_n = \frac{2Ld}{n^2\pi^2} \qquad (2.258)$$

Comparing these results with those for the shorted line, the roles of C_n and L_n in the model are interchanged and, since this is a parallel resonator (high impedance at resonance), the model now is as shown in Figure 2.36. The first section corresponds to $n = 0$ ($\omega = 0$ or DC), the second to the fundamental resonant frequency ($n = 1$), and the third and up to the higher order resonances ($n = 2, 3, \ldots$). The higher order resonant frequencies are $n\omega_0$, where $n = 2, 3, \ldots$ and ω_0 is the fundamental frequency as calculated in (2.178).

2.11.4 Tapped transmission line resonators

A somewhat different approach to the design of transmission line resonators is to use the capacitive and inductive behavior of shorted and open transmission line resonators. From Figure 2.12, we note that a shorted transmission line of length $0 < d_1 < \lambda/4$ [or $n\lambda/2 < d_1 < (\lambda/4 + n\lambda/2)$, $n = 0, 1, 2, \ldots$] behaves as an inductor. Similarly, from Figure 2.11, an open line of length $0 < d_2 < \lambda/4$ [or $n\lambda/2 < d_2 < (\lambda/4 + n\lambda/2)$, $n = 0, 1, 2, \ldots$] behaves as a capacitor. Thus, a resonator can be built as a $\lambda/4$ line as shown in Figure 2.37. Note that selecting the line to be of length $\lambda/4$ guarantees that d_1 and d_2 are each shorter than one quarter wavelength and hence guaranteed to resonate. The circuit shown is a parallel resonator and at resonance the impedance must be real and high; hence, we will calculate the admittance at the connection point.

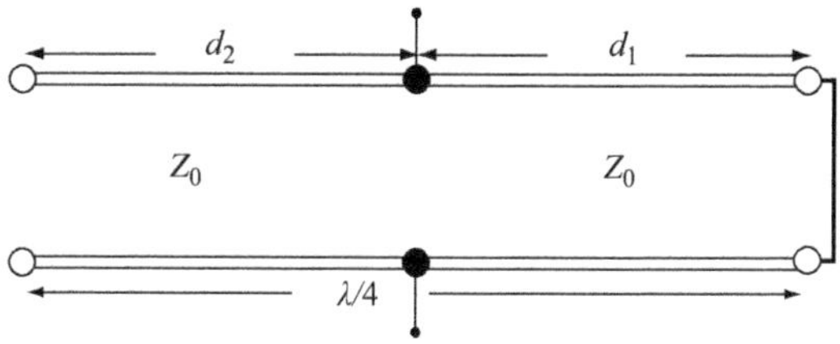

Figure 2.37 A λ/4 tapped transmission line resonator

The input admittance of the shorted section is

$$Z_{\text{in}1} = Z_1 \tanh(\alpha_1 + j\beta_1)d_1 \;\rightarrow\; Y_{\text{in}1} = Y_1 \coth(\alpha_1 + j\beta_1)d_1 \tag{2.259}$$

The input admittance of the open section is

$$Z_{\text{in}2} = Z_2 \coth(\alpha_2 + j\beta_2)d_2 \;\rightarrow\; Y_{\text{in}2} = Y_2 \tanh(\alpha_2 + j\beta_2)d_2 \tag{2.260}$$

Assuming the properties of the two sections are the same and $d_1 = d$, $d_2 = \lambda/4 - d$, we have

$$Y_d = Y_0[\coth(\alpha + j\beta)d + \tanh(\alpha + j\beta)(\lambda/4 - d)] \tag{2.261}$$

Expanding this using the properties of half-angles, we write

$$Y_d = Y_0\left[\frac{\sinh 2\alpha d - j\sin 2\beta d}{\cosh 2\alpha d - \cos 2\beta d} + \frac{\sinh 2\alpha(\lambda/4 - d) + j\sin 2\beta(\lambda/4 - d)}{\cosh 2\alpha(\lambda/4 - d) + \cos 2\beta(\lambda/4 - d)}\right] \tag{2.262}$$

Now, since the lines are low loss, the attenuation constant is small, and we can replace $\sinh 2\alpha d \simeq 2\alpha d$, $\sinh 2\alpha(\lambda/4 - d) \simeq 2\alpha(\lambda/4 - d)$
Also

$$\sin 2\beta\left(\frac{\lambda}{4} - d\right) = \sin 2\beta\left(\frac{\lambda}{4}\right)\cos 2\beta d - \cos 2\beta\left(\frac{\lambda}{4}\right)\sin 2\beta d = \sin 2\beta d \tag{2.263}$$

$$\cos 2\beta\left(\frac{\lambda}{4} - d\right) = \cos 2\beta\left(\frac{\lambda}{4}\right)\cos 2\beta d - \sin 2\beta\left(\frac{\lambda}{4}\right)\sin 2\beta d = -\cos 2\beta d \tag{2.264}$$

Substituting these

$$Y_d = Y_0\left[\frac{2\alpha d - j\sin 2\beta d}{1 - \cos 2\beta d} + \frac{2\alpha(\lambda/4 - d) + j\sin 2\beta d}{1 - \cos 2\beta d}\right] = \frac{2\alpha(\lambda/4)}{1 - \cos 2\beta d} \tag{2.265}$$

This shows that the device must resonate at any value of d since the admittance is real regardless of d and the imaginary part cancels. The resonant frequency is defined by the length of the line (see below). The quality factor of the line is the same as calculated for the parallel resonator above. As a consequence, the bandwidth and half-power points are also the same.

One can also build a tapped transmission line resonator using two shorted segments as in Figure 2.38. In this case, one segment, say the left, must be shorter than a quarter wavelength to act as an inductor. The second must be longer than a quarter wavelength but shorter than a half wavelength to act as a capacitor (see Figure 2.11). Under these conditions, the line becomes a half-wavelength resonator. Although we do not show it here, the analysis is very similar to the analysis for the $\lambda/4$ transmission line resonator.

The tapped resonators in Figures 2.37 and 2.39(b) are guaranteed to resonate for any position of the taps, and they are very specific as to the dimensions d_1 and d_2. The resonant frequency can be calculated from the properties of the shorted and open lines. Since in most practical applications the lines are either lossless or can be approximated as lossless lines, we calculate the resonant frequency for lossless stubs. Considering Figure 2.39, suppose that the left branch in Figure 2.39(b) is made so that it is equivalent to a capacitance C. The input impedance of this segment must be

$$Z_{in1} = jZ_{01} \tan \beta_1 d_1 = \frac{1}{j\omega C} \ (\Omega) \tag{2.266}$$

where Z_{01} is the characteristic impedance of the shorted line forming this segment, and β_1 is the phase constant of the segment.

Line (2), which is also shorted, must behave as an equivalent inductor of inductance L. Its input impedance is

$$Z_{in2} = jZ_{02} \tan \beta_2 d_2 = j\omega L \ (\Omega) \tag{2.267}$$

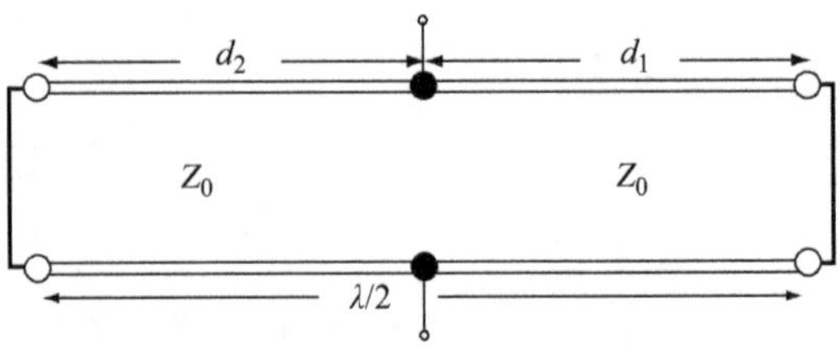

Figure 2.38 A $\lambda/2$ tapped transmission line resonator made with shorted segments

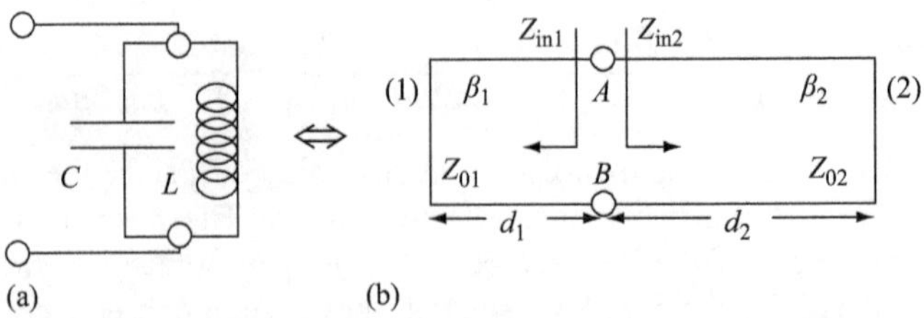

Figure 2.39 (a) A parallel resonant circuit and (b) the transmission line implementation of the parallel resonant circuit

At resonance, the impedance of the circuit is ∞ or, alternatively, the admittance of the parallel circuit is zero. Using the latter, we can write

$$\frac{1}{Z_{\text{in1}}} + \frac{1}{Z_{\text{in2}}} = \frac{1}{jZ_{01}\tan\beta_1 d_1} + \frac{1}{jZ_{02}\tan\beta_2 d_2} = 0 \tag{2.268}$$

Rearranging terms, we get the required condition for resonance:

$$Z_{01}\tan\beta_1 d_1 + Z_{02}\tan\beta_2 d_2 = 0 \tag{2.269}$$

or, in terms of the frequency itself ($\beta = \omega/v_p = 2\pi f/v_p$), we can write

$$Z_{01}\tan\frac{2\pi f d_1}{v_{p1}} + Z_{02}\tan\frac{2\pi f d_2}{v_{p2}} = 0 \tag{2.270}$$

This is a transcendental equation and we cannot solve it explicitly. However, since β_1, β_2, Z_{01}, and Z_{02} are known from the line parameters, all that remains to be defined are d_1 and d_2. This can be done in two ways: if the frequency is given, then a relation between d_1 and d_2 is obtained. We fix one value and find the second such that it satisfies the relation. Alternatively, we can fix both d_1 and d_2 and find the frequencies at which the resulting circuit resonates. Any method of solving the transcendental equation in (2.269) or (2.270) is acceptable for solution.

Note that we should expect multiple solutions from the periodic nature of the tangent functions. The resonant circuit resonates at an infinite number of discrete frequencies.

Series resonant circuits can also be built, at least in principle, using the same approach, although not with all types of transmission lines. A simple series resonant circuit is shown in Figure 2.40 together with its equivalent implementation in terms of shorted, lossless transmission lines. The line shown is a two-wire line, but other transmission lines may be used. Note, however, that there is no proper mechanism to connect two coaxial lines in series. This circuit is limited to open lines such as the parallel plate transmission line or the two-conductor open line.

Following the same process as for the parallel resonant circuit, one segment, say segment (1), must be of length d_1 to make it capacitive and therefore will have the impedance in (2.266). The second segment is made to be inductive and will

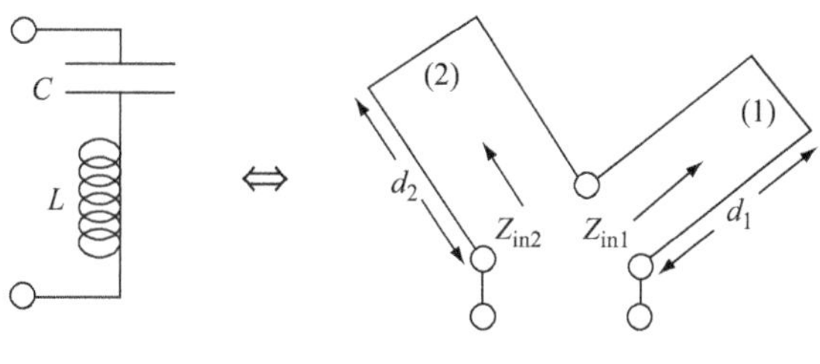

Figure 2.40 A series resonant circuit and its transmission line implementation

have the impedance in (2.267). Now, the total impedance of the line is zero and we write

$$Z_{in1} + Z_{in2} = jZ_{01} \tan \beta_1 d_1 + jZ_{02} \tan \beta_2 d_2 = 0 \tag{2.271}$$

Thus, the equation that must be satisfied is, again,

$$Z_{01} \tan \beta_1 d_1 + Z_{02} \tan \beta_2 d_2 = 0 \tag{2.272}$$

Also, instead of using shorted transmission lines, open transmission lines may be used as well since the impedances of shorted and open transmission lines are the same if one line is shortened or lengthened by one-quarter wavelength. Thus, once a resonator is designed with either type of line, it is a simple matter to find a resonator made of the second type or a combination of the two.

In general, shorted transmission lines are preferred for a variety of reasons, including noise, but sometimes, especially when the line is used to measure external conditions, an open line is more practical. For example, resonant coaxial transmission lines are often used to measure water content in snow to evaluate runoff levels and water reserves. An open resonant circuit is most useful since it can then simply be pushed into the snow pack for measurement purposes. Similarly, a resonator that measures pollutants in air must be open in one way or another. If coaxial lines are used, the resonator must be made of open segments. If, on the other hand, parallel plates are used, shorted lines may be used because the structure itself is open.

This method also allows for the calculation of the Q-factor of the resonator. Assuming a loss resistance R as in Figure 2.28(c) and (d), the Q-factor is

$$Q = 2\pi f_0 \frac{\text{average energy stored}}{\text{power loss}} \tag{2.273}$$

Since the resonant circuit stores both electric and magnetic energies, which are equal at resonance, we have

$$Q = 2\pi f_0 \frac{W_e + W_m}{P_l} = 2\pi f_0 \frac{2W_e}{P_l} = 2\pi f_0 \frac{2W_m}{P_l} \tag{2.274}$$

The stored magnetic energy is that stored in an equivalent inductor L is $LI^2/2$ and the electric energy is that stored in an equivalent capacitor is $CV^2/2$ where I is the current through the inductor and V the voltage across the capacitor. The power loss is I^2R. For the series circuit in Figure 2.28(c), this gives

$$Q = 2\pi f_0 \frac{2(LI^2/2)}{I^2 R} = 2\pi f_0 \frac{L}{R} \tag{2.275}$$

This is the same as that obtained in (2.172) or in (2.225). The bandwidth and the half power frequencies are also the same and given in (2.226)–(2.228).

Similarly, starting with the stored electric energy as $CV^2/2$ and the power loss as I^2/R, where I is the current through the capacitor. The latter can be calculated from the impedance of the capacitor. The Q factor can then be written as

$$Q = 2\pi f_0 \frac{2W_e}{P_l} = 2\pi f_0 \frac{2(CV_c{}^2/2)}{I_c{}^2 R} = 2\pi f_0 \frac{CV_c{}^2}{(V_c{}^2/(j\omega C)(-j\omega C))/R} = \frac{1}{2\pi f_0 RC} \tag{2.276}$$

Identical relations can be obtained for the parallel circuit in Figure 2.28(d).

The limitation of this simplistic method of looking at transmission line resonance is obvious, especially when considering the series resonant circuit in Figure 2.40. In fact, a circuit of this type would be almost impossible to construct with many physical transmission lines. The methods of design and analysis in the previous sections, especially for series transmission line resonators are much more general. The advantage of the analysis in this section is its simplicity and the intuitive equivalence to circuit theory approach to resonators.

2.12 The Smith chat

The dominant feature in transmission line analysis is the use of the reflection coefficient. The reflection coefficient was used to find the conditions on the line, to calculate the line impedance, and to calculate the SWR. Voltage, current, and power were all related to the reflection coefficient. The reflection coefficient, in turn, was defined in terms of the load and line impedances (or any equivalent load impedances such as at a discontinuity). The calculations themselves can be rather complex because of use of complex numbers and harmonic functions. Much of this tedium can be simplified through the use of the Smith chart. The Smith chart is a chart of normalized impedances (or admittances) in the reflection coefficient plane. As such, it allows calculations of all parameters related to transmission lines, waveguides, as well as impedances in open space, circuits, and the like. The Smith chart is a common design tool in electromagnetics. Some measuring instruments such as network analyzers actually use a Smith chart to display conditions on lines and networks. Naturally, any chart can also be implemented in a computer program, and the Smith chart has. A computerized Smith chart can then be used to analyze conditions on lines.

The Smith chart is an impedance chart. As such it does not provide for direct calculations of voltages, currents or power. Nevertheless, it is a useful tool in the calculation of voltages and currents as well as power since it provides important information such as the generalized reflection coefficient, SWR, and the location of voltage and current maxima and minima. One of the most important uses of the chart is as an aid in matching impedances.

To better understand the Smith chart and to gain some insight in its use, we will "build" a Smith chart, gradually, based on the definitions of the reflection coefficient. Consider the circuit in Figure 2.41. The line impedance is real and

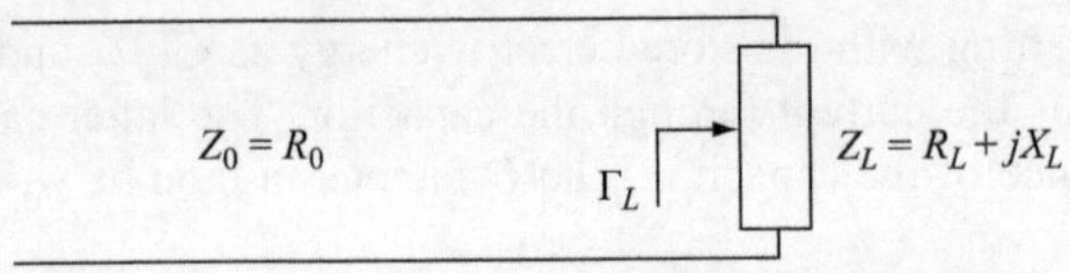

Figure 2.41 A simple transmission line used to introduce the Smith chart

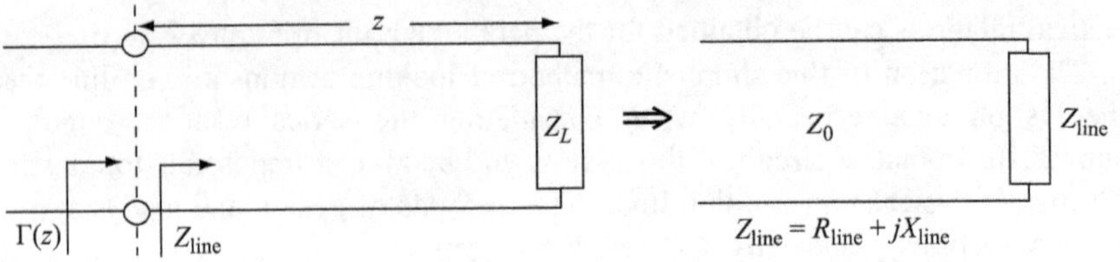

Figure 2.42 Use of an equivalent transmission line to describe the line impedance at a distance z from the load

equals Z_0, but the load is a complex impedance $Z_L = R_L + jX_L$, where R_L is the load resistance and X_L the load reactance. The reflection coefficient [see (2.72) and (2.73)] may be written in one of two forms. The first is a rectangular form (i.e., written in complex variables):

$$\Gamma_L = \frac{Z_L - Z_0}{Z_L - Z_0} = \frac{(R_L - Z_0) + jX_L}{(R_L + Z_0) + jX_L} = \Gamma_r + \Gamma_i \tag{2.277}$$

The reflection coefficient is not modified by normalizing the numerator and denominator by Z_0:

$$\Gamma_L = \frac{(Z_L - Z_0)/Z_0}{(Z_L + Z_0)/Z_0} = \frac{(R_L/Z_0 - 1) + jX_L/Z_0}{(R_L/Z_0 + 1) + jX_L/Z_0} = \frac{(r-1) + jx}{(r+1) + jx} = \Gamma_r + \Gamma_i \tag{2.278}$$

To obtain this result, we substituted $r = R_L/Z_0$ and $x = X_L/Z_0$ as the normalized resistance and reactance. For much of the remainder of this section, we will drop the specific notation for load partly to simplify notation but mostly because the magnitude of the reflection coefficient remains constant along the line and, therefore, the results we obtain apply equally well for any impedance on the line (see Figure 2.42). In the latter case, the generalized reflection coefficient is obtained, and this can be written in exactly the same form as (2.277) or (2.278) by replacing Z_L with $Z(z)$. Equation (2.278) defines a complex plane for the reflection coefficient as shown in Figure 2.43(a). Any normalized impedance (load impedance or line impedance) is represented by a point on this diagram.

The second form of the reflection coefficient is the polar form. This may be written as

$$\Gamma_L = |\Gamma|e^{j\theta_\Gamma} = |\Gamma|(\cos\theta_\Gamma + j\sin\theta_\Gamma) \tag{2.279}$$

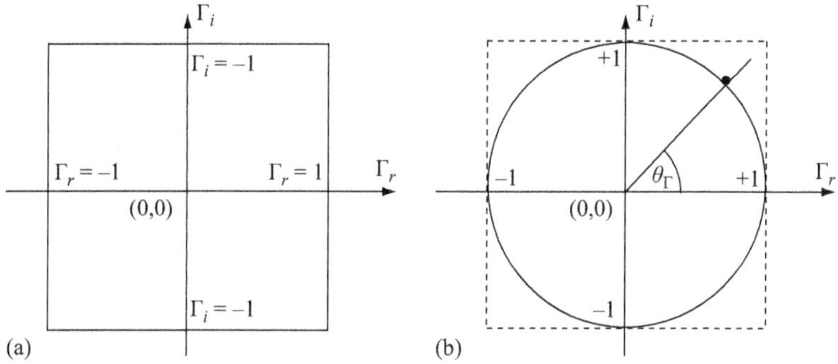

Figure 2.43 The complex plane representation of the reflection coefficient: (a) in rectangular form and (b) in polar form

where θ_Γ is the phase angle of the load reflection coefficient as discussed in Section 2.7.1. For a given magnitude of the reflection coefficient, the phase angle defines a point on the circle of radius $|\Gamma_L|$. Thus, since $|\Gamma_L| \leq 1$, only that section of the rectangular diagram enclosed by the circle of radius 1 is used, as shown in Figure 2.43(b). The polar form is more convenient to use than the rectangular form but we will, for the moment, retain both.

We now go back to the rectangular representation and calculate the real and imaginary parts of the reflection coefficient in terms of the normalized impedance. The starting point is (2.278):

$$\Gamma_r + \Gamma_i = \frac{(r-1) + jx}{(r+1) + jx} \tag{2.280}$$

Crossmultiplying gives

$$(r+1)\Gamma_r - x\Gamma_i + j\Gamma_i(r+1) + jx\Gamma_r = (r-1) + jx \tag{2.281}$$

Separating the real and imaginary parts and rearranging terms, we get two equations:

$$(\Gamma_r - 1)r - \Gamma_i x = -(\Gamma_r + 1) \tag{2.282}$$

$$(\Gamma_r - 1)x + \Gamma_i r = -\Gamma_i \tag{2.283}$$

We now write two equations: one for r and one for x, by first eliminating x and then, separately, r.

From (2.283), we write

$$x = -\frac{\Gamma_i(r+1)}{\Gamma_r - 1} \tag{2.284}$$

Substituting this into (2.282), we get

$$(\Gamma_r - 1)r + \frac{\Gamma_i^2(r+1)}{\Gamma_r - 1} = -(\Gamma_r + 1) \tag{2.285}$$

Multiplying both sides by $\Gamma_r - 1$ and rearranging terms, this gives

$$\Gamma_r^2(r+1) - 2\Gamma_r r + \Gamma_i^2(r+1) = 1 - r \tag{2.286}$$

Dividing by the common term $(r+1)$,

$$\Gamma_r^2 - \frac{2\Gamma_r r}{r+1} + \Gamma_i^2 = \frac{1-r}{r+1} \tag{2.287}$$

Adding $r^2/(r+1)^2$ to both sides of the equation and rearranging terms, we get

$$\left(\Gamma_r - \frac{r}{r+1}\right)^2 + \Gamma_i^2 = \frac{1}{(r+1)^2} \tag{2.288}$$

Repeating the process, we now eliminate r in (2.283) by first writing from (2.282):

$$r = -\frac{(\Gamma_r + 1) - \Gamma_i x}{(\Gamma_r - 1)} \tag{2.289}$$

Substituting this back into (2.283):

$$\Gamma_i \frac{(\Gamma_r + 1) - \Gamma_i x}{\Gamma_r - 1} + (\Gamma_r - 1)x = -\Gamma_i \tag{2.290}$$

Multiplying both sides of (2.290) by $\Gamma_r - 1$ and rearranging terms, we get

$$(\Gamma_r - 1)^2 x + \Gamma_i^2 x - 2\Gamma_i = 0 \tag{2.291}$$

The equation now is divided by x:

$$(\Gamma_r - 1)^2 + \Gamma_i^2 - 2\Gamma_i\left(\frac{1}{x}\right) = 0 \tag{2.292}$$

To bring this into a useful form, we add $1/x^2$ to both sides of the equation:

$$(\Gamma_r - 1)^2 + \Gamma_i^2 - 2\Gamma_i\left(\frac{1}{x}\right) + \left(\frac{1}{x}\right)^2 = \left(\frac{1}{x}\right)^2 \tag{2.293}$$

Rearranging terms, we get

$$(\Gamma_r - 1)^2 + \left(\Gamma_i - \frac{1}{x}\right)^2 = \left(\frac{1}{x}\right)^2 \tag{2.294}$$

Both (2.288) and (2.294) describe circles in the complex Γ plane.

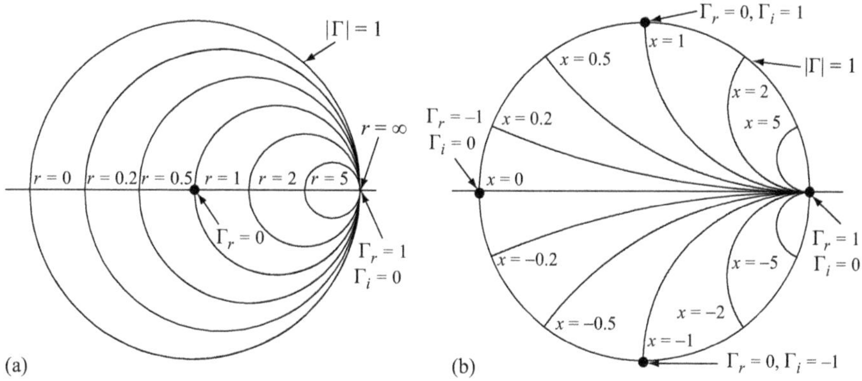

Figure 2.44 The basic components of the Smith chart: (a) circles of constant values of r and (b) circles of constant values of x or −x

Equation (2.288) is the equation of a circle, with its center at $\Gamma_r = r/(r+1)$, $\Gamma_i = 0$, and radius $1/(r+1)$. The center of any circle is on the real axis and can be anywhere between $\Gamma_r = 0$ for $r = 0$ to $\Gamma_r = 1$ for $r \to \infty$. For example, for $r = 1$, the center of the circle is at $\Gamma_r = 0.5$, and its radius equals 0.5. A number of these circles are drawn in Figure 2.44(a). The larger the normalized resistance, the smaller the circle. All circles pass through $\Gamma_r = 1$, $\Gamma_i = 0$. The normalized resistance r can only be positive. Should there ever be a need to describe normalized impedances with negative real part, these must be multiplied by -1 before analysis using the Smith chart can commence.

From (2.294), we obtain a second set of circles for x. Since x can be positive or negative, the circles are centered at $\Gamma_r = 1$, $\Gamma_i = 1/x$ for positive values of x and at $\Gamma_r = 1$, $\Gamma_i = -1/x$ for x negative. These circles are shown in Figure 2.44(b) for a number of values of the normalized reactance x. Figure 2.45 shows the r and x circles on the Γ plane, truncated at the circle $|\Gamma| = 1$. This is the basic Smith chart. A number of properties of the two sets of circles are immediately apparent:

1. The circles are loci of constant r or constant x.
2. x and r circles are orthogonal to each other.
3. There is an infinite number of circles for r and for x.
4. All circles pass through the point $\Gamma_r = 1$, $\Gamma_i = 0$.
5. The circles for x and $-x$ are images of each other, reflected about the real axis.
6. The center of the chart is at $\Gamma_r = 0$, $\Gamma_i = 0$.
7. The intersections of the r circles with the real axis, for $r = r_0$ and $r = 1/r_0$, occur at points symmetric about the center of the chart ($\Gamma_r = 0$, $\Gamma_i = 0$).
8. The intersections of the x circles with the outer circle ($|\Gamma| = 1$) for $x = x_0$ and $x = 1/x_0$ occur at points symmetrically opposite each other.
9. The intersection of any r circle with any x circle represents a normalized impedance point.

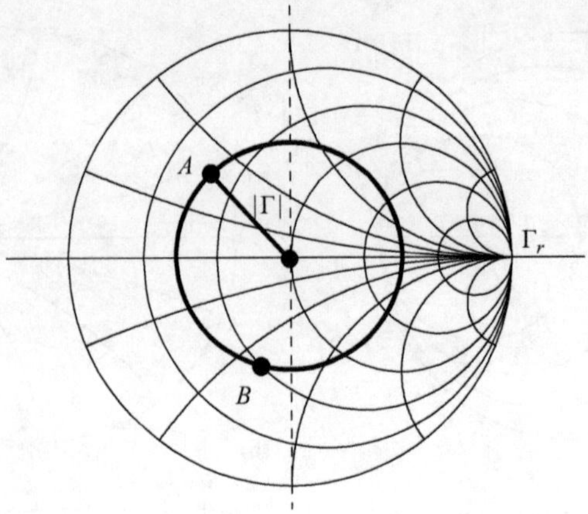

Figure 2.45 The Smith chart. A normalized impedance is a point on the Smith chart defined by the intersection of a circle of constant normalized resistance r and a circle of constant normalized reactance x

10. The real part of the normalized impedance, r, can only be positive but x can be negative or positive.

    The chart as described above is an impedance chart since we defined all points in terms of normalized impedance. Its use as an admittance chart is described in item 14 below.

    In addition to the properties of the r and x circles given above, we note the following:
11. The point $\Gamma_r = 1$, $\Gamma_i = 0$ (rightmost point in Figure 2.45) represents $r = \infty$, $x = \infty$. This is the impedance of an open transmission line. This point is therefore the *open-circuit point*.
12. The diametrically opposite point, at $\Gamma_r = -1$, $\Gamma_i = 0$ represents $r = 0$, $x = 0$. This is the impedance of a short circuit and is called the *short-circuit point*.
13. The outer circle represents $|\Gamma| = 1$. The center of the diagram represents $|\Gamma| = 0$. Any circle centered at the center of the diagram ($\Gamma_r = 0$, $\Gamma_i = 0$) with radius a is a circle on which the magnitude of the reflection coefficient is constant, $|\Gamma| = a$. Moreover, if we take the intersection between any r and x circles, the distance between this point to the center of the diagram is the magnitude of the reflection coefficient for this normalized impedance. A circle drawn through this point represents the generalized reflection coefficient at different locations on the line for this normalized load impedance. The intersection of the reflection coefficient circle with r and x circles represents line impedances at various locations. These aspects of the use of transmission lines are shown in Figure 2.45. For example, point A represents an impedance $r_A + jx_A$ and point B represents an impedance $r_B + jx_B$, but the magnitude of the reflection coefficient is the same. This will later be used to calculate the line impedance as well as voltages and currents on the line.

14. Any point on the chart represents a normalized impedance, say, $z = r + jx$. The admittance of this point is $y = 1/(r + jx) = (r - jx)/(r^2 + x^2)$. The admittance point corresponding to an impedance point lies on the reflection coefficient circle that passes through the impedance point, diametrically opposite to the impedance point. Thus, if we mark a normalized impedance as z and draw the reflection coefficient circle through point z, this circle passes through the admittance point $y = 1/z$. The admittance point y is found by passing a line through z and the center of the diagram. The intersection of this line with the reflection coefficient circle is point y. These steps are shown in Figure 2.46(a). These considerations allow the calculation of admittances instead of impedances. The chart then becomes an admittance chart.

    The Smith chart also provides for calculation of phase angles and lengths of transmission lines. For this purpose, the Smith chart is equipped with a number of scales, marked on the outer periphery of the diagram. These are defined as follows:

15. For a given impedance, a point on the chart is found. The distance from the center of the chart to the point is the magnitude of the line reflection coefficient. If the line connecting the center of the chart with the impedance point is continued until it intersects the outer ($\Gamma = 1$) circle, the location of intersection gives the phase angle of the reflection coefficient in degrees. This is the first set of values given on the circumference of the Smith chart and is shown in Figure 2.46(b). Note that the open-circuit point has zero phase angle ($\Gamma = +1$) and the short-circuit point has either a 180° or $-180°$ phase angle. The difference is in the sign of the imaginary part of the load impedance (below or above the real axis). Intermediate points will vary in phase depending on the distance from the load. For example, for point A in Figure 2.46(b), the phase angle of the reflection coefficient is 104°, whereas for point B, it is $-120°$.

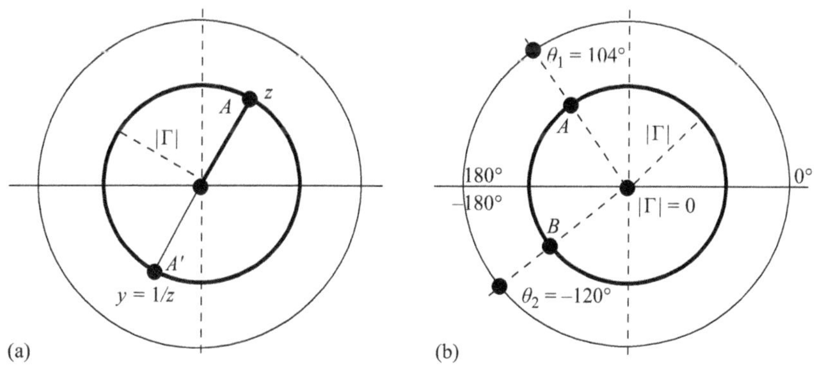

Figure 2.46 (a) Normalized impedance, reflection coefficient, and normalized admittance and (b) indication of phase angle of the reflection coefficient on the Smith chart

16. We recall that the distance between a point of maximum voltage and a point of minimum voltage was found to be $\lambda/4$ in Section 2.7.3. In particular, the impedance of a shorted transmission line changes from zero to infinity (or negative infinity) if we move a distance $\lambda/4$ from the short. Thus, the distance between the short-circuit and open-circuit points is $\lambda/4$. This fact is indicated on the outer circle of the chart, starting at the short-circuit point. Since the short (or any other load) can be anywhere on a line, we may wish to move either toward the generator or toward the load to evaluate the line behavior. These two possibilities are indicated with arrows showing the direction toward load and toward generator (Figure 2.47). Although the distance is marked from the short-circuit point, the distance is always relative: If a point is given at any location on the chart, movement on the chart a distance $\lambda/4$ represents half the circumference of the chart.
17. The direction toward the generator is the clockwise direction. If we wish to calculate the line impedance starting from the load, we move in the clockwise direction toward the generator. If, on the other hand, we wish to calculate the line impedance starting from the generator going toward the load or, starting at the load and going away from the generator, we must move in the counterclockwise direction and use the appropriate distance charts (see Figure 2.47).
18. The whole Smith chart encompasses one-half wavelength. This, of course, is due to the fact that all conditions on lines repeat at intervals of $\lambda/2$ regardless of loading or any other effect that may happen on the line. If we need to analyze lines longer that $\lambda/2$, we simply move around the chart as many half-wavelengths as are necessary. Only the remainder length (length beyond any integer numbers of half-wavelengths) need be analyzed.

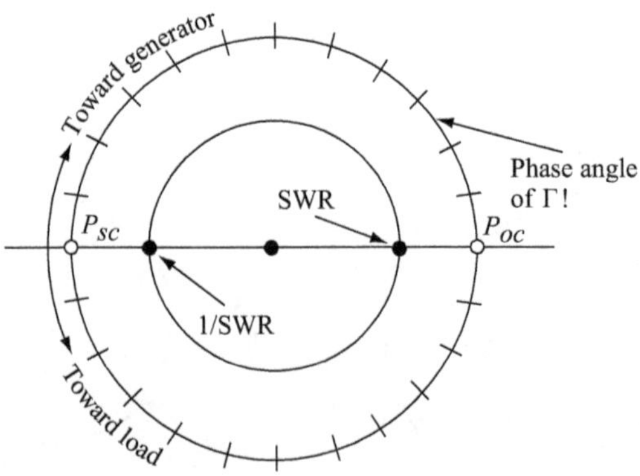

Figure 2.47 Directions on the Smith chart and indication of SWR. The distance between short- and open-circuit points is $\lambda/4$

The Smith chart also allows for the calculation of SWRs. The SWR is calculated from the reflection coefficient as

$$\text{SWR} = \frac{1 + |\Gamma|}{1 - |\Gamma|} \text{ (dimensionless)} \tag{2.295}$$

We note that the circle of radius $|\Gamma|$ intersects the positive real axis at $x = 0$. At this point, the normalized impedance is equal to r and the reflection coefficient is given as $\Gamma = (r - 1)/(r + 1)$. Substituting this into the relation for SWR, we get

$$\text{SWR} = r \tag{2.296}$$

Thus, the SWR equals the value of normalized resistance at the location of intersection of the reflection coefficient circle and the real axis, right of the center of the Smith chart. From property (7) above, the intersection of the reflection coefficient circle with the real axis, left of the center of the chart is at point $1/r$. Thus, this point gives the value 1/SWR. The two points are shown for the reflection coefficient in Figure 2.47.

Now that we discussed the individual parts making up the Smith chart, it is time to put it all together. The result is the Smith chart shown in Figure 2.48. You will immediately recognize the r and x circles as well as the scales discussed. There are, however, a number of other scales given at the bottom of the chart as well as a number of indications on the chart itself, which we have not discussed. These have to do with losses on the line (which we have neglected) and the use of the chart as an admittance rather than impedance chart. The scales at the bottom of the figures are merely for convenience and are not necessarily available in all implementations of the Smith chart. They are usually present in printed charts but not in displayed charts or when the chart is used for output such as in a network analyzer.

Although the chart is relatively simple, it contains considerable information and can be used in many different ways and for purposes other than transmission lines. The main difference in the Smith chart solution and the analytic solution is that the Smith chart uses normalized impedances, whereas in analytic calculations, we tend to use the actual values of the impedance. Also, because it is a graphical chart, the results are approximate and depend on our ability to accurately read the values off the chart. The Smith chart is available commercially as a paper chart as well as computer software. The advantage of a software-based Smith chart is that calculations are exact in addition to the ease of analysis and display of results. It should be remembered that distances on the chart are in wavelengths. Physical distances can be calculated provided the frequency and speed of propagation on the line are known.

Figure 2.49 shows the screen display of a vector network analyzer in the process of matching an antenna. It shows the impedance as the frequency changes from 850 to 950 MHz. The marker shows the normalized impedance at 915 MHz.

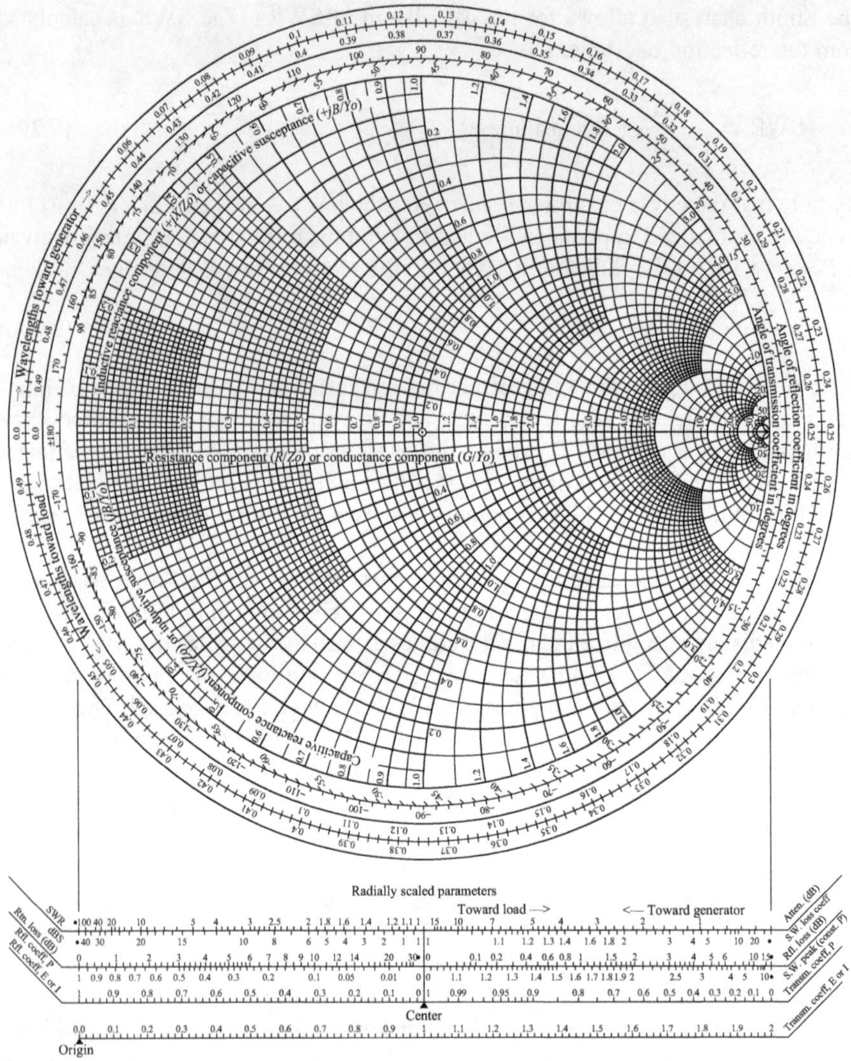

Figure 2.48 The complete Smith chart

The latter equals $0.23 + j2.31$. For a 50-Ω system that equals $11.5 + j115$ Ω. In this case, matching means moving the real part of the impedance until it intersects the $r = 1$ circle and the imaginary part until it intersects the $x = 0$ circle. Although matching is only of cursory interest in this work, it is done by adding series/parallel inductors and/or capacitors to affect the required change in impedance. Note however the options on the network analyzer to set the scan range and the types of displays. The network analyzer can obviously calculate the *S*-parameters, in this case for a two-port network as indicated at the bottom of the screen.

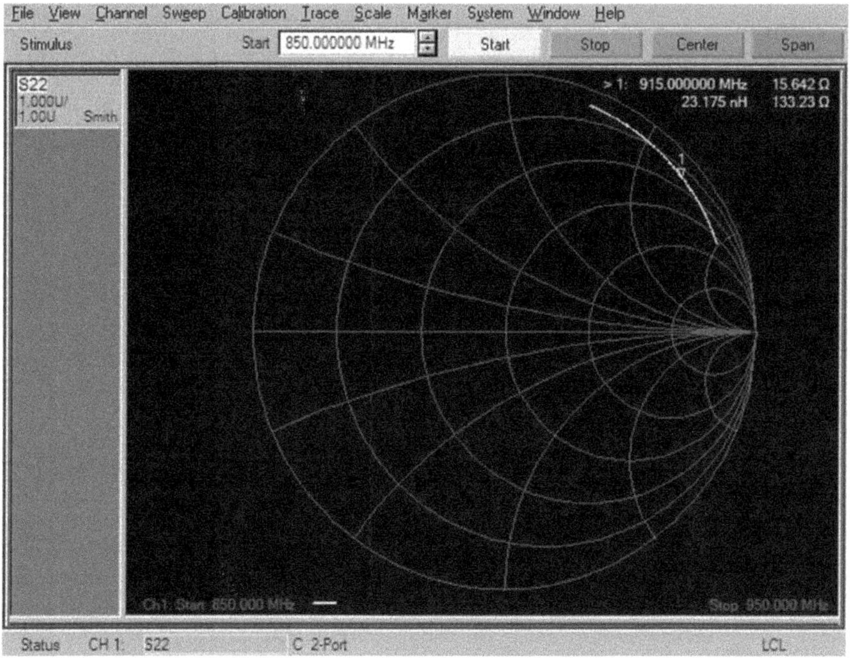

Figure 2.49 Screenshot from a network analyzer display showing an impedance point Z = 0.23 + j2.31 at the marked point

Bibliography

The material in this chapter is based on many sources, including monographs, textbooks, and published articles and reports. As a rule I only listed books when possible with only a few key articles when necessary. There are, of course, many articles on the various subjects as well as much material in web pages.

The theory of transmission lines and that of waveguides can be found in many books including most college-level textbooks on electromagnetics including the following. The material in this chapter follows mostly the exposition in [1] with additions and expansions from [2, 6, 10].

[1] N. Ida, "Engineering Electromagnetics," 3rd edition, Springer, New York, NY, 2015.
[2] G. F. Miner, "Lines and Electromagnetic Fields for Engineers," Oxford University Press, New York, NY, 1996.
[3] D. K. Cheng, "Field and Wave Electromagnetics," 2nd edition, Addison-Wesley, Reading, MA, 1992.
[4] J. C. Freeman, "Fundamentals of Microwave Transmission Lines," Wiley, New York, 1996.

More advanced treatment of transmission lines and waveguides can be found in

[5] R. E. Collin, "Foundations for Microwave Engineering," 2nd edition, McGraw-Hill, New York, NY, 1992.
[6] S. R. Seshadri, "Fundamentals of Transmission Lines and Electromagnetic Fields," Addison-Wesley, Reading, MA, 1971.
[7] F. A. Benson and T. M. Benson, "Fields, Waves and Transmission Lines," Chapman and Hall, 1990.
[8] R. Collier, "Transmission Lines: Equivalent Circuits, Electromagnetic Theory, and Photons," Cambridge University Press, 2013.
[9] F. Olyslager, "Electromagnetic Waveguides and Transmission Lines," Clarendon Press, Oxford, 1999.
[10] H. A Wheeler, "Electrical Characteristics of Transmission Lines," Artech House, Norwood, MA, 1979.

The material on general resonant circuits follows more or less that in [11] with modifications and extensions. Resonant circuits can also be found in circuits-oriented books. See, for example, [12].

[11] D. M. Pozar, "Microwave Engineering," 2nd edition, John Wiley & Sons, Inc., New York, NY, 1998.
[12] C. K. Alexander and M. N. O. Sadiku, "Fundamentals of Electric Circuits," 2nd edition, McGraw-Hill, Boston, MA, 2002.
[13] P. R. Karmel, G. D. Colef and R. L. Camisa, "Introduction to Electromagnetic and Microwave Engineering," John Wiley and Sons, Inc., New York, NY, 1998.

The subject of transmission line circuits is extensive and covered very well in the literature both in research papers and in books. Many of the more recent texts and publications deal with planar structures rather than with classical transmission line. Even though we discuss planar structures in Chapter 3, the basic passive circuits needed in this work were described in this chapter. The references below should be viewed in this context and, because the principles are based on classical transmission line theory, the references given here are appropriate.

[14] M. W. Medley, "Microwave and RF Circuits: Analysis, Synthesis and Design," Artech House, Norwood, MA, 1992.
[15] J. A. G. Malherbe, "Microwave Transmission Line Filters," Artech House, Norwood, MA, 1979.
[16] J. Helszajn, "Waveguide Junction Circulators," Wiley, Chichester, 1998.
[17] R. J. Cameron, R. R. Mansour and C. M. Kudsia, "Microwave Filters for Communication Systems: Fundamentals, Design and Applications," Wiley, New York, 2007.
[18] S. A. Maas, "Microwave Mixers," Artech House, Norwood, MA, 1993.
[19] C. G. Montgomery, R. H. Dicke and E. M. Purcell, "Principles of Microwave Circuits," Peter Peregrinus, London, 1987.
[20] R. J. Weber, "Introduction to Microwave Circuits: Radio Frequency and Design Applications," IEEE Press, Piscataway, NJ, 2001.

[21] C. A. Lee and G. C. Dalman, "Microwave Devices, Circuits and Their Interaction," Wiley, New York, 1994.
[22] T. Itoh, "Numerical Techniques for Microwave and Millimeter-Wave Passive Structures," Wiley, New York, 1989.
[23] F. Giannini and G. Leuzzi, "Non-Linear Microwave Circuit Design," Wiley, Chichester, 2004.
[24] S. A. Maas, "Practical Microwave Circuits," Artech House, Norwood, MA, 2014.
[25] A. A. Behagi, "RF and Microwave Circuit Design: A Design Approach Using (Ads)," TechnoSearch, Ladera Ranch, CA, 2015.
[26] S. Y. Yao, "Microwave Devices and Circuits, 3rd edition" Prentice Hall, Englewood Cliffs, NJ, 1990.
[27] P. A. Rizzi, "Microwave Engineering: Passive Circuits," Prentice Hall, Englewood Cliffs, NJ, 1988.

The issue of active microwave circuits was only mentioned in this chapter in passing because it is not important in the context of this work. But it should be mentioned that these circuits are extremely important for practical work. For example, internal oscillators and amplifiers in network analyzers are based on active circuits. The following references are general sources on some of the active circuits encountered in microwave work:

[28] C. Poole and I. Darwazeh, "Microwave Active Circuit Analysis and Design," Elsevier, Amsterdam, 2015.
[29] S. Voinigescu, "High-Frequency Integrated Circuits," Cambridge University Press, Cambridge, 2013.
[30] G. Ghione and M. Pirola, "Microwave Electronics," Cambridge University Press, Cambridge, 2017.
[31] A. S. Gilmour, "Principles of Klystrons, Traveling Wave Tubes, Magnetrons, Cross-Field Ampliers, and Gyrotrons," Artech House, Norwood, MA, 2011.
[32] J. Whitaker, "Power Vacuum Tubes Handbook," 3rd edition, CRC Press, BVoca Raton, FL, 2012.

Needless to say that there are many good publications on all aspects of transmission lines and on resonators. The available literature gives particular attention to planar structure because of their importance in integrated devices. These are discussed in Chapter 3. The reader is encouraged to search for some of these but I have not included them here as the material needed for understanding of the present and following chapters is "classical" and is given in sufficient detail.

Chapter 3

Planar transmission lines and coupled structures

3.1 Introduction

In Chapter 2, we discussed the properties of the classical transmission lines as well as circuits, including resonators, based on these lines. The basis of the theory and circuits is the two-conductor transmission line (see Figure 2.3). This led to the distributed circuit model in Figure 2.4, and all subsequent properties rose from this model. However, transmission lines are not limited to two conductors. One can easily envisage multiconductor transmission lines as well as other methods of constructing transmission lines including one-conductor waveguides. For example, a three-phase power distribution line consists of three conductors or, if a ground conductor is included, of four conductors. Another example is that of two or more two-conductor lines in close proximity so that they affect each other. More often, however, multiconductor transmission lines are designed for specific purposes. Multiconductor transmission lines have, necessarily, properties that can be very different than the classical two-conductor transmission line. The fields within and around the lines and the modes of propagation can be vastly different, and these lead to properties and applications that are not possible with the classical two-conductor transmission lines.

Of particular interest in this work are the so-called planar transmission lines. Planar structures are extensions of the parallel plate transmission line in Figure 2.3(e), which is considered a planar transmission line. The term planar simply refers to the fact that in these transmission lines, the characteristics of the structure can be determined by the dimensions in a plane—the cross section of the structure. The term is often narrowed to those structures that can be conveniently produced by lithographic methods with particular applications to microwave integrated circuits although, in this work, we will use planar structures that are relatively large and their fabrication is entirely different.

Planar structures can be made of two conductors, and in that form, they resemble the classical two-conductor transmission lines. In fact, they are simple extension of the classical two-wire transmission line. Multiconductor planar structures are also common, and these are often said to be coupled transmission lines. Coupling means that power from one line couples to another, and by doing so, it affects the circuit model and hence affects both the properties of the lines and their applications. A related structure is the waveguide, especially the parallel plate waveguide. Waveguides will also be discussed here because the methods of analysis we use for the coupled transmission line resonators are similar and to an extent, the resonators used in this work resemble microwave cavity resonators.

3.2 Planar transmission lines: the stripline

As was mentioned in the introduction, the term transmission line is rather generic and includes a variety of structures, almost all are multiconductor structures of which the most common is the two-conductor transmission line. Within this narrower definition, in addition to the regular lines such as coaxial, two parallel wires and parallel plate or strip transmission lines, there are certain structures that have found particularly useful applications. The parallel plate transmission line [Figure 2.3(e)] has become particularly useful in printed circuit structures although, in general, the strips do not have to be parallel. Some other transmission lines that can be viewed as extensions to the parallel plate lines are shown in Figure 3.1. Figure 3.1(a) shows the so-called stripline. These structures have become important because of their properties at high frequencies but also because they lend themselves to construction by lithographic means, similar to methods used in integrated circuits. This is a planar structure consisting of a thin conducting strip of width w between two wide conducting plates, separated a distance d apart. The space between the plates is a dielectric of permittivity ε and, in general, a loss factor related to its conductivity σ_c. The strip does not have to be at the center between the plates, and the plates can be of various widths (although, in most cases we will assume the plates are infinite for analysis purpose). The stripline in Figure 3.1(a) is a two-conductor structure since the two outer plates are at the same potential (usually serve as ground planes). Figure 3.1(b) shows another structure, called microstrip. The thin strip of width w is separated from a ground plane by a dielectric of thickness d which supports the stripline. The popularity of this structure is again the ease with which it can be fabricated either by etching or by deposition and hence is easily built into integrated circuits. But it can be as simple as a strip on a printed circuit substrate above a ground plane. One can easily see the relation between this structure and the parallel plate transmission line although, the fact that the dielectric is not present everywhere can complicate analysis. Figure 3.1(c) shows another common planar structure called a slotline. Here, a ground plane is interrupted by a slot. Although one can clearly see

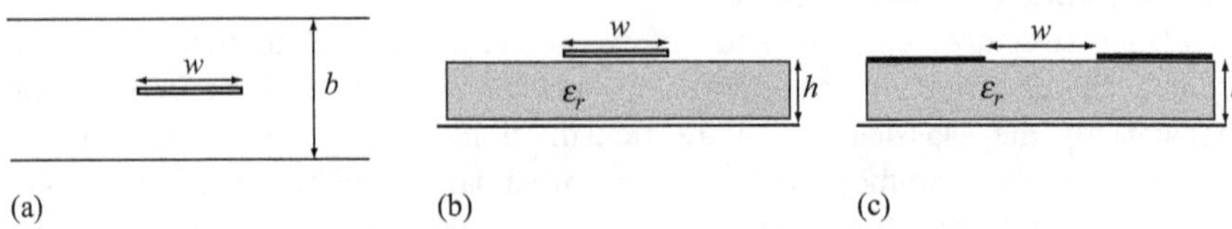

Figure 3.1 Planar structures: (a) stripline, (b) microstrip line, and (c) slotline

similarities between the lines, they are quite different both in properties and, particularly, in the way waves propagate along the line. For example, the stripline propagates transverse electromagnetic (TEM) modes (i.e., the fields along the line behave as plane waves in much the same way as on parallel plate transmission lines), whereas slotted lines support non-TEM modes. The microstrip supports what are called quasi-TEM modes of propagation.

The analysis of planar transmission lines is, in general, more complex than that for the common transmission lines (say the coaxial or even the two wire transmission line) because the field distributions are much more complex, and hence, the parameters of the lines often cannot be calculated analytically. Often properties such as characteristic impedance, attenuation constant, or resonant frequency are calculated using approximations or are calculated numerically. Often too, these may be evaluated experimentally. Approximate solutions exist for common lines.

Both the stripline and the microstrip come in many variants in constructions and hence in properties and performance. Some of the common variations are shown in Figure 3.2. The structures in Figure 3.2(a)–(c) still qualify as two-conductor lines. However, those in Figure 3.2(d) and (e) do not. Figure 3.2(e) is a three-conductor line if the ground plane is present as shown. One can then drive each stripline separately with respect to ground or the source can be applied

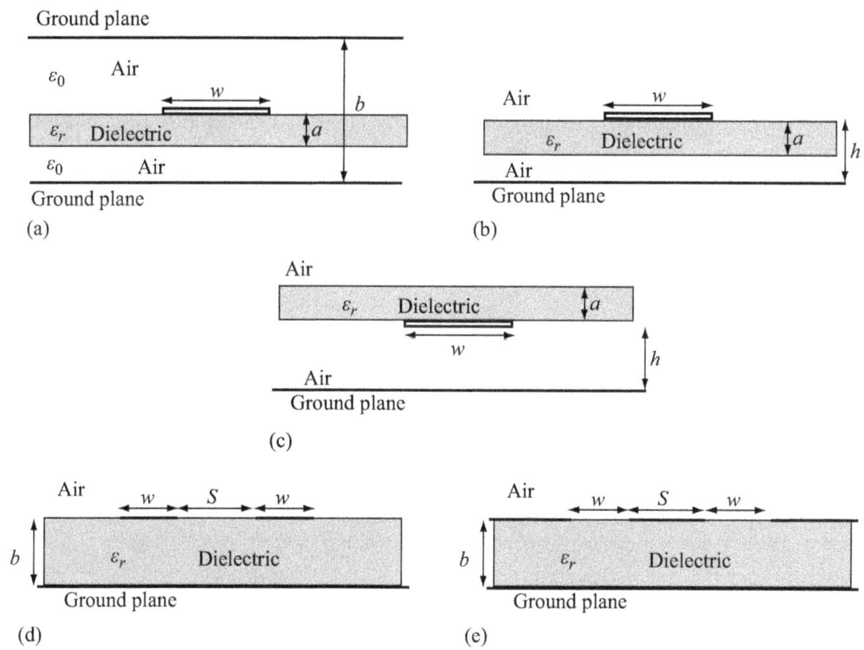

Figure 3.2 Variants of the basic structures in Figure 2.29: (a) suspended stripline, (b) suspended microstrip, (c) inverted microstrip, (d) coplanar waveguide, and (e) coplanar stripline

between the striplines. If the two strips are driven each with respect to the ground plane, they form two separate parallel lines and, because they are in close proximity, power from one line couples with the second line. If, on the other hand, the ground plane is removed, the structure becomes a two-conductor transmission line rather than a coupled line. The structure in Figure 3.2(e) can usually be viewed as a two-conductor line since the two side strips are normally connected to the ground plane. As a whole, these structures will not be discussed here except for the coupled transmission lines and their use in directional couplers.

3.2.1 Coupled transmission lines

As was indicated above, the structure in Figure 3.2(e) is unique in that it forms a coupled stripline. In fact, any three-conductor stripline will behave as a couple transmission line. Of course, this is not limited to three conductors but we will limit ourselves here to three conductors. The coupled structure in Figure 3.2(e) is a coplanar stripline, but it should be obvious that other configuration of striplines, microstrips, or even nonuniform transmission lines can be realized. We will limit discussion here to coupled striplines because these are used in this work and, to an extent, because coupled striplines support TEM modes and hence are easiest to analyze in terms of effective capacitance and phase velocities on the lines.

Given a three-conductor transmission line in any configuration, its equivalent capacitance model is shown in Figure 3.3. In this model, C_{11} represents the capacitance between strip 1 and ground in the absence of strip 2, C_{22} represents the capacitance between strip 2, and ground in the absence of strip 1, and C_{12} represents the capacitance between the two strips in the absence of the ground plane. There are three basic structures that come into consideration in the context of coupled transmission line. The first is that of Figure 3.2(e). The other two are shown in Figure 3.4. The striplines in Figure 3.4(a) are said to be edge-coupled, whereas in Figure 3.4(b), they are broadside coupled. Both consist of a pair of

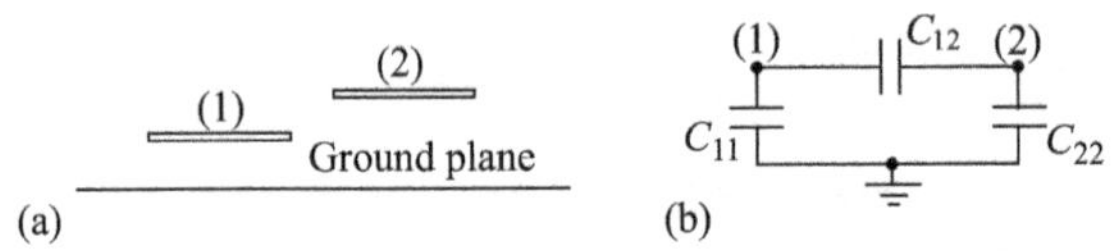

Figure 3.3 Capacitance model for the two striplines above a conducting ground plane

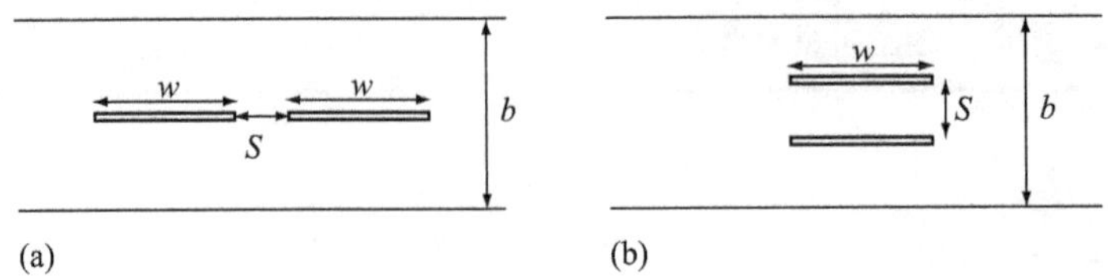

Figure 3.4 (a) Edge-coupled striplines and (b) broadside coupled striplines

striplines in a homogeneous dielectric (usually air) and both are placed between two ground planes. Of particular interest in this work is the structure in Figure 3.4(b), but it should be noted that any of the structures can be modified in many ways. One can use strips of different widths, the strips can be offset and they do not have to be symmetric relative to the ground planes. Edge and broadside coupling can also be combined by duplicating the broadside coupled transmission line as in Figure 3.5. In this case, each two pair of striplines forms either an edge-coupled stripline or a broadside coupled stripline. The capacitance model for this configuration is more complex than the one in Figure 3.3, but the method of analysis is essentially the same. It should be noticed as well that the ground plane does not have to be a "ground" conductor—it is simply the third conductor of the line. In practice, however, it is most often connected to ground, hence the designation.

As for any stripline, the coupled stripline can support TEM modes of propagation (for a homogeneous dielectric like air) but, unlike the classical stripline, the coupled stripline supports two fundamental, orthogonal TEM modes. These are referred to as "even" and "odd" modes indicating two distinct ways the striplines can be excited. The even mode is characterized by the two striplines being excited with equal amplitudes and in phase (i.e., they are both positive with respect to the ground planes) as shown in Figure 3.6(a). As a consequence of this excitation, the electric field intensity has zero normal component on the line of symmetry (the symmetry line A–A' is therefore a "magnetic wall"). The odd mode is excited with the two lines having equal amplitudes but with opposite phases (one positive, the other negative with respect to the ground plane) as shown in Figure 3.6(b). Here, the electric field intensity only has a normal component on the line of symmetry. (The symmetry line A–A' is an electric wall.)

In terms of the capacitance model in Figure 3.3, the even mode now takes the form in Figure 3.7(a). In effect, the two lines are decoupled.

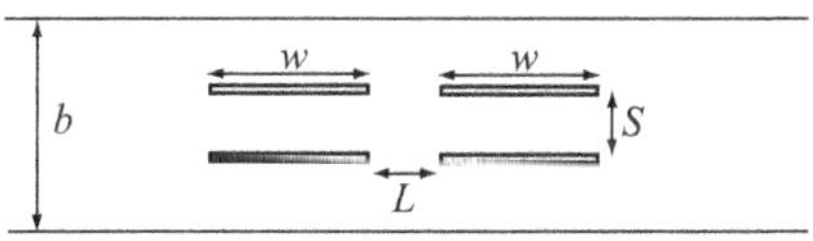

Figure 3.5 A 4 stripline coupled structure as a combination of edge and broadside coupling

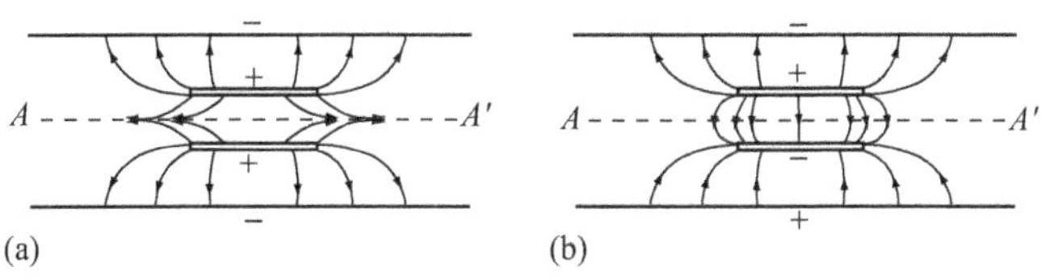

Figure 3.6 Fields in broadside coupled striplines: (a) even and (b) odd modes

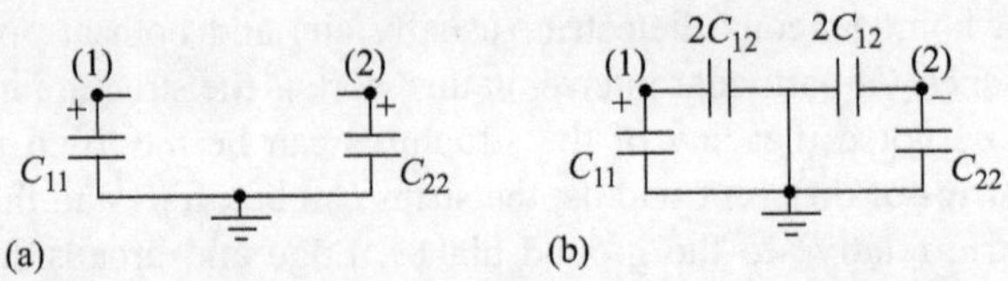

Figure 3.7 (a) The model for even-mode propagation and (b) the model for odd-mode propagation

If the coupling is symmetric, that is, if the two strips are identical in size and at identical positions with respect to the ground planes, the two capacitances C_{11} and C_{22} are identical, and we can write the even-mode capacitance as

$$C_e = C_{11} = C_{22} \tag{3.1}$$

The characteristic impedance of the line for even-mode propagation can now be written as

$$Z_{0e} = \sqrt{\frac{L}{C_e}}\ (\Omega) \tag{3.2}$$

or

$$Z_{0e} = \frac{\sqrt{LC_e}}{C_e} = \frac{1}{vC_e}\ (\Omega) \tag{3.3}$$

Clearly, only the capacitance and speed of propagation are necessary hence the simple model in Figure 3.3.

The odd-mode impedance is calculated based on the model in Figure 3.7(b). This model takes into account the fact that the centerline between the strips is an electric wall, and hence, the capacitance between each strip and the electric wall is twice that between the strips, and the electric wall is at ground potential. The capacitance in the odd mode therefore is

$$C_o = C_{11} + 2C_{12} = C_{22} + 2C_{12} \tag{3.4}$$

C_{11} was calculated above and C_{12} is the capacitance per unit length between the two strips. Equation (3.4) assumes that the coupling is symmetric. The impedance now is written as in (3.3):

$$Z_{0o} = \frac{\sqrt{LC_o}}{C_o} = \frac{1}{vC_o}\ (\Omega) \tag{3.5}$$

The even- and odd-mode impedances can be calculated provided that the even- and odd-mode capacitances in (3.1) and (3.4) can be calculated. This is, in general, an arduous task since the fields within the line, as indicated schematically in Figure 3.6, are rather complex and calculation of capacitances is not trivial. The capacitances and hence impedances are normally calculated using conformal transformation methods or, in more complex configurations, using numerical techniques.

For example, the characteristic impedance for the even and odd mode in the structure shown in Figure 3.4(b) are given by

$$Z_{0e} = \frac{60\pi K(k')}{\sqrt{\varepsilon_r}\,K(k)} \text{(even mode)} \tag{3.6}$$

$$Z_{0o} = \frac{293.9S}{b\sqrt{\varepsilon_r}\tanh^{-1}(k)} \text{(odd mode)} \tag{3.7}$$

k is the complete elliptic function of the first kind, S is the separation between the transmission lines, w their width, and b the distance between the ground planes [Figure 3.4(b)]. The relative permittivity between the striplines is ε_r. The parameter k may be calculated as the solution to the following transcendental equation:

$$w = \frac{b}{\pi}\left\{\ln\left(\frac{1+R}{1-R}\right) - \frac{S}{b}\ln\left(\frac{1+R/k}{1-R/k}\right)\right\} \tag{3.8}$$

with

$$R = \sqrt{\frac{k(S/b) - 1}{(b/kS) - 1}} \tag{3.9}$$

Then, k' is calculated as

$$k' = \sqrt{1 - k^2} \tag{3.10}$$

Similar results exist for edge-coupled striplines and for many other configurations. These expressions were calculated using conformal mapping and although fairly complex, they are only approximate, increasing in accuracy as w/S increases. There are also approximations available that avoid the calculation of the elliptic functions as well as online resources for calculations. It should also be noted that the thickness of the striplines were neglected and that the assumption of pure TEM propagation implies lossless media, a condition that can only be approximately satisfied. Approximations for finite thickness striplines exist but, in general, the more complex configurations, including lossy dielectrics, must be treated numerically.

To better understand the behavior of the broadside coupled transmission lines, it is useful to look at a very simple configuration, which can be treated analytically in terms of parallel plate capacitances.

We use the configuration in Figure 3.4(b) assuming the two strips are identical, centered between the ground planes and that there are no fringing effects (i.e., $w \gg d$ and $w \gg S$). Under these conditions, the capacitances are calculated as those of parallel plate capacitors. In this case, $C_{11} = C_{22}$. The capacitance, say C_{11}, is the sum of the capacitance between one of the strips and the upper plane and the capacitance between the same strip and the lower plane. Using the dimensions in Figure 3.4(b), the capacitance is

$$C_{11} = C_{22} = \frac{\varepsilon w}{(b-S)/2} + \frac{\varepsilon w}{S + (b-S)/2} = \frac{4\varepsilon w b}{b^2 - S^2} \text{ (F/m)} \tag{3.11}$$

This is the even-mode capacitance as indicated in (3.1). Using (3.3), the characteristic impedance of the line for even-mode propagation can now be written from (3.5) as

$$Z_{0e} = \frac{1}{vC_e}\ (\Omega) \tag{3.12}$$

With $\varepsilon = \varepsilon_0\varepsilon_r$, $v = 1/\sqrt{\varepsilon_0\varepsilon_r\mu_0}$, and C_e from (3.11), the impedance becomes

$$Z_{0e} = \frac{1}{vC_o} = \frac{b^2 - S^2}{4\varepsilon_0\varepsilon_r wb}\sqrt{\varepsilon_0\varepsilon_r\mu_0} = \sqrt{\frac{\mu_0}{\varepsilon_0}}\frac{b^2 - S^2}{4wb\sqrt{\varepsilon_r}} = \eta_0\frac{b^2 - S^2}{4wb\sqrt{\varepsilon_r}}\ (\Omega) \tag{3.13}$$

where ε_r is the relative permittivity of the space between the ground planes and η_0 the intrinsic impedance of vacuum. This is the characteristic impedance of either of the strips relative to the ground plane when the coupled line operates in the even mode.

The odd-mode impedance is calculated in a similar fashion using (3.4) and (3.5). C_{11} was calculated in (3.11), and C_{12} is simply the capacitance per unit length between the two strips:

$$C_{12} = \frac{\varepsilon_0\varepsilon_r w}{S} \tag{3.14}$$

The odd-mode capacitance in (3.4) is

$$C_o = C_{11} + 2C_{12} = \frac{4\varepsilon_0\varepsilon_r wb}{b^2 - S^2} + 2\frac{\varepsilon_0\varepsilon_r w}{S} = 2\varepsilon_0\varepsilon_r w\left[\frac{2b}{(b^2 - S^2)} + \frac{1}{S}\right]\ (\text{F/m}) \tag{3.15}$$

The odd-mode impedance is therefore

$$\begin{aligned} Z_{0o} &= \frac{1}{vC_o} = \frac{\sqrt{\varepsilon_0\varepsilon_r\mu_0}}{(4\varepsilon_0\varepsilon_r wb/(b^2 - S^2)) + (2(\varepsilon_0\varepsilon_r w/S))} \\ &= \eta_0\frac{1}{2w\sqrt{\varepsilon_r}[2b/(b^2 - S^2) + 1/S]}\ (\Omega) \end{aligned} \tag{3.16}$$

This is the characteristic impedance of either of the strips relative to the ground plane when the coupled line operates in the odd mode and is clearly different than the even-mode characteristic impedance.

The results in (3.13) and (3.16) are fairly poor approximations, but they show the general behavior of the coupled striplines. First we note that the odd-mode impedance is lower than the even-mode impedance. The expressions also show explicit relations between the dimensions and impedances and hence can serve for initial design of coupled transmission lines.

Although the characteristic impedance for the two modes is different, the phase velocity is the same and is entirely dependent on the permittivity of the dielectric in the line.

These relations are given here to demonstrate the fact that calculation of parameters for this type of lines, especially when the dielectric between the lines is nonhomogeneous, can be challenging. In fact, often, the line properties are evaluated using numerical methods.

Although approximate expressions for impedance and resonance for many planar transmission lines exist, these expressions become more complex as the structure becomes more complex. In particular, introduction of discontinuities in the lines or in materials between the lines complicates the expressions considerably. When we come to calculation of resonant frequencies of structures based on these types of lines, it will be much more profitable to use a numerical method rather than analytical techniques because the relation between the length of the line and its resonant frequency is not as simple as in the simple two-conductor uniform transmission line we discussed in Chapter 2. The effects of the ground planes, proximity to external objects, and presence of supporting structures within the resonator all complicate the analysis. Even when the structure is simple, introduction of materials or objects for testing purposes complicates the analysis beyond what can be done analytically. These all combine to favor numerical analysis of the structures.

Coupled transmission lines form the basis of a variety of microwave circuits, perhaps the most obvious of them being the class of circuits called dividers and couplers. Dividers are passive devices capable of power division or power combination, while couplers rely on the power coupling capabilities of transmission lines in close proximity. Another common application is in design of filters. Coupled transmission lines can also serve as microwave resonators by introducing appropriate discontinuities in the lines. In that role, they exhibit two modes of resonance—even and odd resonance—in parallel to the two modes of propagation in continuous lines.

3.3 Waveguides and cavity resonators

In Chapter 2, the discussion centered around transmission lines that were defined as structures made of two or more conductors with the stated purpose of transferring power. The difference between mere "circuits" and transmission lines was shown to be the length of the lines in relation to the wavelength, with short lines being classical circuits, treated by lumped circuit methods and long lines being treated by distributer circuit methods. The latter meant that transmission lines are, necessarily, lines operating at high frequencies.

Analysis of transmission lines, including the planar structures in the present chapter, followed circuit theory methods by which we obtained voltages and currents on the lines as well as all other properties including propagation parameters, impedance, and the behavior of voltages and currents at loads and discontinuities. In effect, the quantities of interest were circuit quantities. But it was also indicated in Section 2.6 that a completely equivalent analysis in terms of the electric and magnetic field is possible and leads to the same behavior. That was based on the

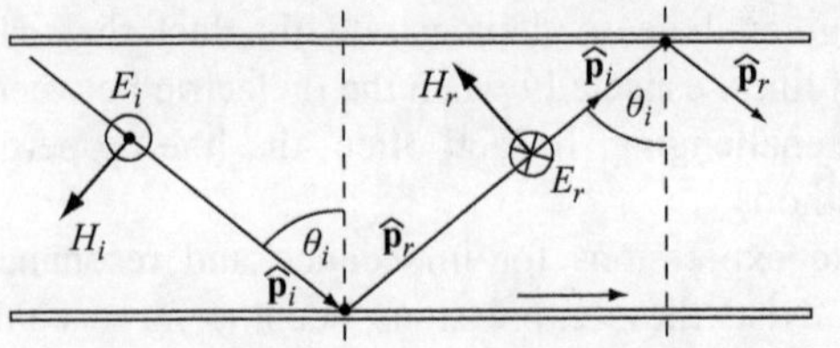

Figure 3.8 Reflection of the incident wave off the lower and upper plates in a parallel plate waveguide, transverse electric field

assumption that the electric and magnetic fields are perpendicular to each other (see Figure 2.5) and to the direction of propagation (see also Appendix D), that is, we have used the concepts of plane waves or, alternatively, that of TEM propagation. Because all two-wire transmission lines support TEM waves, that analysis was deemed correct and useful.

However, that is not the whole story of transmission lines. One can take a different approach with some transmission lines that lead to the conclusion that TEM propagation is not the only possibility and, more significantly, that non-TEM propagation has certain advantages. To see that, consider the parallel plate transmission line in Figure 3.8. Instead of currents and voltages, we can start by allowing a TEM wave to propagate at an angle to, say, the lower surface. In this wave, the electric and magnetic fields are perpendicular to each other and to the direction of propagation. The wave will be reflected off the surface as shown in Figure 3.8 and propagate toward the upper conductor. The wave reflected off the lower surface is also a TEM wave. This wave will now be reflected off the upper surface as shown in Figure 3.8. The process repeats indefinitely. Without calculating the fields, the important effects to notice are

1. The wave propagates to the right—that is, power propagates to the right along the transmission line. We say that power is guided by the two conductors.
2. The larger the angle θ, the longer the path of the wave and hence the slower the propagation.
3. The electric field intensity is perpendicular to the direction of propagation—that is, the electric field intensity is a transverse field.
4. The magnetic field intensity has two components—one is transverse to the direction of propagation (pointing along the x-axis), the other is in the direction of propagation (along the z-axis).

In other words, the reflections off the conductors cause an incident TEM wave to produce a wave in which only the electric field intensity is transverse, whereas the magnetic field intensity has a component in the direction of propagation. We call this wave a transverse electric (TE) wave.

Suppose now that the electric and magnetic field intensities are interchanged as shown in Figure 3.9. Again without much difficulty, we conclude that the magnetic field intensity remains perpendicular to the direction of propagation and hence is transverse, whereas the electric field intensity has a component in the direction of

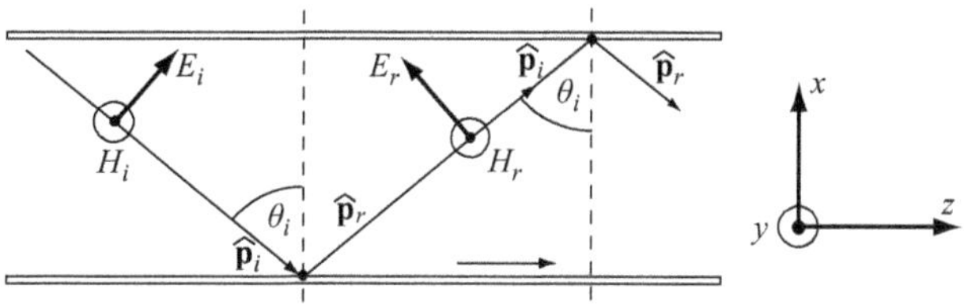

Figure 3.9 Reflection of the incident wave off the lower plate in a parallel plate waveguide, transverse magnetic field

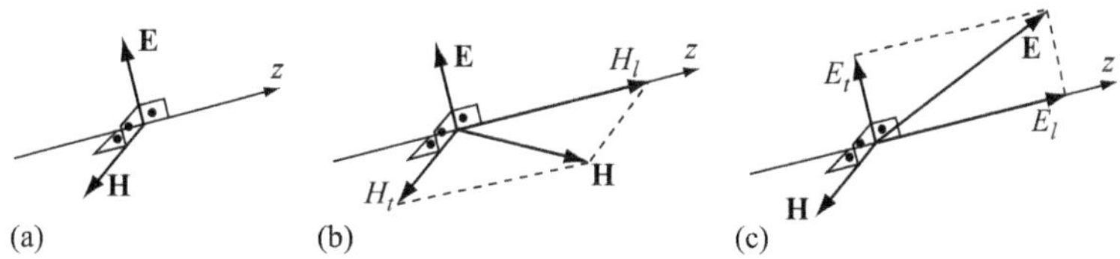

Figure 3.10 Types of waves: (a) TEM wave, (b) TE wave, and (c) TM wave. The index l stands for longitudinal and t for transverse component

propagation. For this reason, the wave now is called a transverse magnetic (TM) wave.

The conclusion so far is that the parallel plate transmission line can support TEM, TE, and TM waves depending on the angle of incidence of the wave and the relation of the electric field intensity to the surface of the plates. If $\theta = 90°$, the propagation is TEM since then both **E** and **H** are perpendicular to the direction of propagation. For any other angle, if **E** is parallel to the surface of the conductors, the propagation is TE, whereas if the magnetic field intensity is parallel to the conducting surfaces, the propagation is TM.

Figure 3.10 shows schematically the three types of waves that may exist in the parallel plate waveguide and the relations between the electric and magnetic field components in each mode of propagation. The TEM mode can only exist in two-conductor structures, whereas the TE and TM modes can exist in one-conductor structures as well.

The parallel plate transmission line has now become a parallel plate waveguide simply by virtue of how we use it. In this case, a wave is launched between the plates, whereas as a classical transmission line, we connect the two conductors to a generator or circuit, which results in a current in and voltage on the line. But that seemingly small difference is far reaching in consequences.

It should be intuitively obvious that the three types of waves have different properties and these can be deduced from the fields themselves. Also, the parallel plate waveguide is only one of a number of structures that can serve as waveguides. Some of the more common are rectangular cross-section waveguides and circular cross-section waveguides. In both of these, one cannot distinguish two separate conductors; hence, in the sense of classical transmission lines, they cannot be considered as transmission lines. However, in the sense that they transfer and guide power they are. In fact, dielectric waveguides have no conductors at all. The main reason these structures are discussed here is that sections of waveguides can be transformed into resonators, and these resonators, called cavity resonators, have much in common with transmission line resonators. We will use some of the concepts of cavity resonators in this chapter and in the following chapters to understand and analyze the behavior of the open transmission line resonators that are at the center of this work.

In lieu of a full theoretical development, something that would take much space, the following sections give the important relations in waveguides and in cavity resonators without giving their full justification. We also restrict ourselves to rectangular waveguides and resonators, simply because these are more useful here.

3.3.1 TE propagation in parallel plate waveguides

The fields in waveguides are usually written as longitudinal and transverse components. With reference to Figure 3.8, the fields in the parallel plate waveguide are

Longitudinal component:

$$H_z(x,z) = -\frac{E_0}{\eta}\frac{\lambda}{\lambda_{\mathrm{cm}}}\cos\left(\frac{m\pi x}{d}\right)e^{-j\beta_g z}\ (\mathrm{A/m}) \tag{3.17}$$

Transverse components:

$$E_y(x,z) = jE_0\sin\left(\frac{m\pi x}{d}\right)e^{-j\beta_g z}\ (\mathrm{V/m}) \tag{3.18}$$

$$H_x(x,z) = -j\frac{E_0}{\eta}\frac{\lambda}{\lambda_g}\sin\left(\frac{m\pi x}{d}\right)e^{-j\beta_g z}\ (\mathrm{A/m}) \tag{3.19}$$

In these relations, d is the separation between the two plates, λ_g is the guide wavelength, λ is the wavelength in the medium between the plates, and λ_{cm} is the cutoff wavelength for mode m. λ_g and λ_{cm} are

$$\lambda_g = \frac{\lambda}{\sqrt{1 - f_{\mathrm{cm}}^2/f^2}}\ (\mathrm{m}),\ \lambda_{\mathrm{cm}} = \frac{2d}{m}\ (\mathrm{m}),\ m = 1, 2, 3, \ldots \tag{3.20}$$

The cutoff wavelength and hence the cutoff frequency depend on the mode and the dimension of the waveguide. The cutoff frequency is

$$f_{\mathrm{cm}} = \frac{m}{2d\sqrt{\mu\varepsilon}}\ (\mathrm{Hz}),\ m = 1, 2, 3, \ldots \tag{3.21}$$

f_{cm} and λ_{cm} are the frequency below which and the wavelength above which TE propagation cannot occur in the waveguide and m is the mode number. A cutoff frequency exists for each mode, and the fields in (3.17)–(3.19) are also mode dependent. Note as well that the TEM mode has no cutoff frequency—it can exist at any frequency.

The power propagated in the wave is calculated using the Poynting vector. The time-averaged power density propagated in the z direction is

$$\mathcal{P}_{\text{av}} = \frac{E_0^2}{2\eta}\frac{\lambda}{\lambda_g}\sin^2\left(\frac{m\pi x}{d}\right)\ (\text{W/m}^2) \tag{3.22}$$

Finally, since the propagation is in the z direction, we can also define the characteristic impedance of the waveguide by dividing the transverse component of the electric field intensity (E_y) by the transverse component of the magnetic field intensity (H_x):

$$Z_{\text{TE}} = \frac{\eta}{\sqrt{1 - f_{\text{cm}}^2/f^2}}\ (\Omega) \tag{3.23}$$

The wave impedance is, in fact, larger than the impedance for TEM waves. Also, unlike the wave impedance η for TEM waves, the wave impedance for TE waves is frequency dependent. The wave impedance tends to infinity at cutoff ($f = f_{\text{cm}}$) and to η as the frequency approaches infinity ($f \gg f_{\text{cm}}$).

3.3.2 TM propagation in parallel plate waveguides

Longitudinal component:

$$E_z(x,z) = jE_0\frac{\lambda}{\lambda_{\text{cm}}}\sin\left(\frac{m\pi x}{d}\right)e^{-j\beta_g z}\ (\text{V/m}) \tag{3.24}$$

Transverse components:

$$E_x(x,z) = E_0\frac{\lambda}{\lambda_g}\cos\left(\frac{m\pi x}{d}\right)e^{-j\beta_g z}\ (\text{V/m}) \tag{3.25}$$

$$H_y(x,z) = \frac{E_0}{\eta}\cos\left(\frac{m\pi x}{d}\right)e^{-j\beta_g z}\ (\text{A/m}) \tag{3.26}$$

Power propagated by the m mode is

$$\mathcal{P}_{\text{av}} = \frac{E_0^2\lambda}{2\eta\lambda_g}\cos^2\left(\frac{m\pi x}{d}\right)\ (\text{W/m}^2) \tag{3.27}$$

We can also calculate the wave impedance of the waveguide as

$$Z_{\text{TM}} = \frac{E_x}{H_y} = \eta\sqrt{1 - f_{\text{cm}}^2/f^2}\ (\Omega) \tag{3.28}$$

Note that the wave impedance for TM waves is always smaller than the intrinsic impedance η, whereas at cutoff, $Z_{\text{TM}} = 0$.

3.3.3 Rectangular waveguides

In the parallel plates waveguide in the previous section, the fields only varied in one transverse direction, whereas propagation was along the plates. Because of that, analysis of the fields was simple. True parallel plate waveguides are not practical since all dimensions must be finite. Although structures resembling the parallel plate waveguide can be built (such as the striplines in Section 3.2) and are quite common in microwave integrated circuits, most waveguides are closed structures. A rectangular or cylindrical tube or some other type of enclosed conductor may be used. In the most general sense, the conductor is not a condition of existence of guided waves; only total reflection from a boundary is required. However, to simplify the discussion, we will restrict ourselves to waveguides defined by highly conducting surfaces.

One of the most common and simple waveguide structures is the rectangular waveguide built with two intersection parallel plates as shown in Figure 3.11. This view has the advantage of defining the waveguide in terms of the parallel plate waveguides we have already discussed and produces both TE and TM waves but TEM waves cannot exist since now there is only one conductor.

A rectangular waveguide is shown in Figure 3.12. The dimensions of the waveguide are the internal dimensions, and the walls are assumed to be perfectly conducting. We will also restrict discussion to lossless media within the waveguide. The longitudinal and transverse components of the fields are calculated directly from Maxwell's equations, the details of which are not shown here (but see the "Bibliography" section).

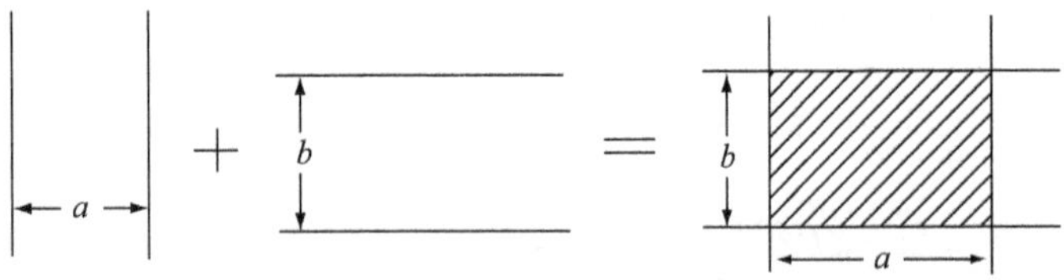

Figure 3.11 A rectangular waveguide (shown in cross section) as a combination of two parallel plate waveguides

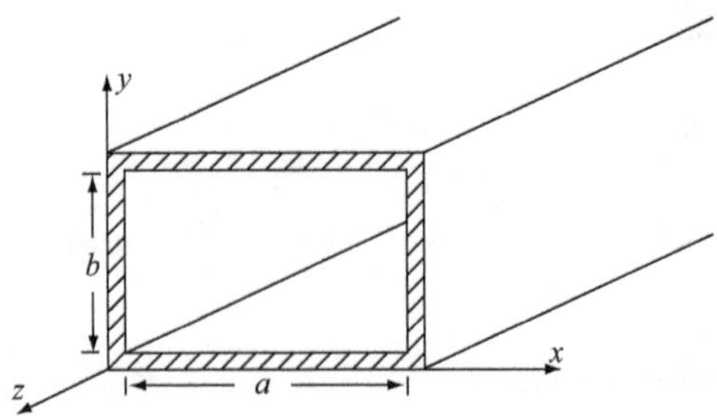

Figure 3.12 Structure and dimensions of a rectangular waveguide

3.3.4 TM modes in rectangular waveguides

For TM modes, the longitudinal component is the electric field intensity:

$$E_z(x, y, z) = E_0 \sin\left(\frac{m\pi}{a}x\right)\sin\left(\frac{n\pi}{b}y\right)e^{-j\beta_g z} \text{ (V/m)} \tag{3.29}$$

The transverse components are

$$E_x(x, y, z) = \frac{-j\beta_g}{k_{\text{cmn}}^2}E_0\frac{m\pi}{a}\cos\left(\frac{m\pi x}{a}\right)\sin\left(\frac{n\pi y}{b}\right)e^{-j\beta_g z} \text{ (V/m)} \tag{3.30}$$

$$E_y(x, y, z) = \frac{-j\beta_g}{k_{\text{cmn}}^2}E_0\frac{n\pi}{b}\sin\left(\frac{m\pi x}{a}\right)\cos\left(\frac{n\pi y}{b}\right)e^{-j\beta_g z} \text{ (V/m)} \tag{3.31}$$

$$H_x(x, y, z) = \frac{j\omega\varepsilon}{k_{\text{cmn}}^2}E_0\frac{n\pi}{b}\sin\left(\frac{m\pi x}{a}\right)\cos\left(\frac{n\pi y}{b}\right)e^{-j\beta_g z} \text{ (A/m)} \tag{3.32}$$

$$H_y(x, y, z) = \frac{-j\omega\varepsilon}{k_{\text{cmn}}^2}E_0\frac{m\pi}{a}\cos\left(\frac{m\pi x}{a}\right)\sin\left(\frac{n\pi y}{b}\right)e^{-j\beta_g z} \text{ (A/m)} \tag{3.33}$$

In these relations, a is typically the larger dimension. The guide phase constant β_g and the cutoff wavenumber k_{cmn} are

$$\beta_g = \sqrt{\omega^2\mu\varepsilon - \left(\frac{m\pi}{a}\right)^2 - \left(\frac{n\pi}{b}\right)^2} \text{ (rad/m)} \tag{3.34}$$

$$k_{\text{cmn}} = \sqrt{\left(\frac{m\pi}{a}\right)^2 + \left(\frac{n\pi}{b}\right)^2} \text{ (rad/m)} \tag{3.35}$$

The cutoff frequency may be obtained from the cutoff wavenumber by writing $k_{\text{cmn}} = 2\pi f_{\text{cmn}}\sqrt{\mu\varepsilon}$:

$$f_{\text{cmn}} = \frac{1}{2\sqrt{\mu\varepsilon}}\sqrt{\left(\frac{m}{a}\right)^2 + \left(\frac{n}{b}\right)^2} \text{ (Hz)} \tag{3.36}$$

The wave impedance is obtained by taking the ratio between the transverse components of the electric and magnetic field intensities. Since the wave must propagate in the z direction (i.e., the Poynting vector must be in the z direction), we can take either the ratio between E_x and H_y or the negative ratio of E_y and H_x:

$$Z_{\text{TM}} = \frac{E_x}{H_y} = -\frac{E_y}{H_x} = \frac{\beta_g}{\omega\varepsilon} = \sqrt{\frac{\mu}{\varepsilon}\left(1 - \frac{f_{\text{cmn}}^2}{f^2}\right)} \ (\Omega) \tag{3.37}$$

The total power in the waveguide cross section is the power density, integrated over the waveguide cross section. This gives

$$P = \frac{\omega\varepsilon\beta_g E_0^2 ab}{8k_{\text{cmn}}^2} \text{ (W)} \tag{3.38}$$

where $k_{\mathrm{cmn}}^2 = (m\pi/a)^2 + (n\pi/b)^2$. The total power is directly proportional to the cross-sectional area of the wave guide (ab). In any given waveguide, the total power may be increased by using larger fields (electric and magnetic) or increasing the physical dimensions of the waveguide. Also, the power is proportional to frequency and the dielectric constant in the waveguide. Most waveguides use air as the dielectric, but increasing the frequency is feasible up to certain limits, imposed by the circuits used to generate the fields. The modes, m, n, are any integer values including zero. However, the obvious $m=0$, $n=0$ mode does not exist. Also, observing the transverse fields in (3.30)–(3.33), modes with indices $m=0$, $n\neq 0$ and $n=0$, $m\neq 0$ cannot exist because with these indices, all transverse fields are identically zero.

3.3.5 TE modes in rectangular waveguides

In TE modes, the longitudinal component is the magnetic field intensity:

$$H_z(x,y) = H_0 \cos\left(\frac{m\pi x}{a}\right)\cos\left(\frac{n\pi y}{b}\right)e^{-\gamma z}\ (\mathrm{A/m}) \tag{3.39}$$

The transverse components of the electric and magnetic field intensities are

$$E_x(x,y,z) = \frac{j\omega\mu}{k_{\mathrm{cmn}}^2} H_0 \frac{n\pi}{b}\cos\left(\frac{m\pi x}{a}\right)\sin\left(\frac{n\pi y}{b}\right)e^{-j\beta_g z}\ (\mathrm{V/m}) \tag{3.40}$$

$$E_y(x,y,z) = \frac{-j\omega\mu}{k_{\mathrm{cmn}}^2} H_0 \frac{m\pi}{b}\sin\left(\frac{m\pi x}{a}\right)\cos\left(\frac{n\pi y}{b}\right)e^{-j\beta_g z}\ (\mathrm{V/m}) \tag{3.41}$$

$$H_x(x,y,z) = \frac{j\beta_g}{k_{\mathrm{cmn}}^2} H_0 \frac{m\pi}{a}\sin\left(\frac{m\pi x}{a}\right)\cos\left(\frac{n\pi y}{b}\right)e^{-j\beta_g z}\ (\mathrm{A/m}) \tag{3.42}$$

$$H_y(x,y,z) = \frac{j\beta_g}{k_{\mathrm{cmn}}^2} H_0 \frac{n\pi}{b}\cos\left(\frac{m\pi x}{a}\right)\sin\left(\frac{n\pi y}{b}\right)e^{-j\beta_g z}\ (\mathrm{A/m}) \tag{3.43}$$

The propagation properties for TE modes in the waveguide are identical to those for TM modes except for the wave impedance. This can be seen by direct calculation of the various properties (f_{cmn}, β_g, v_g, etc.). The wave impedance, however, is different for TE modes and is given by the ratio of the transverse components of the electric and magnetic fields as follows:

$$Z_{\mathrm{TE}} = \frac{E_x}{H_y} = -\frac{E_y}{H_x} = \frac{\omega\mu}{\beta_g} = \sqrt{\frac{\mu}{\varepsilon}\left(\frac{1}{1-f_{\mathrm{cmn}}^2/f^2}\right)}\ (\Omega) \tag{3.44}$$

TE_{mn} modes are obtained for all possible pairs of the integers m and n, except for $m=0$ and $n=0$. Unlike TM modes, in TE modes, either m or n can be zero but not both. This indicates that the lowest propagating mode is a TE_{0n} or TE_{m0}, depending on the dimensions a and b of the waveguide. If $a > b$, the lowest cutoff frequency is for a TE_{10} mode. Also to be noted is that TM and TE modes with the same indices have the same cutoff frequency, as can be seen from (3.36).

The power density in a waveguide propagating a TE mode may be calculated using steps identical to those for TM modes, using the Poynting vector for the transverse components of **E** and **H**. This gives

$$P = \frac{\omega\mu\beta_g H_0^2 ab}{8k_{cmn}^2} \text{ (W)} \tag{3.45}$$

The lowest cutoff frequency mode in any waveguide is called *dominant mode*. In rectangular waveguides, this is a TE mode and, usually, the TE_{10} mode. Different modes which have the same cutoff frequency are called *degenerate modes*.

3.3.6 Cavity resonators

A rectangular cavity resonator is built out of a rectangular waveguide by adding two conducting walls at $z=0$ and $z=d$, as shown in Figure 3.13(a). The cavity resonator may be viewed as being made of three parallel plate waveguides, as shown in Figure 3.13(b). The cavity is a modified waveguide, in which there are standing waves in the z direction as well as in the x and y directions. The cavity acts as a resonant structure in which there is exchange of energy between the electric and magnetic field at given (resonant) frequencies. This is equivalent to resonant *LC* circuits in the case of lossless cavities and to *RLC* circuits in the case of lossy cavities.

The analysis of fields in a cavity requires the solution of the full three-dimensional wave equation with the required boundary conditions. The procedure here will be to take the TM and TE waves we have defined for waveguides and to modify them to satisfy the additional boundary conditions imposed by the additional conducting walls. However, the TE and TM equations for waveguides cannot be used directly. The main reason is that in waveguides, we assumed explicitly that the wave propagates in the z direction and that the transverse directions are the directions perpendicular to the direction of propagation (z direction). In cavities,

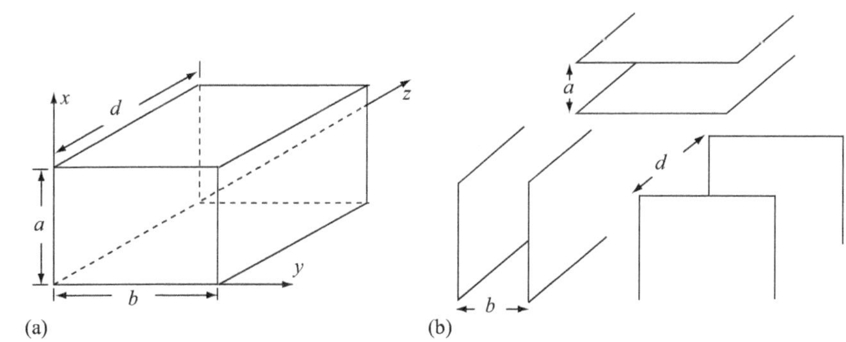

Figure 3.13 (a) Structure and dimensions of a rectangular cavity resonator and (b) construction of the cavity resonator as the intersection of three parallel plate waveguides

there is no clear direction we can take as a transverse direction. The approach here is to take the z direction (usually the long dimension of the cavity) as a reference direction, allow the waves to propagate along this direction, and calculate the total waves as the sum of the forward- and backward-propagating waves reflected off the shorting walls. We define the TE and TM modes as

1. A TM mode in a cavity resonator is any mode which has no magnetic field component in the z direction of the cavity.
2. A TE mode in a cavity resonator is any mode which has no electric field component in the z direction of the cavity.

3.3.7 TM modes in cavity resonators

The longitudinal component is

$$E_z(x,y,z) = E_0 \sin\left(\frac{m\pi x}{a}\right)\sin\left(\frac{n\pi y}{b}\right)\cos\left(\frac{p\pi z}{d}\right) \text{ (V/m)} \tag{3.46}$$

The transverse components in the cavity resonator are as follows

$$E_x(x,y,z) = \frac{-1}{\gamma^2+k^2}\frac{p\pi}{d}\frac{m\pi}{a}E_0\cos\left(\frac{m\pi x}{a}\right)\sin\left(\frac{n\pi y}{b}\right)\sin\left(\frac{p\pi z}{d}\right) \text{ (V/m)} \tag{3.47}$$

$$E_y(x,y,z) = \frac{-1}{\gamma^2+k^2}E_0\frac{n\pi}{b}\frac{p\pi}{d}\sin\left(\frac{m\pi x}{a}\right)\cos\left(\frac{n\pi y}{b}\right)\sin\left(\frac{p\pi z}{d}\right) \text{ (V/m)} \tag{3.48}$$

$$H_x(x,y,z) = \frac{j\omega\varepsilon}{\gamma^2+k^2}E_0\frac{n\pi}{b}\sin\left(\frac{m\pi x}{a}\right)\cos\left(\frac{n\pi y}{b}\right)\cos\left(\frac{p\pi z}{d}\right) \text{ (A/m)} \tag{3.49}$$

$$H_y(x,y,z) = \frac{-j\omega\varepsilon}{\gamma^2+k^2}E_0\frac{m\pi}{a}\cos\left(\frac{m\pi x}{a}\right)\sin\left(\frac{n\pi y}{b}\right)\cos\left(\frac{p\pi z}{d}\right) \text{ (A/m)} \tag{3.50}$$

Note that these are standing waves—they do not propagate, as required. The wavenumber is defined as

$$k^2 = \left(\frac{m\pi}{a}\right)^2 + \left(\frac{n\pi}{b}\right)^2 + \left(\frac{p\pi}{d}\right)^2 = \omega^2\mu\varepsilon \tag{3.51}$$

From this, the resonant frequency is calculated as

$$f_{mnp} = \frac{1}{2\sqrt{\mu\varepsilon}}\sqrt{\left(\frac{m}{a}\right)^2 + \left(\frac{n}{b}\right)^2 + \left(\frac{p}{d}\right)^2} \text{ (Hz)} \tag{3.52}$$

where the indices m, n, and p indicate the mode in which the cavity resonates. In resonant cavities, the concept of cutoff is different than in waveguides. Since there is no propagation in a cavity, these are called *resonant frequencies* or *resonant modes* rather than cutoff frequencies.

Any combination of mode indices m, n, and p results in a resonant frequency of the cavity except for those with $m = 0$ or $n = 0$ [for which the longitudinal component of the field in (3.46) becomes zero]. If m or n or both are zero, all field components become zero. However, p can be zero. The lowest TM resonant mode (assuming $a > b > c$) is TM_{110}.

3.3.8 TE modes in cavity resonators

The longitudinal component is the magnetic field intensity:

$$H_z(x, y, z) = H_0 \cos\left(\frac{m\pi x}{a}\right)\cos\left(\frac{n\pi y}{b}\right)\sin\left(\frac{p\pi z}{d}\right) \text{ (A/m)} \tag{3.53}$$

The transverse components are

$$E_x(x, y, z) = \frac{j\omega\mu}{\gamma^2 + k^2} H_0 \frac{n\pi}{b} \cos\left(\frac{m\pi x}{a}\right)\sin\left(\frac{n\pi y}{b}\right)\sin\left(\frac{p\pi z}{d}\right) \text{ (V/m)} \tag{3.54}$$

$$E_y(x, y, z) = \frac{-j\omega\mu}{\gamma^2 + k^2} H_0 \frac{m\pi}{a} \sin\left(\frac{m\pi x}{a}\right)\cos\left(\frac{n\pi y}{b}\right)\sin\left(\frac{p\pi z}{d}\right) \text{ (V/m)} \tag{3.55}$$

$$H_x(x, y, z) = -\frac{1}{\gamma^2 + k^2} H_0 \frac{m\pi}{a}\frac{p\pi}{d} \sin\left(\frac{m\pi x}{a}\right)\cos\left(\frac{n\pi y}{b}\right)\cos\left(\frac{p\pi z}{d}\right) \text{ (A/m)} \tag{3.56}$$

$$H_y(x, y, z) = -\frac{1}{\gamma^2 + k^2} H_0 \frac{n\pi}{b}\frac{p\pi}{d} \cos\left(\frac{m\pi x}{a}\right)\sin\left(\frac{n\pi y}{b}\right)\cos\left(\frac{p\pi z}{d}\right) \text{ (A/m)} \tag{3.57}$$

From the fields in (3.53)–(3.57), we see that for TE modes, either m or n can be zero (but not both), while p must be nonzero (otherwise the longitudinal component of the field is zero). For $p = 0$ or for $m = n = 0$, all components of the field are zero. The lowest resonant mode is therefore either the TE_{101} or TE_{011}, depending on the dimensions a, b, c.

The resonant frequencies for TE modes are the same as for the TM modes:

$$f_{mnp} = \frac{1}{2\pi\sqrt{\mu\varepsilon}}\sqrt{\left(\frac{m}{a}\right)^2 + \left(\frac{n}{b}\right)^2 + \left(\frac{p}{d}\right)^2} \text{ (Hz)} \tag{3.58}$$

Some of the modes may have the same resonant frequency even though they are different modes. As an example, for a cubic cavity ($a = b = d$), TE_{011} and TE_{101} have the same frequency. These are called degenerate modes, as in waveguides.

3.3.9 Energy relations in a cavity resonator

Power and energy relations in a cavity are defined by the Poynting theorem. Since there is a certain amount of energy stored in the fields of a cavity, the calculation of this energy is an important aspect of analysis. This is particularly obvious if we recall that in a resonant device, these relations change dramatically at or near resonance. This was true with resonant circuits or resonant transmission lines and is certainly true with resonant cavities. The stored and dissipated power in a cavity define the basic qualities of the cavity. A lossless cavity is not practically realizable; therefore, we also defined the quality factor of the cavity, which is a measure of losses in the cavity. A shift in the resonant frequency of the cavity can also be described in terms of energy as we have done in Chapter 2 for general resonant circuits.

To define the energy relations in the cavity, we need to calculate the Poynting vector ($\mathcal{P} = \mathbf{E} \times \mathbf{H}$) in the cavity. From (3.46)–(3.50) or (3.53)–(3.57), we note that the Poynting vector for lossless cavities is purely imaginary, that is, the time-averaged power density in the cavity is zero:

$$\mathcal{P}_{\mathrm{av}} = \frac{1}{2}\mathrm{Re}(\mathbf{E} \times \mathbf{H}^*) = 0 \tag{3.59}$$

This means that there is no real power transferred in or out of the cavity, but there is stored energy in the magnetic and electric fields inside the cavity. From the complex Poynting vector [(D.135) in Appendix D], we have

$$S = j2\omega\int_v \left(\frac{1}{4}\varepsilon\mathbf{E}\cdot\mathbf{E}^* - \frac{1}{4}\mu\mathbf{H}\cdot\mathbf{H}^*\right)dv\ (\mathrm{W}) \tag{3.60}$$

The total time-averaged stored electric and magnetic energy in the cavity can now be written as

$$W_0 = \int_v \left(\frac{\varepsilon\mathbf{E}\cdot\mathbf{E}^*}{4} - \frac{\mu\mathbf{H}\cdot\mathbf{H}^*}{4}\right)dv\ (\mathrm{J}) \tag{3.61}$$

where **E** and **H** are the fields in the cavity, and v the volume of the cavity. This relation is correct at any frequency regardless of resonance.

If the cavity also has wall losses, the time-averaged dissipated power in the cavity walls is

$$P_{\mathrm{loss}} = \frac{R_s}{2}\int_s J_s^2 ds = \frac{R_s}{2}\int_s |H_t|^2 ds\ (\mathrm{W}) \tag{3.62}$$

where R_s is the surface resistance of the cavity walls, H_t is the tangential magnetic field intensity at the walls surface, and s is the internal surface of the cavity walls. In addition, there may also be losses due to the dielectric inside the cavity, and these must be added to (3.62).

The stored energy and the power loss in the cavity define the two most important properties of cavity resonators. These are the resonant frequency and the Q-factor in the cavity. These were discussed at length in Chapter 2 and are taken up again in the following section where analysis of resonant frequency and quality factor is done through the intermediary of energy and power. However, before we leave the discussion of cavity resonators, it is useful to indicate that any power loss in the cavity must be supplied from outside the cavity to maintain resonance. One can easily imagine that if power lost in the cavity is not replaced, the fields will diminish and eventually dissipate. Similarly, if too much power is supplied to the cavity, the fields will increase over time, usually associated with an increase in losses. Thus, the issue of coupling energy to the cavity is important and must be taken care of for the cavity to resonate.

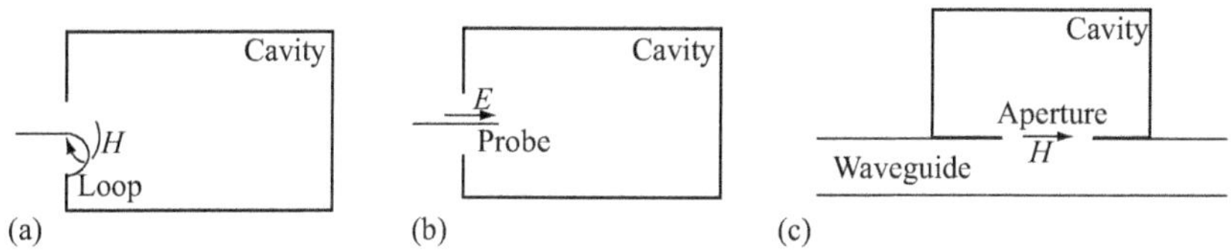

Figure 3.14 (a) Coupling to a cavity by a small loop in the cavity, (b) coupling to a cavity by a small probe in the cavity, and (c) coupling to a cavity by a small aperture in the wall of the cavity

The introduction of energy into the cavity can be done in a number of ways. The most obvious of these is to have a source within the cavity that generates the necessary fields. A small loop [Figure 3.14(a)] or a simple probe excitation [Figure 3.14(b)] can be used. A loop generates a magnetic field intensity, and this magnetic field intensity excites a mode with magnetic field intensity parallel to that generated by the loop. Different modes can be generated by simply locating the probe or the loop at different locations in the cavity, although, for obvious reasons, these must be close to the outer surfaces of the cavity. The same applies to the small probe except that now the excitation is through the electric field intensity of the probe. Similarly, the cavity can be coupled through a small aperture through which a small amount of energy "leaks" into the cavity [Figure 3.14(c)]. In this case, the modes excited are those that have fields parallel to those in the waveguide at the location of the aperture. The three coupling methods in Figure 3.14 excite different modes. These are shown for a cavity, but identical considerations apply to waveguides. We will see in the following chapter that the transmission line resonator used in the fabric sensors or the rubber thickness sensor use the small probe to couple energy into the sensor and the sensor itself is a modified cavity resonator.

3.4 Coupled stripline resonators

Coupled striplines are used extensively for transmission line couplers, filters, transformers, and other applications. However, one important application in the context of this work is their use in open transmission line resonators. Particularly useful for this purpose is the broadside coupled structure of Figure 3.4(b) because of its field distribution (see Figure 3.6), but resonators based on the edge-coupled stripline in Figure 3.4(a) can also be built and have their own application. In either type of line, the ground planes serve as a shield, partially isolating the resonator from external influences (see Figure 3.15). The planes are relatively large compared to the classical striplines. In fact, in most methods of analysis, they are considered as infinite planes. The planes can also be bent, to either entirely enclose the striplines or, as required by an open resonator, to partially enclose the striplines (Figure 3.16). This modification also helps improve the Q-factor of the resonator but, naturally, complicates analysis. A distinct advantage of coupled stripline resonators is that just like the dual mode of propagation, there are also two resonant

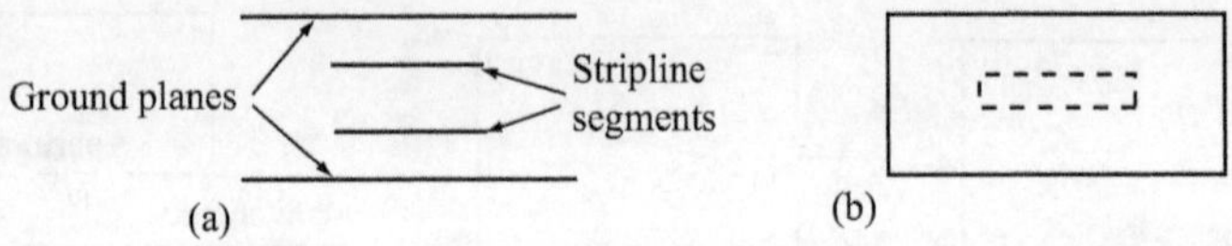

Figure 3.15 A broadside coupled stripline resonator: (a) side and (b) top views

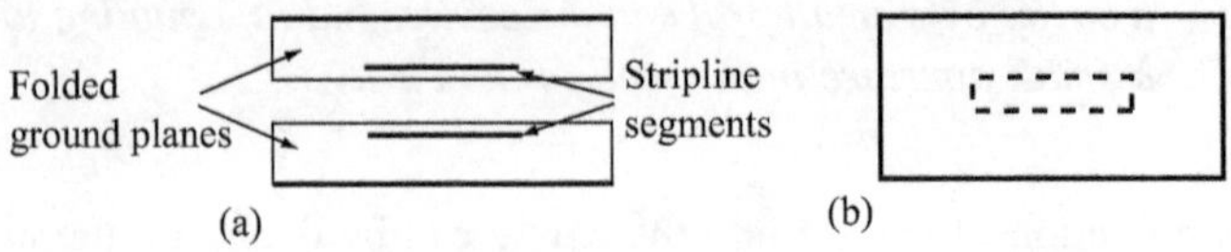

Figure 3.16 Partially shielded, broadside-coupled stripline resonator: (a) side view with striplines recessed from the edges of the folded ground planes and (b) top view

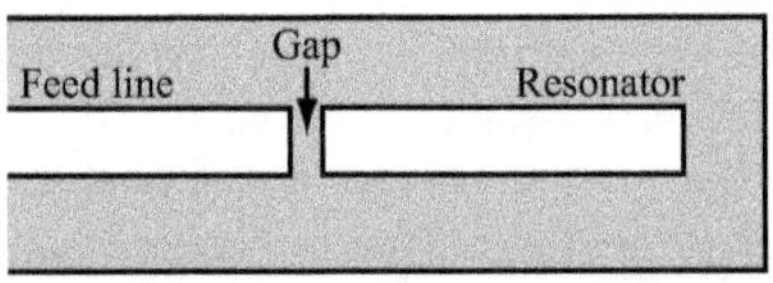

Figure 3.17 Capacitive coupling to a stripline resonator. The gap between the feed line and the resonator serves as a capacitor (top view)

modes. These are also called the odd and even mode of resonance. As can be seen from the field distributions in Figure 3.6, the odd mode produces fields perpendicular to the centerline of the broadside-coupled structure, whereas the even mode produces electric fields parallel to the center line. One can, for example, expect that a thin dielectric sheet on the center line will interact with the electric field in either mode, but its vertical position will affect the even mode much more than the odd mode. Hence, if one is interested in the thickness (or alternatively, the permittivity) of the material, an even-mode excitation is more useful. Bulk effects, such as humidity in the air or temperature, will tend to affect the odd mode and even mode roughly equally so that these effects can be used for compensation purposes.

Resonators, of any type, also require that energy be coupled into the resonator to compensate for power losses. In circuits, that is done through a feedback mechanism, whereas in the tapped transmission line resonator (see e.g., Figure 2.37), the coupling is direct through a transmission line feed. In the microwave cavity resonators in Figure 3.14, the coupling is through an electric probe, a magnetic loop, or through an aperture shared by the waveguide and the cavity. In stripline resonators, the coupling can also be capacitive thorough a gap between the stripline and the feeding transmission line (also a stripline) as shown in Figure 3.17. A direct resistive connection is also possible although not as commonly used. A third possibility, one that is used in this work, is coupling through a probe. This is

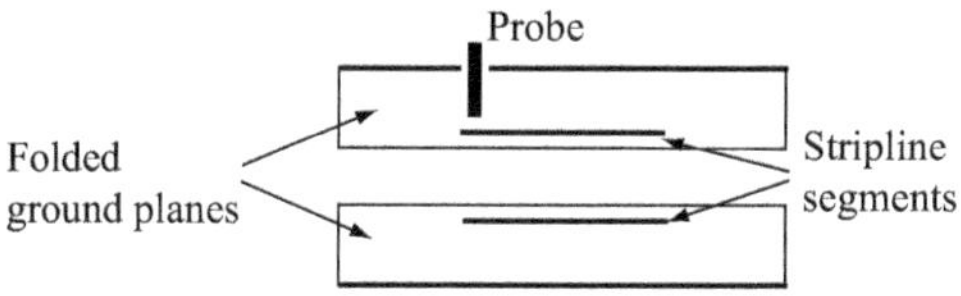

Figure 3.18 Probe coupling to a stripline resonator

depicted in Figure 3.18 for the resonator in Figure 3.16 and is similar to the cavity resonator feed in Figure 3.14(b). The optimization of this coupling is discussed in the following chapter as part of the overall optimization of the sensors but, as in the cavity resonator, the function of the probe is to excite the proper fields at the location it is placed so as to reinforce the resonator field distribution.

There are other properties that we will discuss in the following chapter when we undertake the design of such a resonator.

The resonator in Figure 3.15 is made as a transmission line section between two ground planes but is open on four sides. As a first approximation, one may assume that this structure will behave similar to that in Figure 2.33 or that we could start the analysis using Section 2.11.2. This approach, which looks at resonance in terms of the input impedance of the open line, can supply basic information on the properties of the resonator. Thus, one could assume that a half-wavelength long broadside coupled stripline will produce a resonant frequency equal to $f = c/\lambda$. However, that would imply that we neglect the ground planes entirely and by so doing disregard the issue of even and odd modes. This, of course, is inappropriate since this type of line will resonate at distinct even and odd modes with different resonant frequencies. It is this dual resonance that makes this type of resonator attractive, and in the present work, the dual resonance is critical to the successful development of the sensors and sensing strategies. On the other hand, exact analysis is difficult because of the nonuniform distribution of the fields, especially around the edges. Therefore, we will opt for a numerical evaluation of the resonant frequency and the Q-factor of the resonator as well as the effect of dielectrics in the resonator. Unlike analytical modeling, numerical analysis requires specific numerical dimensions and properties. Thus, we delay this until a resonator has been designed based on specific requirements. We shall do so in the following chapter. However, it is useful to look first at an analytical method of analysis called the perturbation method because it provides a way of understanding how the shift in resonant frequency relates to the properties (geometric and electric) of the space within the resonator and any dielectric inclusion such as when testing materials. It can also provide examples that are easy to calculate and therefore can be used to verify numerical models. The perturbation method is discussed next.

3.5 Resonant cavity perturbation

The discussion in this chapter focuses mostly on transmission lines and transmission line resonators. There is, however, a parallel between transmission lines and waveguides and between transmission line resonators and microwave cavity resonators. In particular, the work reported in this book is based on a partially enclosed transmission line resonator that resembles both in structure and in function a partially open-cavity resonator. This is fortunate because some of the tools for analysis and measurements developed for cavity resonators can be adapted to transmission line resonators either exactly or, often, as an approximation. One such tool that will be used extensively is the idea of cavity perturbation.

In general terms, by measuring the shift in resonant frequency and of the Q-factor of a cavity due to a sample inserted in the cavity or due to changes in material properties of the cavity as a whole, one can determine the complex permittivity of the sample or the material filling the cavity. Alternatively, the properties of materials, introduced into a cavity, change the resonant frequency and the Q factor of the cavity, and these changes can then be related to material properties. If the material is lossless, the shift in resonant frequency alone is sufficient to determine the dielectric constant of the material. In general, however, both the permittivity and the permeability must be considered as complex values. To understand the process by which the resonant frequency shifts, we consider first a closed (cavity) resonator. The extension to open transmission line resonators is then discussed.

3.5.1 Whole cavity perturbation, lossless media

The shift in resonant frequency of a cavity is due to the change in energy stored in the cavity, whereas the change in the Q-factor is due to power loss in the cavity (and hence related to the loss factor of the material within the cavity as well as in the cavity walls). The following discussion starts with the fields in the cavity and calculates the energy relations using the Poynting theorem in complex form, but we assume for now that the medium in the cavity is lossless. To properly account for energy and power, we must account for the fields everywhere in the space of interest. In terms of resonant structures, we assume that the space is bounded by conducting surfaces so that the space is clearly defined and the fields on the boundary are known. Hence, the discussion below assumes a cavity bounded by perfect conductors on which the tangential electric field intensity and the normal magnetic flux density are zero.

A closed resonator with lossless walls (perfect conductors) and a lossless medium with permittivity ε and μ filling the resonator is shown in Figure 3.19(a).

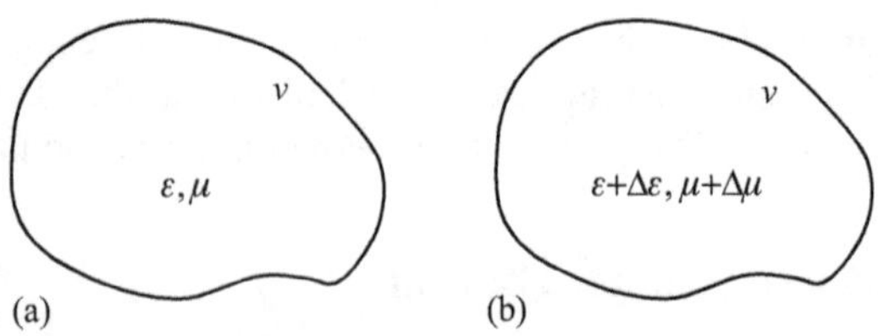

Figure 3.19 The concept of cavity perturbation: (a) unperturbed cavity and (b) whole cavity perturbation

The resonator resonates at a frequency which we denote as ω_0 and is viewed here as the reference frequency of resonance. Assuming the fields within the cavity are $\mathbf{E}_0$ and $\mathbf{H}_0$, we have from Maxwell's equations (see Sections D.1 and D.4) and

$$\nabla \times \mathbf{E}_0 = -j\omega_0\mu\mathbf{H}_0 \tag{3.63}$$

$$\nabla \times \mathbf{H}_0 = j\omega_0\varepsilon\mathbf{E}_0 \tag{3.64}$$

If the cavity is perturbed, that is, if the material properties in the cavity change by small values $\Delta\varepsilon$ and $\Delta\mu$ [Figure 3.19(b)], the fields change to $\mathbf{E}$ and $\mathbf{H}$ and the relations become

$$\nabla \times \mathbf{E} = -j\omega(\mu + \Delta\mu)\mathbf{H} \tag{3.65}$$

$$\nabla \times \mathbf{H} = j\omega(\varepsilon + \Delta\varepsilon)\mathbf{E} \tag{3.66}$$

In these equations, $\mathbf{E}$ and $\mathbf{H}$ are phasors, and ω is the perturbed resonant frequency, distinct from the reference resonant frequency.

To calculate the shift in resonant frequency, we take the scalar product of (3.66) with $\mathbf{E}_0^*$ and the scalar product of the conjugate of (3.63) with $\mathbf{H}$ and add the two together:

$$(\nabla \times \mathbf{H}) \cdot \mathbf{E}_0^* - (\nabla \times \mathbf{E}_0) \cdot \mathbf{H} = j\omega(\varepsilon + \Delta\varepsilon)\mathbf{E} \cdot \mathbf{E}_0^* - j\omega_0\mu\mathbf{H}_0^* \cdot \mathbf{H} \tag{3.67}$$

From the following vector identity

$$\mathbf{H} \cdot (\nabla \times \mathbf{E}_0) - \mathbf{E}_0^* \cdot (\nabla \times \mathbf{H}) = \nabla \cdot \left(\mathbf{E}_0^* \times \mathbf{H}\right) \tag{3.68}$$

we get

$$\nabla \cdot \left(\mathbf{H} \times \mathbf{E}_0^*\right) = j\omega(\varepsilon + \Delta\varepsilon)\mathbf{E} \cdot \mathbf{E}_0^* - j\omega_0\mu\mathbf{H}_0^* \cdot \mathbf{H} \tag{3.69}$$

Similarly, taking the scalar product of (3.65) with $\mathbf{H}_0^*$, the conjugate of (3.64) with $\mathbf{E}$, adding them together and using again the identity in (3.68) we get

$$\nabla \cdot \left(\mathbf{H}_0^* \times \mathbf{E}\right) = j\omega(\mu + \Delta\mu)\mathbf{H} \cdot \mathbf{H}_0^* - j\omega_0\varepsilon\mathbf{E}_0^* \cdot \mathbf{E} \tag{3.70}$$

The sum of these two equations is

$$\begin{aligned}\nabla \cdot \left(\mathbf{H}_0^* \times \mathbf{E}\right) + \nabla \cdot \left(\mathbf{H} \times \mathbf{E}_0^*\right) &= j\omega(\mu + \Delta\mu)\mathbf{H} \cdot \mathbf{H}_0^* - j\omega_0\varepsilon\mathbf{E}_0^* \cdot \mathbf{E} \\ &\quad + j\omega(\varepsilon + \Delta\varepsilon)\mathbf{E} \cdot \mathbf{E}_0^* - j\omega_0\mu\mathbf{H}_0^* \cdot \mathbf{H}\end{aligned} \tag{3.71}$$

We now integrate the expression over the volume of the cavity:

$$\begin{aligned}&\int_v \left[\nabla \cdot \left(\mathbf{H}_0^* \times \mathbf{E}\right) + \nabla \cdot \left(\mathbf{H} \times \mathbf{E}_0^*\right)\right] dv \\ &= \int_v \left[j\omega(\mu + \Delta\mu)\mathbf{H} \cdot \mathbf{H}_0^* - j\omega_0\varepsilon\mathbf{E}_0^* \cdot \mathbf{E} + j\omega(\varepsilon + \Delta\varepsilon)\mathbf{E} \cdot \mathbf{E}_0^* - j\omega_0\mu\mathbf{H}_0^* \cdot \mathbf{H}\right] dv\end{aligned} \tag{3.72}$$

Applying the divergence theorem to the left-hand side

$$\int_v [\nabla \cdot (\mathbf{H}_0^* \times \mathbf{E}) + \nabla \cdot (\mathbf{H} \times \mathbf{E}_0^*)] dv = \oint_s (\mathbf{H}_0^* \times \mathbf{E}) \cdot d\mathbf{s} + \oint_s (\mathbf{H} \times \mathbf{E}_0^*) \cdot d\mathbf{s} \tag{3.73}$$

This integral vanishes since **E** and $\mathbf{E}_0$ are normal to the surface of the perfectly conducting cavity wall, whereas **H** and $\mathbf{H}_0$ must be tangential. Thus, we have

$$\int_v [j\omega(\mu + \Delta\mu)\mathbf{H} \cdot \mathbf{H}_0^* - j\omega_0\varepsilon\mathbf{E}_0^* \cdot \mathbf{E} + j\omega(\varepsilon + \Delta\varepsilon)\mathbf{E} \cdot \mathbf{E}_0^* - j\omega_0\mu\mathbf{H}_0^* \cdot \mathbf{H}] dv = 0 \tag{3.74}$$

or, expanding the expression after dividing both sides by j:

$$\omega \int_v \mu\mathbf{H} \cdot \mathbf{H}_0^* dv + \omega \int_v \Delta\mu\mathbf{H} \cdot \mathbf{H}_0^* dv - \omega_0 \int_v \varepsilon\mathbf{E}_0^* \cdot \mathbf{E} dv + \omega \int_v \varepsilon\mathbf{E} \cdot \mathbf{E}_0^* dv$$
$$+ \omega \int_v \Delta\varepsilon\mathbf{E} \cdot \mathbf{E}_0^* dv - \omega_0 \int_v \mu\mathbf{H}_0^* \cdot \mathbf{H} dv = 0 \tag{3.75}$$

or

$$(\omega - \omega_0)\left[\int_v \mu\mathbf{H} \cdot \mathbf{H}_0^* dv + \int_v \varepsilon\mathbf{E}_0^* \cdot \mathbf{E} dv\right] = -\omega \int_v \Delta\varepsilon\mathbf{E} \cdot \mathbf{E}_0^* dv - \omega \int_v \Delta\mu\mathbf{H} \cdot \mathbf{H}_0^* dv \tag{3.76}$$

This gives

$$\frac{\omega - \omega_0}{\omega} = -\frac{\int_v \Delta\varepsilon\mathbf{E} \cdot \mathbf{E}_0^* dv + \int_v \Delta\mu\mathbf{H} \cdot \mathbf{H}_0^* dv}{\int_v \mu\mathbf{H} \cdot \mathbf{H}_0^* dv + \int_v \varepsilon\mathbf{E}_0^* \cdot \mathbf{E} dv} \tag{3.77}$$

The result in (3.77) is exact under the assumptions it was developed. If we can assume that the change in permittivity and permeability is small, then **E** is approximately equal to $\mathbf{E}_0$ and **H** to $\mathbf{H}_0$. Similarly, because the change in frequency is small, we can approximate ω in the denominator by ω_0, and we have

$$\frac{\omega - \omega_0}{\omega_0} \approx -\frac{\int_v \Delta\varepsilon|E_0|^2 dv + \int_v \Delta\mu|H_0|^2 dv}{\int_v \varepsilon|E_0|^2 dv + \int_v \mu|H_0|^2 dv} \tag{3.78}$$

The exact relation in (3.77) or the approximation in (3.78) may be used to calculate the resonant frequency due to changes in permeability and/or permittivity. It should be emphasized again that this can only be done if we know the electric and magnetic field intensities throughout the cavity. In general, this is not the case, but in practice, the fields of specific cavities are known and may be used in (3.78).

In most cases of practical interest, the media in cavities are dielectrics with permeability equal to μ_0. Under these conditions, (3.77) and (3.78) become

$$\frac{\omega - \omega_0}{\omega} = -\frac{\int_v \Delta\varepsilon \mathbf{E} \cdot \mathbf{E}_0^* dv}{\int_v \mu_0 \mathbf{H} \cdot \mathbf{H}_0^* dv + \int_v \varepsilon \mathbf{E}_0^* \cdot \mathbf{E} dv} \tag{3.79}$$

$$\frac{\omega - \omega_0}{\omega_0} \approx -\frac{\int_v \Delta\varepsilon |E_0|^2 dv}{\int_v \varepsilon |E_0|^2 dv + \int_v \mu_0 |H_0|^2 dv} \tag{3.80}$$

3.5.2 Cavity perturbation by small, lossless material samples

The discussion above assumes that the whole interior of the cavity is uniformly perturbed. However, more often, only part of the cavity will be perturbed as in, for example, testing a small sample in the cavity (see Figure 3.20). Now, of course, the change in resonant frequency depends on the volume of the sample and, significantly, where in the sample within the cavity it is located. For example, a small dielectric sample placed at a location of zero electric field intensity will produce little or no change in resonant frequency. To calculate the change in resonant frequency, one must be able to relate the electric and magnetic field intensities in the sample to those in the unperturbed part of the cavity. This is not possible in general although it is relatively simple for some samples, depending on their shape and is based on interface conditions at the sample's boundaries. This method simply assumes that the fields within the sample and the cavity relate to each other as in the static case (a quasistatic approximation). To obtain a general expression for the shift in resonant frequency, we note that the denominator in (3.77) or (3.78) is the total energy stored in the cavity, whereas the nominator is the energy due to the change in material properties. Since a sample introduced in a cavity is small, we may safely approximate the fields in the denominator as in (3.77) and (3.78) because the change in total energy stored in the cavity is small. These are $\mathbf{E}_0$

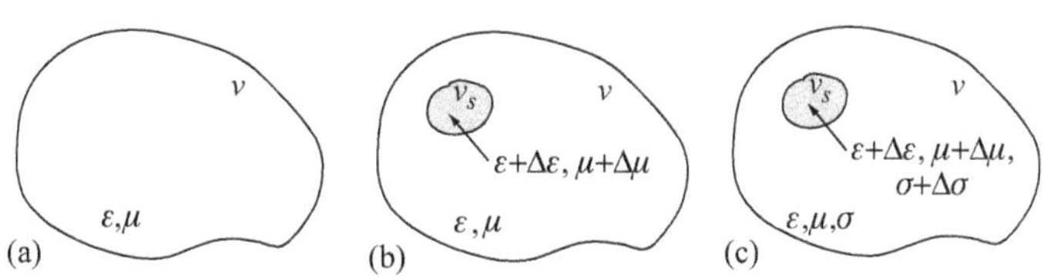

Figure 3.20 Cavity perturbation by small sample: (a) empty cavity,(b) small lossless sample in the cavity, and (c) small lossy sample in the cavity

and $\mathbf{H}_0$. In the nominator, we must use the fields internal to the sample for **E** and **H**. We denote these as $\mathbf{E}_i$ and $\mathbf{H}_i$. Thus, we can write

$$\frac{\omega - \omega_0}{\omega_0} = -\frac{\int_v \Delta\varepsilon \mathbf{E}_i \cdot \mathbf{E}_0^* dv + \int_v \Delta\mu \mathbf{H}_i \cdot \mathbf{H}_0^* dv}{\int_v \mu \mathbf{H}_0 \cdot \mathbf{H}_0^* dv + \int_v \varepsilon \mathbf{E}_0^* \cdot \mathbf{E}_0 dv} \tag{3.81}$$

or

$$\frac{\omega - \omega_0}{\omega_0} \approx -\frac{\int_v \Delta\varepsilon \mathbf{E}_i \cdot \mathbf{E}_0^* dv + \int_v \Delta\mu \mathbf{H}_i \cdot \mathbf{H}_0^* dv}{\int_v \varepsilon |E_0|^2 dv + \int_v \mu |H_0|^2 dv} \tag{3.82}$$

In most cases of interest, the materials in the cavity are dielectrics. In these cases, $\Delta\mu$ is zero and the stored magnetic energy equals the stored electric energy:

$$\frac{\omega - \omega_0}{\omega_0} \approx -\frac{\int_v \Delta\varepsilon \mathbf{E}_i \cdot \mathbf{E}_0^* dv}{2\int_v \varepsilon |E_0|^2 dv} \tag{3.83}$$

Most cavities are air-filled with permittivity ε_0. For a sample of isotropic (complex) permittivity ε, $\Delta\varepsilon = \varepsilon - \varepsilon_0$ and the approximation may be written as

$$\frac{\omega - \omega_0}{\omega_0} \approx -(\varepsilon_r - 1)\frac{\int_{v_s} \mathbf{E}_i \cdot \mathbf{E}_0^* dv}{2\int_v |E_0|^2 dv} \tag{3.84}$$

where ε_r is the relative permittivity of the perturbing sample and v_s the volume of the perturbation. The latter can be the whole volume of the cavity (in the case of perturbation in the permittivity of the space in the cavity, such as due to moisture in air) or the volume of the sample (such as in testing of permittivity of dielectrics introduced into the cavity). In essence, the shift in resonant frequency is proportional to the volume of the sample and its relative permittivity.

In this case, we have also assumed that the electric and magnetic energy in the cavity remain equal because the perturbation is small.

3.5.3 *Cavity perturbation, lossy media*

If the perturbing medium is lossy, the permittivity is complex. We start by denoting the perturbation in material conductivity, permittivity, and permeability as $\sigma + \Delta\sigma$, $\varepsilon + \Delta\varepsilon$, $\mu + \Delta\mu$ where σ, ε, and μ are the properties of the unperturbed cavity [see Figure 3.20(c)].

Starting again with Maxwell's equations, we write for the unperturbed cavity:

$$\nabla \times \mathbf{E}_0 = -j\omega_0\mu\mathbf{H}_0 \tag{3.85}$$

$$\nabla \times \mathbf{H}_0 = (\sigma + j\omega_0\varepsilon)\mathbf{E}_0 \tag{3.86}$$

For the perturbed cavity, we have

$$\nabla \times \mathbf{E} = -j\omega(\mu + \Delta\mu)\mathbf{H} \tag{3.87}$$

$$\nabla \times \mathbf{H} = (\sigma + j\omega\varepsilon)\mathbf{E} = (\sigma + \Delta\sigma + j\omega(\varepsilon + \Delta\varepsilon))\mathbf{E} \tag{3.88}$$

The following vector identity applies in this case

$$\nabla \cdot (\mathbf{H} \times \mathbf{E}_0) = \mathbf{E}_0 \cdot (\nabla \times \mathbf{H}) - \mathbf{H} \cdot (\nabla \times \mathbf{E}_0) \tag{3.89}$$

Substituting from (3.41) and (3.39)

$$\nabla \cdot (\mathbf{H} \times \mathbf{E}_0) = (\sigma + \Delta\sigma + j\omega(\varepsilon + \Delta\varepsilon))\mathbf{E} \cdot \mathbf{E}_0 + j\omega_0\mu\mathbf{H}_0 \cdot \mathbf{H} \tag{3.90}$$

and substituting from (3.88) and (3.86):

$$\nabla \cdot (\mathbf{H} \times \mathbf{E}) = (\sigma + j\omega\varepsilon)\mathbf{E} \cdot \mathbf{E}_0 + j\omega_0(\mu + \Delta\mu)\mathbf{H}_0 \cdot \mathbf{H} \tag{3.91}$$

Subtracting the former from the latter:

$$\begin{aligned}\nabla \cdot (\mathbf{H} \times \mathbf{E}) - \nabla \cdot (\mathbf{H} \times \mathbf{E}_0) &= (\sigma + j\omega\varepsilon)\mathbf{E} \cdot \mathbf{E}_0 + j\omega_0(\mu + \Delta\mu)\mathbf{H}_0 \cdot \mathbf{H} \\ &\quad - (\sigma + \Delta\sigma + j\omega(\varepsilon + \Delta\varepsilon))\mathbf{E} \cdot \mathbf{E}_0 + j\omega_0\mu\mathbf{H}_0 \cdot \mathbf{H} \\ &= \nabla \cdot (\mathbf{H} \times \mathbf{E}_0) = (\sigma + \Delta\sigma + j\omega(\varepsilon + \Delta\varepsilon))\mathbf{E} \cdot \mathbf{E}_0 + j\omega_0\mu\mathbf{H}_0 \cdot \mathbf{H} \\ &= j(\omega - \omega_0)[\varepsilon\mathbf{E} \cdot \mathbf{E}_0 - \mu\mathbf{H} \cdot \mathbf{H}_0] - j\omega\Delta\mu\mathbf{H} \cdot \mathbf{H}_0 + j\omega\left(\Delta\varepsilon - j\frac{\Delta\sigma}{\omega}\right)\mathbf{E} \cdot \mathbf{E}_0\end{aligned} \tag{3.92}$$

Integrating over the volume on both sides:

$$\begin{aligned}\int_v [\nabla \cdot (\mathbf{H} \times \mathbf{E}) - \nabla \cdot (\mathbf{H} \times \mathbf{E}_0)]dv &= \oint_s [(\mathbf{H} \times \mathbf{E}) - (\mathbf{H} \times \mathbf{E}_0)] \cdot d\mathbf{s} \\ &= \int_v \left[j(\omega - \omega_0)[\varepsilon\mathbf{E} \cdot \mathbf{E}_0 - \mu\mathbf{H} \cdot \mathbf{H}_0] - j\omega\Delta\mu\mathbf{H} \cdot \mathbf{H}_0 + j\omega\left(\Delta\varepsilon \quad j\frac{\Delta\sigma}{\omega}\right)\mathbf{E} \cdot \mathbf{E}_0\right]dv\end{aligned} \tag{3.93}$$

where we used the divergence theorem to convert the volume to a closed surface integral on the left-hand side. Because s is a conducting surface, $\hat{\mathbf{n}} \times \mathbf{E}_0 = 0$ and $\hat{\mathbf{n}} \times \mathbf{E} = 0$ so that the left-hand side is zero (no power propagation across the boundary). Hence,

$$\int_v \left[j(\omega - \omega_0)[\varepsilon\mathbf{E} \cdot \mathbf{E}_0 - \mu\mathbf{H} \cdot \mathbf{H}_0] - j\omega\Delta\mu\mathbf{H} \cdot \mathbf{H}_0 + j\omega\left(\Delta\varepsilon - j\frac{\Delta\sigma}{\omega}\right)\mathbf{E} \cdot \mathbf{E}_0\right]dv = 0 \tag{3.94}$$

Rearranging terms

$$\frac{\omega - \omega_0}{\omega} = -\frac{\int_v [(\Delta\varepsilon - j(\Delta\sigma/\omega))\mathbf{E} \cdot \mathbf{E}_0 - \Delta\mu\mathbf{H} \cdot \mathbf{H}_0]dv}{\int_v [\varepsilon\mathbf{E} \cdot \mathbf{E}_0 - \mu\mathbf{H} \cdot \mathbf{H}_0]dv} \tag{3.95}$$

or, approximating ω in the denominator with ω_0:

$$\frac{\omega - \omega_0}{\omega_0} = -\frac{\int_v [(\Delta\varepsilon - j(\Delta\sigma/\omega))\mathbf{E} \cdot \mathbf{E}_0 - \Delta\mu\mathbf{H} \cdot \mathbf{H}_0]dv}{\int_v [\varepsilon\mathbf{E} \cdot \mathbf{E}_0 - \mu\mathbf{H} \cdot \mathbf{H}_0]dv} \tag{3.96}$$

Note that this is completely equivalent to (3.82) since **H** and **E** are 90° out of phase (in space); hence, the energy terms in the brackets add up. This means that we can still use the form in (3.83) for the case where $\Delta\mu = 0$:

$$\frac{\omega - \omega_0}{\omega_0} = -\frac{\int_v [(\Delta\varepsilon - j(\Delta\sigma/\omega))\mathbf{E} \cdot \mathbf{E}_0]dv}{2\int_v [\varepsilon\mathbf{E} \cdot \mathbf{E}_0]dv} \tag{3.97}$$

Because the perturbation is small, we can usually assume in the denominator: $E \sim E_0$

$$\frac{\omega - \omega_0}{\omega_0} = -\frac{\int_{v_s} [(\Delta\varepsilon - j(\Delta\sigma/\omega))\mathbf{E} \cdot \mathbf{E}_0]dv}{2\int_v \varepsilon E_0^2 dv} \tag{3.98}$$

As in the lossless media case, the integration in the nominator is only on the perturbed section of the cavity, whereas in the denominator, it is over the whole cavity.

It should be noted here that both ω and ω_0 must be complex, whereas in the lossless case, both ω and ω_0 were real.

In the lossless (ideal) cavity, the quality factor is infinite, but in the lossy cavity, it is finite and is related to the losses in the cavity. In (3.97) or (3.98), the losses are due to the change in conductivity caused by the perturbation. In particular, if the change is due to the introduction of a lossy sample into the cavity, then $\Delta\varepsilon$ is the change in the dielectric constant of the sample, σ_s is the conductivity of the sample, and the quantity in the round brackets can be written in terms of the complex permittivity of the sample, ε_c:

$$\varepsilon_c = \varepsilon' + j\varepsilon'' = \varepsilon' - j\frac{\sigma_s}{\omega} \tag{3.99}$$

Equations (3.97) and (3.98) may now be written as

$$\frac{\omega - \omega_0}{\omega_0} = -\frac{\int_v [(\varepsilon' - j\varepsilon'' - \varepsilon_0)\mathbf{E} \cdot \mathbf{E}_0]dv}{2\int_v [\varepsilon_0\mathbf{E} \cdot \mathbf{E}_0]dv} \tag{3.100}$$

$$\frac{\omega - \omega_0}{\omega_0} = -\frac{\int_v [(\varepsilon' - j\varepsilon'' - \varepsilon_0)\mathbf{E} \cdot \mathbf{E}_0] dv}{2\int_v \varepsilon_0 {E_0}^2 dv} \tag{3.101}$$

Using the idea of the complex resonant frequency in (2.190) where we replaced the resonant frequency with a complex resonant frequency, we can write for the cavity without the inclusion:

$$\omega = \omega_0 + j\frac{\omega_0}{2Q_0} \tag{3.102}$$

where ω_0 is clearly the resonant frequency of the empty cavity and Q_0 its loaded Q-factor.

We can do the same for the perturbed cavity and write

$$\omega = \omega_r + j\frac{\omega_r}{2Q_s} \tag{3.103}$$

where ω_r is the resonant frequency of the perturbed cavity and Q_s the loaded Q-factor with the sample present:

$$\omega_r + j\frac{\omega_r}{2Q_s} - \omega_0 = -\frac{\omega_0 \int_v [(\varepsilon' - j\varepsilon'' - \varepsilon_0)\mathbf{E} \cdot \mathbf{E}_0] dv}{2\int_v \varepsilon_0 {E_0}^2 dv} \tag{3.104}$$

Equating real terms on the two sides of (3.104):

$$\omega_r - \omega_0 = -\frac{\omega_0 \int_v [(\varepsilon' - \varepsilon_0)\mathbf{E} \cdot \mathbf{E}_0] dv}{2\int_v \varepsilon_0 {E_0}^2 dv} \tag{3.105}$$

This equation is identical to (3.84) indicating that the losses do not affect the resonant frequency.

Equating the imaginary terms, we get

$$\frac{\omega_r}{Q_s} = \frac{\omega_0 \int_v [\varepsilon'' \mathbf{E} \cdot \mathbf{E}_0] dv}{\int_v \varepsilon_0 {E_0}^2 dv} \tag{3.106}$$

Since the change in frequency is small, we can write $\omega_r \approx \omega_0$ to obtain

$$\frac{1}{Q_s} = \frac{\int_v [\varepsilon'' \mathbf{E} \cdot \mathbf{E}_0] dv}{\int_v \varepsilon_0 E_0^2 dv} \tag{3.107}$$

This assumes that the losses in the unperturbed cavity are negligible. If, however, they are not, then (3.107) is modified as

$$\frac{1}{Q} = \frac{\int_v [\varepsilon'' \mathbf{E} \cdot \mathbf{E}_0] dv}{\int_v \varepsilon_0 E_0^2 dv} \tag{3.108}$$

where

$$\frac{1}{Q} = \frac{1}{Q_s} - \frac{1}{Q_0} \tag{3.109}$$

and Q_0 is the Q-factor of the unperturbed cavity.

The use of the perturbation formulas above requires that the fields within the whole of the cavity and within the perturbed space be known so that the various volume integrals may be calculated. In many cases, this is, at best, an arduous task. In others, such as in the case of small material samples, one can assume that $\mathbf{E} \sim \mathbf{E}_0$, that is, that the perturbation changes the fields within the sample very little, to simplify he integrals. Yet, from these formulas, it is clear that the shift in the resonant frequency depends on these fields and, as a consequence, on where in the cavity the perturbation occurs. If, for example, a sample is placed in a place where the fields are low (such as at the walls of the cavity or close to corners), the shift in resonant frequency will be minimal, whereas the same sample placed at a locations of very high field will result in a large shift in the frequency. Certain relatively simple configurations can be treated analytically. These include dielectric sheets and disks, spherical and cylindrical dielectrics, and ellipsoids. Other, more complex configurations must be analyzed numerically both for the shift in resonant frequency and for the calculation of the Q-factor. Even for the simple configurations indicated here, the calculation is only valid within enclosed cavities since we assumed the tangential electric field intensity and the normal magnetic field intensity vanish on the boundary. Open or partially open cavities must, again, be treated numerically in the majority of cases. Nevertheless, the perturbation method is valuable as a tool in design because it allows one to approximate a solution and reach a preliminary design that can then be followed by a full numerical analysis for more accurate results, should that be needed.

The perturbation method can also be used to calculate the change in Q-factor due to geometry or dielectric perturbations throughout the cavity. However, this aspect of the method is not discussed here because it has limited relevance to the present work. The discussion above should at least show how this can be done in general.

Bibliography

The theory of transmission lines was discussed in Chapter 2. The extension to coupled transmission lines and in particular to broadside coupled transmission lines (but not resonators) are discussed in Chapter 7 of [1]. The reference delves into other types of planar structures. A more concise description of broadside coupled transmission lines can be found in [2] and pages 156–160 in [3]. A particularly clear and concise approach to coupled transmission lines and their use in couplers can be found in [4]. Reference [10] discusses many aspects and applications of coupled microstrip lines, and although microstrip lines have properties that differ significantly from striplines, the source is useful in its treatment of methods of analysis as well as applications. Properties of coupled striplines and calculation of impedances can be found in [7,10].

[1] B. Bhat and S. K. Koul, "Stripline-Like Transmission Lines for Microwave Integrated Circuits," John Wiley & Sons, New York, 1989.

[2] E. Nyfors and P. Vainikainen, "Industrial Microwave Sensors," 1991 IEEE MTT-S Digest, 1991, pp. 1009–1012.

[3] E. Nyfors and P. Vainikainen, "Industrial Microwave Sensors," Artech House, Norwood, MA. 1989.

[4] D. M. Pozar, "Microwave Engineering," 2nd edition, John Wiley & Sons, Inc., New York, 1998.

[5] T. H. Lee, "Planar Microwave Engineering: A Practical Guide to Theory, Measurement, and Circuits," Cambridge University Press, Cambridge, 2004.

[6] C. Nguyen, "Analysis Methods for RF, Microwave and Millimeter-Wave Planar Transmission Line Structures," Wiley, New York, 2000.

Other sources for broadside coupled transmission lines as well as other planar structures are

[7] R. Mongia, I. Bahl and P. Bhartia, "RF and Microwave Coupled-Line Circuits," Artech House, Boston, 1999.

[8] R. Garg, I. Bahl and M. Bozzi, "Microstrip Lines and Slotlines," Artech House, Boston, 2013.

[9] T. Itoh, "Planar Transmission Line Structures," IEEE Press, New York, 1987.

[10] S. B. Cohn, "Characteristic impedances of broadside coupled strip transmission lines," IEEE Transactions on Microwave Theory and Techniques, Vol. MTT-8, June 1960, pp. 633–637.

[11] H. Howe, "Stripline Circuit Design," Artech House, Norwood, MA, 1974.

[12] P. Bhartia and P. Pramanick, "Computer-aided design models for broadside-coupled striplines and millimeter-wave suspended substrate microstrip lines," IEEE Transactions on Microwave Theory and Techniques, Vol. 36, No. 11, November 1988, pp. 1476–1481.

[13] S. Iezekiel, "Microwave Photonics: Devices and Applications," Wiley, New York, 2009.

[14] R. Mongia, I. J. Bahl and P. Bhartia, "RF and MW Coupled-Line Circuits," Artech House, Norwood, MA, 1999.

[15] "Design Equations for Broadside and Edgewise Stripline Couplers," Rogers Corporation, 1998.

[16] L. Young, "Parallel Coupled Lines and Directional Couplers," Artech House, Norwood, MA, 1972.

[17] J. Helszajn, "The Stripline Circulators: Theory and Practice," Wiley, New York, 2008.

The short section on waveguides and cavity resonators in this work is intended simply to show the connection between the type of transmission line resonator used here and the classical cavity resonators. It also serves as introduction to the following sections on stripline resonators and, more importantly, on cavity perturbation. For further information on waveguides and cavity resonators, including nonrectangular structures, one can consult any source on microwave and waveguides including [4,18–21]. Almost any textbook on electromagnetics will have the fundamentals; some more advanced texts will also discuss nonrectangular structures and optical as well as dielectric waveguides and resonators.

[18] R. E. Collin, "Foundations for Microwave Engineering," 2nd edition, McGraw-Hill, New York, NY, 1992.

[19] C. A. Balanis, "Advanced Engineering Electromagnetics," 2nd edition, Wiley, 2012.

[20] N. Marcuvitz, "Waveguide Handbook," IET, London, 1986.

[21] A. Das and S. K. Das, "Microwave Engineering," McGraw Hill, Boston, MA, 2000.

[22] K. Chang, "Encyclopedia of RF and Microwave Engineering," Wiley, New York, 2005.

[23] T. Rozzi and M. Mongiardo, "Open Electromagnetic Waveguides," IET, London, 1987.

Perturbation theory is discussed in a number of good references. The sections on perturbation methods in this work follows that in [24] which gives the general formulae as well as examples of calculation in layered structures and in spherical geometries. Additional examples can be found in [25–28].

[24] R. F. Harrington, "Time-Harmonic Electromagnetic Fields," McGraw-Hill, NY, 1961.

[25] W. R. Smith, "Static and Dynamic Electricity," McGraw-Hill, New York, 1950, pp. 67–68.

[26] J. Van Bladel, "Electromagnetic Fields," Hemisphere Publishing, New York, 1985.

[27] J. A. Stratton, "Electromagnetic Theory," McGraw-Hill, New York, 1941, pp. 205–213.

[28] M. Fischer, P. Vainikainen and E. Nypors, "Design Aspects of Stripline Resonator Sensors for Industrial Application," Report S 214, Helsinki University of Technology, August 1995.

In addition to the above, there are many good research papers on specific subjects as well as tutorial and calculators that can be found online and used in the design. These were not included here.

Chapter 4

Microwave measurements

4.1 Introduction

Although the frequencies used in the two sensors described in this work are relatively low, below 500 MHz, and microwaves are usually seen as being above that, the term microwave measurements is appropriate for two reasons. First, the methods and the instrumentation used are normally associated with microwaves, and, in fact, the common designation of the microwave range includes these frequencies. Of course, one can get away from this slight difficulty by using the term "high frequency" measurements but that is a less well-defined and broader concept. Second, related reason lies with the fact that the range of frequencies usually associated with microwaves is arbitrary and what matters are the measurements and how they can be performed most accurately.

In general terms, at low frequencies, one can measure quantities such as voltages, currents impedance, frequency, and power in circuits directly using relatively simple instruments, familiar to any engineer. The use of Ohm's and Kirchhoff's laws allows one to analyze circuits quickly and accurately. These are based on a lumped-parameter approach, which, in turn, is based on the fact that at low frequencies, the wavelength is long compared to the dimension of any component of the circuit. Thus, for example, we can safely assume that on a conductor, the voltage is constant, and the current entering and leaving the conductor is equal in magnitude and phase. Under these conditions, a (passive) circuit can be reduced to a collection of components (resistive, capacitive, and inductive) producing a mesh of currents and voltages.

As the frequency increases, the wavelength becomes smaller and, eventually, the lumped approach fails since the phases of currents and voltages in any segment of the circuit cannot be assumed to remain constant over any length of conductors. In effect, we must assume that voltages and currents are a result of propagation along the segment. The circuit itself must be described as a distributed parameter circuit, very different than the low-frequency representation. The common low-frequency circuit approach using loop currents and node voltages cannot be used, and one must then either use field-analysis methods through solution of Maxwell's equations or adapt circuit analysis to high-frequency applications. Clearly, not all microwave problems can be analyzed using circuit theory extensions. If one is interested in the behavior in a space, either inside a medium or in free space, there is no alternative to field analysis through solution of Maxwell's equations. In principle, of course, all problems can be analyzed through Maxwell's equations although the difficulties of doing so can be enormous. Transmission lines can and almost always are analyzed through circuit concepts adapted to high-frequency propagation using methods "borrowed" from field analysis. In Chapter 2, we did exactly that: We used the ideas of propagation, transmission, reflection and attenuation, and the concept of plane waves [transverse electromagnetic modes (TEM) propagation] to analyze the behavior of voltages, currents, and impedances on the line. Since on transmission lines one can measure voltages and currents, this approach is natural. Other microwave problems that are not clearly defined circuits, such as waveguides, are better analyzed by using electric and magnetic fields. They can also be analyzed by use of equivalent voltages and equivalent currents although one cannot measure these directly. In this approach, the equivalent voltages are proportional to the electric-field intensities and the equivalent currents are proportional to the magnetic-field intensities. Because the fields vary from location to location and depend on the mode of propagation in the waveguide, the equivalent voltages and currents must be set so that their product provides the power flow of the mode at any location in the waveguide. Clearly too, their ratio must equal the wave impedance. Beyond transmission lines and waveguides, the only alternative is a full field analysis.

Microwave measurements follow these methods of analysis. In transmission lines and in waveguides, one can use network-analysis methods since one can define proper ports for measurements. Thus, one-, two-, and N-port analysis methods can be applied directly. This chapter will discuss network measurements since the present work deals entirely with transmission lines and transmission line resonators. Also, because the sensors described in this work are intended to be connected directly to a network analyzer, understanding of these measurements is crucial to understanding of the operation and specifications of the sensors. However, in order to understand how a network analyzer can perform its task, it is important to understand how the various microwave quantities are defined and measured. We will therefore discuss some general methods of measurement and then link them to the network analyzer through the use of the S-parameters. We will discuss measurements in terms of the S-parameters since in the present work, we will only have recourse to measurements in transmission lines and cavity resonators and because the instrument we use to perform these measurements is a network analyzer.

Although one would think that network analysis is limited to true networks, that is, systems in which an input connection and an output connection of some sort can be identified and connected to, this is not the case. One can easily connect antennas to the ports of the network analyzer and by doing so convert it into a radar system, which then can be used to analyze waves in open space. By doing so, one can extend the measurements to either enclosed or open spaces. That is not to say that it is easy to do so since issues of noise, external sources, and influences from structures must be taken into account, but with properly controlled measuring system measurements in space, especially in enclosed space, are possible and can be accurate. In some cases, such as measurements in a resonant cavity, where the waves are contained by the walls of the cavity, this is particularly simple and, with some modifications, this is the method used in this work.

Other measurements such as antenna measurements and free-space measurements will not be discussed except as necessary for the present work. In addition, measurements of components properties and responses, although commonly performed, will not be discussed here. These are one-port measurements, whereas we restrict ourselves to two-port measurements, simply because that is the type of measurement needed in the context of this work.

4.2 *N*-Port networks

At low frequencies, a port is a physical connection, a pair of terminals in which one can measure voltages on or currents into the terminals. Thus, one can, as indicated in the introduction, measure voltages, currents, impedance, and power by direct means. This is in general also the case on transmission lines in which the ports are well defined. In general microwave applications, this is not the case and a port is defined somewhat differently, as a plane transverse to the direction of propagation of the wave. In a waveguide or transmission line, this is a plane transverse to the line or waveguide. At these ports, one can measure incident, reflected, and transmitted power and hence fields or equivalent voltages and currents. Clearly then, the analysis of low-frequency networks and microwave networks will be different. At low frequencies, we typically use impedance or admittance parameters, whereas in microwaves the primary parameters are called scattering parameters since they define reflection and transmission at ports. Nevertheless, the methods are similar in that they borrow from circuit analysis and, perhaps more importantly, because the impedance and scattering parameters are related.

In what follows, we start with the impedance parameters, followed by the scattering parameters and then show the relations between them. In this work, we will have recourse to measurements in two-port systems, but it is perhaps more useful and certainly more general to start with an *N*-port network which can then be reduced to a two-port or a one-port network as needed. Consider Figure 4.1, which shows an *N*-port network of transmission lines. On each line, there are incident and backward (reflected) currents and voltages, and on each line there is a reference port (sometimes called termination port or even load port) at which the phases of the current and voltage are known. The reference port is indicated symbolically by the dashed line. Incident waves, denoted with a (+), propagate toward the port, reflected waves, indicated with a (−), propagate away from the port as shown by the thin, horizontal arrowed lines. The network shown in Figure 4.1 can also describe an equivalent network of, say, waveguides in which the voltages and currents shown are the equivalent voltages and currents as described in the introduction.

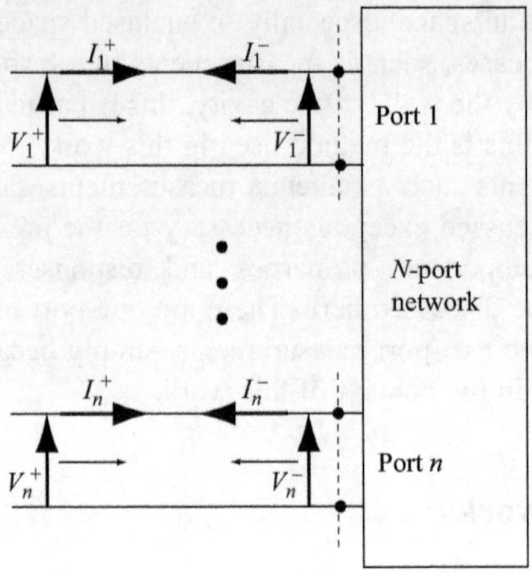

Figure 4.1 An arbitrary N-port network of transmission lines

Starting with the voltages and current on a transmission line [see (2.74)], we write on line n:

$$V_n(z) = V_n^+ e^{-j\beta z} + V_n^- e^{j\beta z} \tag{4.1}$$

$$I_n(z) = I_n^+ e^{-j\beta z} - I_n^- e^{j\beta z} \tag{4.2}$$

where $I_n^+ = V_n^+/Z_{n0}$, $I_n^- = V_n^-/Z_{n0}$, and Z_{n0} is the characteristic impedance of the line. We write the reference voltage and current on the nth port by setting $z = 0$ in (4.1) and (4.2):

$$V_n = V_n^+ + V_n^- \tag{4.3}$$

$$I_n = I_n^+ - I_n^- \tag{4.4}$$

The relation between the voltages and currents on the n ports is then described by an impedance matrix as follows:

$$\begin{Bmatrix} V_1 \\ V_2 \\ \vdots \\ V_N \end{Bmatrix} = \begin{bmatrix} Z_{11} & Z_{12} & \cdots & Z_{1n} \\ Z_{21} & Z_{22} & & \vdots \\ \vdots & & & \vdots \\ Z_{N1} & \cdots & \cdots & Z_{NN} \end{bmatrix} \begin{Bmatrix} I_1 \\ I_2 \\ \vdots \\ I_N \end{Bmatrix} \tag{4.5}$$

The impedance matrix is linear, meaning that $4N$ independent quantities, V^+, V_n^-, I^+, I_n^-, may be selected to describe the behavior of the network. The coefficients of the impedance matrix, Z_{ij}, can be found by driving port j with a current I_j, while all other ports have zero current (open-circuits) and using the open-circuit voltage on port i:

$$Z_{ij} = \left.\frac{V_i}{I_j}\right|_{I_k=0, k\neq j} \tag{4.6}$$

The term Z_{ii} is the impedance of port i when all other ports are open circuited and is clearly the ratio between V_i and I_i. V_i and I_i are given in (4.1) and (4.2) for $n = i$.

This shows as well how the impedance coefficients may be measured either manually or using a computerized instrument such as a network analyzer: starting with one port, while all other ports are open we apply a current to that port (say port i) and measure the voltages on all other ports to get the off diagonal terms Z_{ij}, $j \neq i$. Repeating this for each port, we obtain all off-diagonal terms of the impedance matrix. The diagonal terms are simply the ratio of voltage on a port and the current in the port with all other ports open.

Since the relations between current and voltage can be written in terms of admittances as,

$$I_n^+ = Y_{n0}V_n^+, \quad I_n^- = Y_{n0}V_n^- \tag{4.7}$$

we can define an admittance matrix:

$$\begin{Bmatrix} I_1 \\ I_2 \\ \vdots \\ I_N \end{Bmatrix} = \begin{bmatrix} Y_{11} & Y_{12} & \cdots & Y_{1n} \\ Y_{21} & Y_{22} & & \vdots \\ \vdots & & & \vdots \\ Y_{N1} & \cdots & \cdots & Y_{NN} \end{bmatrix} \begin{Bmatrix} V_1 \\ V_2 \\ \vdots \\ V_N \end{Bmatrix} \tag{4.8}$$

The coefficients are found in a manner analogous to (4.6): Port j is driven with a voltage V_j, while short-circuiting all other ports and measuring the short-circuit current in port i.

$$Y_{ij} = \left.\frac{I_i}{V_j}\right|_{V_k=0, k\neq j} \tag{4.9}$$

A few properties of the impedance and admittance matrices are worth mentioning at this stage:

1. For a general network, the coefficients are complex.
2. For reciprocal networks, that is, networks that do not contain nonreciprocal media such as ferrites or active devices, the impedance and admittance matrices are symmetric.
3. If the network is lossless, the coefficients of the matrices are purely imaginary since the net power into the network must be zero.
4. The impedance and admittance matrices are reciprocal of one another, that is,

$$[Y] = [Z]^{-1} \tag{4.10}$$

5. Although not immediately evident, the coefficients of the matrices are frequency dependent, that is, as the frequency changes, the coefficients change because the impedances are frequency dependent in most cases.

The impedance and admittance matrices in (4.5) and (4.8) were developed from circuit principles and clearly apply to circuit networks including transmission lines. However, they apply equally well to networks in which circuit quantities including voltages and currents are not obvious such as waveguides. In such cases, one measures electric and magnetic fields rather than voltages and currents. However, the concept of impedance is still valid since the ratio of electric-field intensity and magnetic-field intensity is an impedance (wave impedance). Thus, one can define equivalent voltages and equivalent currents in terms of the electric- and magnetic-field intensities and proceed with the use of the impedance or admittance matrix approach as above. We do not show these matrices here because in such cases it is easiest and perhaps more intuitive to use the concept of scattering parameters (described next) and then show that the two approaches are related, that is, one can convert from impedance coefficients to scattering coefficients and vice versa.

Reduction of an N-port network to a two-port network is, perhaps, the most common application since more often than not, we are interested in analysis or measurement of the properties of a device that possesses an input port and an output port. This will be the case used in this work and in particular in the following section. The reduction to a two-port network is shown in Figure 4.2. It should be noted that the incident quantities (voltages and currents) flow into the network, whereas the reflected quantities flow out of the network. The impedance and admittance matrices for the two-port network are

$$\begin{Bmatrix} V_1 \\ V_2 \end{Bmatrix} = \begin{bmatrix} Z_{11} & Z_{12} \\ Z_{21} & Z_{22} \end{bmatrix} \begin{Bmatrix} I_1 \\ I_2 \end{Bmatrix} \tag{4.11}$$

$$\begin{Bmatrix} I_1 \\ I_2 \end{Bmatrix} = \begin{bmatrix} Y_{11} & Y_{12} \\ Y_{21} & Y_{22} \end{bmatrix} \begin{Bmatrix} V_1 \\ V_2 \end{Bmatrix} \tag{4.12}$$

4.2.1 *The scattering matrix and* S*-parameters*

Impedance and admittance networks are useful whenever voltages and currents can be measured, with particularly simple representations for transmission lines. However, in applications where voltages cannot be measured directly, a different approach is needed. Instead of relating the total voltage and total current [Eqs. (4.3)

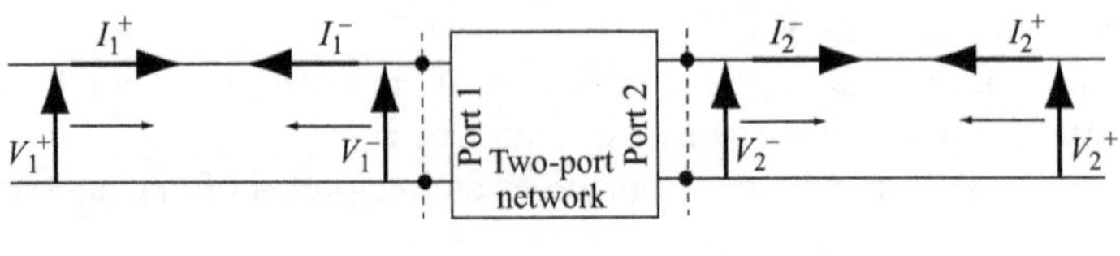

Figure 4.2 A two-port network of transmission lines

and (4.4)], one can use the scattering matrix approach. In this method, the reflected and incident wave amplitudes rather than voltages and currents are related through a scattering matrix. Nevertheless, the scattering matrix is defined through equivalent voltages. Needless to say that the scattering matrix and hence the scattering parameters are related to the impedance and admittance matrices discussed above. The reason that scattering coefficients are more useful for analysis of microwave networks is that they do not rely on measurement of voltages, and the concepts of short and open circuit, which are so vague in microwave analysis, are not required for analysis with *S*-parameters.

In dealing with waves, either in open or closed structures, the quantities that can be measured directly are reflected and transmitted waves through simple measurements such as detection of maxima and minima (both magnitude and location), standing wave ratio (SWR), and reflected and transmitted power. Thus, the scattering parameters deal with reflection and transmission of equivalent voltages. Consider Figure 4.1, which shows a junction of N waveguides. At each port i, there is an incident wave indicated by the equivalent voltage V_i^+ and a reflected wave indicated by the equivalent voltage V_i^-. We can then say that the reflected amplitude due to port i is

$$V_i^- = S_{ii} V_i^+ \tag{4.13}$$

This reflection ignores any waves that may exist due to the other ports in the network. But waves are also transmitted from each port to all other ports, depending on the transmission coefficients between them. The amplitudes of these transmitted (or, one can say scattered) waves on line i can be written as

$$V_n^- = S_{ni} V_i^+, \qquad n = 1, 2, \ldots, N, \quad n \neq i \tag{4.14}$$

Repeating this for all ports produces the scattering matrix:

$$\begin{Bmatrix} V_1^- \\ V_2^- \\ \vdots \\ V_n^- \end{Bmatrix} = \begin{bmatrix} S_{11} & S_{12} & \cdots & S_{1n} \\ S_{21} & S_{22} & & \vdots \\ \vdots & & & \vdots \\ S_{n1} & \cdots & \cdots & S_{nn} \end{bmatrix} \begin{Bmatrix} V_1^+ \\ V_2^+ \\ \vdots \\ V_n^+ \end{Bmatrix} \tag{4.15}$$

The scattering matrix for an N-port network is defined formally in a manner similar to the impedance matrix except, of course, that the coefficients relate forward and reflected waves at the ports. The coefficients of the matrix are found as follows:

$$S_{ij} = \left. \frac{V_i^-}{V_j^+} \right|_{V_k^+ = 0, k \neq j} \tag{4.16}$$

That is, to measure S_{ij}, port j is driven with an incident equivalent voltage, and the reflected voltage is measured at port i, while all other ports are set to zero voltage. The latter can be easily accomplished by terminating all ports other than port i and j

with their characteristic impedance to avoid reflections. Clearly then, S_{ij} is the transmission coefficient from port j to port i, with all other ports terminated in their characteristic impedance (matched loads). If $i=j$, we have

$$S_{ij} = \left.\frac{V_i^-}{V_i^+}\right|_{V_k^+=0, k\neq i} \tag{4.17}$$

This is clearly the reflection coefficient at port i, with all other ports terminated with their characteristic impedance. Note that the reflection coefficient as defined in (2.72) implies that the wave is reflected while there are no other waves transmitted due to, say, internal reflections within the circuit. In that case, S_{ii} is equal to the reflection coefficient. If, however, there are multiple reflections, S_{ii} represents the total reflected wave coefficient rather than the reflection coefficient in (2.72). For this reason, one should not, in general, refer to S_{ii} as the reflection coefficient. That is only true under the clearly defined conditions in (4.17) or in one-port networks in which (4.17) is trivial since there are no other ports.

4.2.2 Generalized scattering parameters

So far, we have implicitly assumed that all ports have the same characteristic impedance by simply not including the characteristic impedance in the definitions. To analyze more general networks, we can define generalized scattering parameters. Rather than viewing the coefficients as relating the amplitudes of the incident and reflected voltages, we normalize these voltages so that the normalized amplitudes squared provide the incident and reflected time averaged power at port n:

$$P_{in} = \frac{1}{2}|a_n|^2 \tag{4.18}$$

$$P_{rn} = \frac{1}{2}|b_n|^2 \tag{4.19}$$

With these, the net time averaged power into port n becomes

$$P_n = P_{in} - P_{rn} = \frac{1}{2}|a_n|^2 - \frac{1}{2}|b_n|^2 \tag{4.20}$$

At least in principle this form is more appealing since it is more general and does not necessarily require voltages. However, to relate to the previous discussion, we can write

$$a_n = \frac{V_n^+}{\sqrt{Z_{0n}}} \tag{4.21}$$

$$b_n = \frac{V_n^-}{\sqrt{Z_{0n}}} \tag{4.22}$$

With these, the generalized scattering matrix becomes

$$\begin{Bmatrix} b_1 \\ b_2 \\ \vdots \\ b_n \end{Bmatrix} = \begin{bmatrix} S_{11} & S_{12} & \cdots & S_{1n} \\ S_{21} & S_{22} & & \vdots \\ \vdots & & & \vdots \\ S_{n1} & \cdots & \cdots & S_{nn} \end{bmatrix} \begin{Bmatrix} a_1 \\ a_2 \\ \vdots \\ a_n \end{Bmatrix} \tag{4.23}$$

The coefficients of the generalized scattering matrix are

$$S_{ij} = \left. \frac{V_i^- \sqrt{Z_{0j}}}{V_i^+ \sqrt{Z_{0i}}} \right|_{V_k^+=0, k\neq i} \tag{4.24}$$

whereas S_{ii} remains unchanged

$$S_{ij} = \left. \frac{V_i^-}{V_i^+} \right|_{V_k^+=0, k\neq i} \tag{4.25}$$

It should be noted that (4.24) reduces to (4.16) if $Z_{0i} = Z_{0j}$ (i.e., if all ports have the same characteristic impedance) and then the generalized scattering matrix in (4.23) reduces to that in (4.15).

4.2.3 *Some properties of* S-*parameters*

1. If ports are perfectly matched, the reflection coefficients are zero, and hence $S_{ii} = 0$. That is, the diagonal coefficients are zero.
2. As with the impedance matrix, reciprocal networks are symmetric, that is, $S_{ij} = S_{ji}$ $(i \neq j)$.
3. For any lossless network, the net power into a port [see (4.20)] is zero. That implies that the sum of the products of each term of any row or any column multiplied by its complex conjugate is equal to 1:

$$\sum_{n=1}^{N} S_{ni} \cdot S_{ni}^* = 1 \tag{4.26}$$

4. The *S*-parameters are easily related to the *Z*- or *Y*-parameters and can be obtained from them.
5. There are other properties that will not be discussed here including properties associated with the phase.

4.2.4 *The* ABCD-*parameters and the transmission matrix*

Another useful set of network parameters is the so-called *ABCD* parameters. These are defined for two-port networks when multiple two-port networks are cascaded to form a more complex network. The reason this is convenient is that then the matrix for the cascaded network is the product of the matrices of the individual two-port networks. To do so, we define the *ABCD* matrix based on Figure 4.3. In this network, the currents and voltages are the same as those in Figure 4.1 except that I_2 points away from the network. This is done so that the output from any network becomes the input to the next two-port network in the cascade (see Figure 4.4). We write

$$V_1 = AV_2 + BI_2 \tag{4.27}$$

$$I_2 = CV_2 + DI_2 \tag{4.28}$$

Or

$$\begin{Bmatrix} V_1 \\ I_1 \end{Bmatrix} = \begin{bmatrix} A & B \\ C & D \end{bmatrix} \begin{Bmatrix} V_2 \\ I_2 \end{Bmatrix} \tag{4.29}$$

If we write the same for the two networks in Figure 4.4:

$$\begin{Bmatrix} V_1 \\ I_1 \end{Bmatrix} = \begin{bmatrix} A_1 & B_1 \\ C_1 & D_1 \end{bmatrix} \begin{Bmatrix} V_2 \\ I_2 \end{Bmatrix} \tag{4.30}$$

$$\begin{Bmatrix} V_2 \\ I_2 \end{Bmatrix} = \begin{bmatrix} A_2 & B_2 \\ C_2 & D_2 \end{bmatrix} \begin{Bmatrix} V_3 \\ I_3 \end{Bmatrix} \tag{4.31}$$

Thus

$$\begin{Bmatrix} V_1 \\ I_1 \end{Bmatrix} = \begin{bmatrix} A_1 & B_1 \\ C_1 & D_1 \end{bmatrix} \begin{bmatrix} A_2 & B_2 \\ C_2 & D_2 \end{bmatrix} \begin{Bmatrix} V_3 \\ I_3 \end{Bmatrix} \tag{4.32}$$

That is, the relation between V_1, I_1 and V_3, I_3 is given by the product of the two matrices.

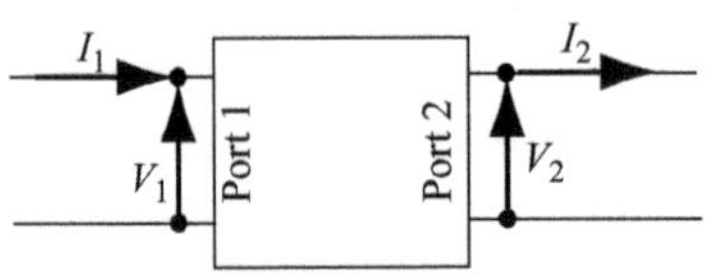

Figure 4.3 The two-port network used to define the ABCD-parameters. Note that I_2 flows out of port 2

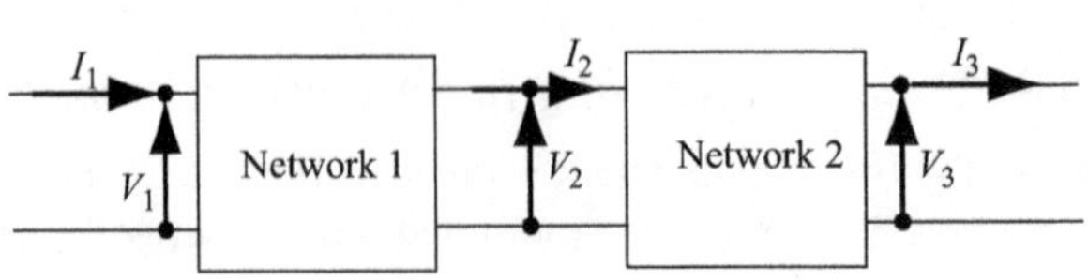

Figure 4.4 Two cascaded two-port networks

The coefficients *ABCD* in (4.29) are measured on the basis of (4.27) and (4.28) and depend of course on the network configuration, as follows:

$$A = \left.\frac{V_1}{V_2}\right|_{I_2=0}, B = \left.\frac{V_1}{I_2}\right|_{V_2=0}, C = \left.\frac{I_1}{V_2}\right|_{I_2=0}, D = \left.\frac{I_1}{I_2}\right|_{V_2=0} \tag{4.33}$$

4.2.5 Relations between the various parameters

The relations between the various parameters can be written down relatively easily from their definitions. This can be done for general N-port networks, but we will limit ourselves here to two-port networks as the most useful for the present work and because it is easier to follow the process.

4.2.6 Shift of reference plane

The voltage and current on a transmission line and, of course, waves (electric field, magnetic field, or power) in space or in a waveguide propagate along a path with an amplitude and phase that change as the waves propagate. Thus, for example, if we were to measure a quantity at a point along the path of propagation and then move to a different point, we would get totally different results. On a lossless transmission line, however, the amplitude remains constant, but the phase changes as the wave propagates along the line. This has important implications on measurement of the S-parameter and, of course, on all other parameters. Consider, for example, an instrument such as an oscilloscope or a network analyzer. The measurement, that is, the quantity that the oscilloscope displays is the quantity that appears at its port (the connector on the front of the instrument). This is the test or reference point for measurements. We call it the reference plane. However, we never actually measure anything at that point—the measurement we need must occur at the end of a transmission line that connects the instrument to the measurement point as shown in Figure 4.5. We use here a network analyzer simply to emphasize that the idea of the reference plane is important in high-frequency applications because only then the shift in frequency as we shift the reference plane becomes significant.

Consider first the transmission line as discussed in Chapter 2. The shift in phase along the line as the wave propagates a distance l between two points z_1

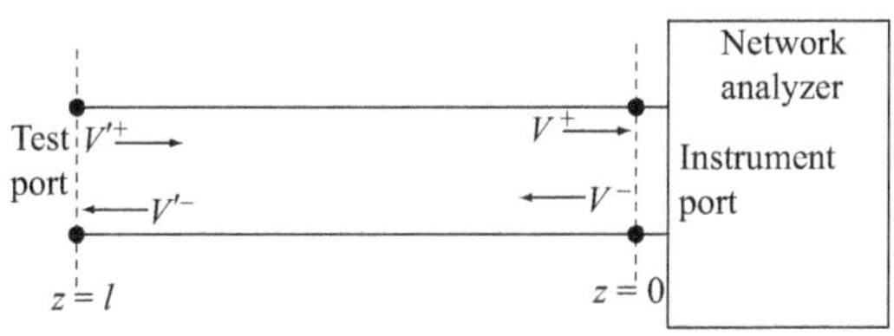

Figure 4.5 The instrument port and test port. The instrument measures the values at $z = 0$

and z_2 can be written as (using the forward voltage wave as an example, see Figure 4.6):

$$V_{z2}^{+} = V_{z1}^{+}e^{-j\beta(z2-z1)} = V_{z1}^{+}e^{-j\beta l} \tag{4.34}$$

where l is the distance between the two points. βl is the electrical length between the two points. If a reflected wave is generated at point z_2, the reflected wave at point z_1 becomes

$$V_{z1}^{-} = \left(V_{z1}^{+}\Gamma e^{-j\beta l}\right)e^{-j\beta l} = V_{z1}^{+}\Gamma e^{-j2\beta l} \tag{4.35}$$

Clearly then there is a shift in the phase proportional to the distance the wave propagates, and this can be understood as a delay in the time domain. The electrical length βl is the phase difference between the two points.

Now, returning to the *S*-parameters and Figure 4.5, we have the instrument port at $z = 0$ and shifted measurement or test port at $z = l$. Assuming the amplitude of the forward wave (voltage, for example) to be V^+, the *S*-parameters can be written at each of the two locations for each of the ports in Figure 4.5 using (4.15) at the reference plane at $z = 0$ as

$$\begin{Bmatrix} V_1^- \\ V_2^- \end{Bmatrix} = \begin{bmatrix} S_{11} & S_{12} \\ S_{21} & S_{22} \end{bmatrix} \begin{Bmatrix} V_1^+ \\ V_2^+ \end{Bmatrix} \tag{4.36}$$

We can do the same at the shifted plane, but now the *S*-parameters must be different:

$$\begin{Bmatrix} V_1'^- \\ V_2'^- \end{Bmatrix} = \begin{bmatrix} S'_{11} & S'_{12} \\ S'_{21} & S'_{22} \end{bmatrix} \begin{Bmatrix} V_1'^+ \\ V_2'^+ \end{Bmatrix} \tag{4.37}$$

However, from the theory of transmission lines (and the discussion above), we have

$$V'^{+} = V^{+}e^{j\beta l} \tag{4.38}$$

$$V'^{-} = V^{-}e^{-j\beta l} \tag{4.39}$$

Substituting (4.38) and (4.39) into (4.36), we get

$$\begin{bmatrix} e^{j\beta l_1} & 0 \\ 0 & e^{j\beta l_2} \end{bmatrix} \begin{Bmatrix} V_1'^- \\ V_2'^- \end{Bmatrix} = \begin{bmatrix} S_{11} & S_{12} \\ S_{21} & S_{22} \end{bmatrix} \begin{bmatrix} e^{-j\beta l_1} & 0 \\ 0 & e^{-j\beta l_2} \end{bmatrix} \begin{Bmatrix} V_1'^+ \\ V_2'^+ \end{Bmatrix} \tag{4.40}$$

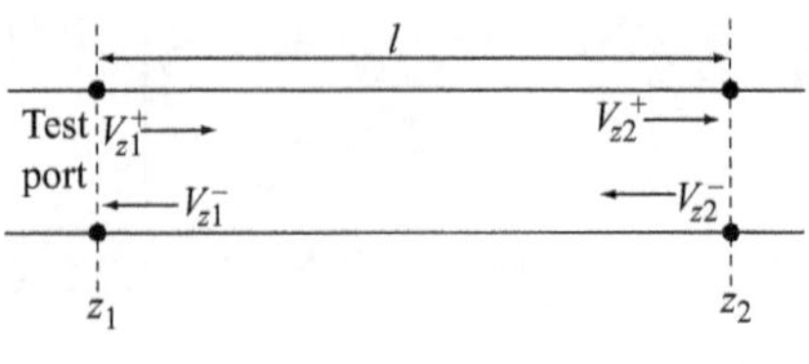

Figure 4.6 Propagation of voltages on a line between two points separated a distance l apart

Multiplying by the inverse of the leftmost matrix

$$\begin{Bmatrix} V_1'^- \\ V_2'^- \end{Bmatrix} = \begin{bmatrix} S_{11} & S_{12} \\ S_{21} & S_{22} \end{bmatrix} \begin{bmatrix} e^{-j2\beta l_1} & 0 \\ 0 & e^{-j2\beta l_2} \end{bmatrix} \begin{Bmatrix} V_1'^+ \\ V_2'^+ \end{Bmatrix}$$

$$= \begin{bmatrix} S_{11}e^{-j2\beta l_1} & S_{12} \\ S_{21} & S_{22}e^{-j2\beta l_2} \end{bmatrix} \begin{Bmatrix} V_1'^+ \\ V_2'^+ \end{Bmatrix} \tag{4.41}$$

That is, the S-parameters at the shifted reference planes is

$$\left[S'\right] = \begin{bmatrix} S_{11}e^{-j2\beta l_1} & S_{12} \\ S_{21} & S_{22}e^{-j2\beta l_2} \end{bmatrix} = \begin{bmatrix} S_{11}e^{-j2\theta_1} & S_{12} \\ S_{21} & S_{22}e^{-j2\theta_2} \end{bmatrix} \tag{4.42}$$

Clearly then the S-parameters of any shifted reference plane are simply those of the test reference plane with the diagonal (S_{ii}) parameters shifted by a phase equivalent to twice the distance between the test reference plane and the shifted reference plane. This phase angle can be either calculated or, more often, measured as part of the overall measurement process. This will be used in Chapter 8 to calibrate the network analyzer to take into account the phase shift due to the insertion of coaxial cables between the ports of the network analyzer and the device under test (DUT).

4.2.7 Transformations between parameters

Although the S-parameters are the most useful parameters in the microwave range, it is important to remember that all the network parameters we defined (and others we have not) are related and can be obtained from each other. This is particularly useful since some coefficients are easier to obtain than others. The following shows a few of these transformations to and from S-parameters and between other parameters for two-port networks:

Transformation from Z- to S-parameters

$$S_{11} = \frac{(Z_{11} - Z_0)(Z_{22} + Z_0) - Z_{12}Z_{21}}{(Z_{11} + Z_0)(Z_{22} + Z_0) - Z_{12}Z_{21}} \tag{4.43}$$

$$S_{12} = \frac{2Z_{12}Z_0}{(Z_{11} + Z_0)(Z_{22} + Z_0) - Z_{12}Z_{21}} \tag{4.44}$$

$$S_{21} = \frac{2Z_{21}Z_0}{(Z_{11} + Z_0)(Z_{22} + Z_0) - Z_{12}Z_{21}} \tag{4.45}$$

$$S_{22} = \frac{(Z_{11} + Z_0)(Z_{22} - Z_0) - Z_{12}Z_{21}}{(Z_{11} + Z_0)(Z_{22} + Z_0) - Z_{12}Z_{21}} \tag{4.46}$$

Note that in these transformations we did not assume symmetric networks (i.e., S_{12} is not assumed to be equal to S_{21}). Of course, if they are equal, it is because the network is symmetric.

Transformation from *ABCD*- to *S*-parameters

$$S_{11} = \frac{A + B/Z_0 - CZ_0 - D}{A + B/Z_0 + CZ_0 + D} \tag{4.47}$$

$$S_{12} = \frac{2(AD - BC)}{A + B/Z_0 + CZ_0 + D} \tag{4.48}$$

$$S_{21} = \frac{2}{A + B/Z_0 + CZ_0 + D} \tag{4.49}$$

$$S_{22} = \frac{-A + B/Z_0 - CZ_0 + D}{A + B/Z_0 + CZ_0 + D} \tag{4.50}$$

Transformation from *S*- to *Z*-parameters

$$Z_{11} = Z_0 \frac{(1 + S_{11})(1 - S_{22}) + S_{12}S_{21}}{(1 - S_{11})(1 - S_{22}) - S_{12}S_{21}} \tag{4.51}$$

$$Z_{12} = Z_0 \frac{2S_{12}}{(1 - S_{11})(1 - S_{22}) - S_{12}S_{21}} \tag{4.52}$$

$$Z_{21} = Z_0 \frac{2S_{21}}{(1 - S_{11})(1 - S_{22}) - S_{12}S_{21}} \tag{4.53}$$

$$Z_{22} = Z_0 \frac{(1 - S_{11})(1 + S_{22}) + S_{12}S_{21}}{(1 - S_{11})(1 - S_{22}) - S_{12}S_{21}} \tag{4.54}$$

Transformation from *S*- to *ABCD*-parameters

$$A = \frac{(1 + S_{11})(1 - S_{22}) - S_{12}S_{21}}{2S_{21}} \tag{4.55}$$

$$B = Z_0 \frac{(1 + S_{11})(1 + S_{22}) - S_{12}S_{21}}{2S_{21}} \tag{4.56}$$

$$C = \frac{1}{Z_0} \frac{(1 - S_{11})(1 - S_{22}) - S_{12}S_{21}}{2S_{21}} \tag{4.57}$$

$$D = \frac{(1 - S_{11})(1 + S_{22}) + S_{12}S_{21}}{2S_{21}} \tag{4.58}$$

Transformation from *ABCD*- to *Z*-parameters

$$Z_{11} = \frac{A}{C} \tag{4.59}$$

$$Z_{12} = \frac{AD - BC}{C} \tag{4.60}$$

$$Z_{12} = \frac{1}{C} \tag{4.61}$$

$$Z_{22} = \frac{D}{C} \tag{4.62}$$

Transformation from Z- to $ABCD$-parameters

$$A = \frac{Z_{11}}{Z_{21}} \tag{4.63}$$

$$B = \frac{Z_{11}Z_{22} - Z_{12}Z_{21}}{Z_{21}} \tag{4.64}$$

$$C = \frac{1}{Z_{21}} \tag{4.65}$$

$$A = \frac{Z_{22}}{Z_{21}} \tag{4.66}$$

The Z-, Y-, $ABCD$-, and S-parameters are the classical parameters in network analysis and all assume linear networks. In essence, one assumes small signal operation and testing. There are, however, instances in which large signal analysis and hence nonlinear analysis is required. Examples are in analysis of power amplifiers, power components, and nonsinusoidal operation in which one must take into account multiple harmonics. Parameters for these applications have been developed in the form of X-parameters, which can be viewed as a superset of S-parameters. These apply to both linear and nonlinear (small- and large-signal) analysis since in the limit of low signal, they reduce to the classical S-parameters as given in this section. We will not pursue these here, but it should be remembered that, in conjunction with nonlinear vector network analyzers, X-parameters are a very important extension of capabilities of parameter analysis.

4.3 Use of the *S*-parameters for practical measurements

The use of S-parameters for measurements in the microwave bands is widespread and is, perhaps, the single most important method for measurement of a wide variety of quantities. It is particularly useful in two-port structures in which the method can be implemented with relative ease. But it can also be used in one-port measurements (such as in evaluation of electronic components) and in multi-port measurements. In the past, measurement of S-parameters required direct measurements of voltages or power-related quantities using discrete instruments that made the testing procedure tedious and prone to errors. However, the method is standard on network analyzers and, therefore, provided that the DUT can be connected to the network analyzer either directly or through the intermediacy of a test fixture, the parameters are measured directly and from these, other important quantities can be obtained and displayed as needed. We will discuss the network analyzer separately in Chapter 8 and, at that point, will reflect back on the

S-parameters. Here we wish to discuss some of the results that can be obtained from the *S*-parameters and explain their usage since these have either direct or indirect bearing on the present work. Although we use a two-port model, as the most common application of *S*-parameters, *N*-ports can be similarly applied if needed.

Consider first the two-port system in Figure 4.2. The *S*-parameters may be calculated from the following:

$$\begin{Bmatrix} b_1 \\ b_2 \end{Bmatrix} = \begin{bmatrix} S_{11} & S_{12} \\ S_{21} & S_{22} \end{bmatrix} \begin{Bmatrix} a_1 \\ a_2 \end{Bmatrix} \tag{4.67}$$

where the *S*-parameters have the following meaning:

S_{11} is the reflection coefficient Γ_1 at port 1 when port 2 is terminated with a matching load (i.e., there is no reflection from port 2 back into port 1). S_{22} is the reflection coefficient Γ_2 at port 2 when port 1 is terminated with a matching load

S_{12} is the transmission coefficient for a wave traveling from port 2 to port 1.

S_{21} is the transmission coefficient for a wave traveling from port 1 to port 2.

These are easy to understand and recalling that the *S*-parameters are defined from incident, reflected, and transmitted power, it is no surprise that all the important properties of microwave systems can be obtained from measurement of *S*-parameters. In the following, we discuss practical measurements with particular emphasis on those that are relevant to the present work.

4.3.1 Matching of loads

Measurement of the S_{11}-parameter provides the reflection coefficient looking into port 1. This has immediate use in matching such as in matching a load (say, an antenna) to a transmission line. In principle, matching means minimization of the S_{11}-parameter. In practice, this is done through minimization of the SWR. SWR is defined directly in terms of the reflection coefficient; hence, by measuring the S_{11}-parameter, one obtains

$$\text{SWR} = \frac{1 + S_{11}}{1 - S_{11}} \tag{4.68}$$

The SWR at matched condition equals 1. In transmission lines, this is obvious, but the same can be done in waveguides. The SWR can be viewed as a measurement of the degree of mismatch and hence is the natural choice of parameter to measure when matching is required. Some practical aspects of matching in the context of measurements will be discussed in Chapter 8.

4.3.2 Detection of resonance

Monitoring of the S_{11}-parameter is also commonly used to measure the resonant frequency of a resonator. The process is simply that of scanning the frequency of the source coupling to the resonator until S_{11} is maximum. For series resonance, the impedance of the resonator is minimum, and hence S_{11} tends to -1 since

$$S_{11} = \Gamma_1 = \frac{Z_{\text{in}} - Z_0}{Z_{\text{in}} + Z_0} \tag{4.69}$$

In general, the S-parameters are complex, but at resonance the input impedance to the resonator is real, and S_{11} becomes real (for a lossless transmission line connecting to the resonator).

For parallel resonance, the impedance of the resonator is high, and S_{11} tends to $+1$, again, under the same conditions. As with matching, resonance can be detected through monitoring of the SWR, but in this case, resonance occurs at maximum SWR. The S-parameter does not measure the resonant frequency but rather detects resonance, and the frequency is deduced from the scan.

4.3.3 Determination of losses

Given an input power P_i fed to port 1, reflected power P_r from port 1, and transmitted power P_o to port 2, one can easily measure losses in the system. In microwaves, one defines a number of losses and quantities associated with losses. The following describe these losses and their measurement through the S-parameters. Losses are usually given in dB.

Given an incident power P_i at the input port of a network, some of that power is reflected back as P_r; hence, the power entering the network is $P_i - P_r$. The network may attenuate this power so that P_o is smaller than $P_i - P_r$. The ratio between the power entering the network and the incident power is

$$\frac{P_{\text{in}}}{P_i} = \frac{P_i - P_r}{P_i} \tag{4.70}$$

whereas the ratio between the output power at the output port and input power at the input port is

$$\frac{P_o}{P_i} = \frac{P_o}{P_i - P_r} \tag{4.71}$$

Therefore

$$\frac{P_o}{P_i} = \frac{P_i - P_r}{P_i} \times \frac{P_o}{P_i - P_r} \tag{4.72}$$

These relations define the fundamental losses in the network as follows.

4.3.3.1 Reflection loss

Reflection loss (RFL) is a measure of the reflected power at the port and is defined by the term in (4.70) as follows:

$$\text{RFL} = 10\log\frac{P_i - P_r}{P_i} = 10\log\left(1 - |S_{11}|^2\right) \tag{4.73}$$

4.3.3.2 Attenuation loss (transmission loss)

Attenuation loss (AL) or transmission loss is defined as the power loss between the input and output ports, that is, given the power entering the input port as $P_i - P_r$ and the power at the output port as P_o, the ratio $P_o/(P_i - P_r)$ defines the AL as follows:

$$\mathrm{AL} = 10\,\log\frac{P_o}{P_i - P_r} = 10\,\log\frac{|S_{12}|^2}{1 - |S_{11}|^2} \tag{4.74}$$

4.3.3.3 Insertion loss

Insertion loss (IL) is defined as the ratio of the power delivered to the load when the load is connected directly to the source and the power when the network is inserted between the source and the load. That is, when the network is inserted, part of the input power is reflected, whereas the transmitted power may be attenuated due to the network. Therefore IL may be viewed as the sum of loses due to reflection and due to attenuation by the inserted network. From (4.72), we have

$$10\,\log\frac{P_o}{P_i} = 10\,\log\frac{P_i - P_r}{P_i} + 10\,\log\frac{P_o}{P_i - P_r} \tag{4.75}$$

The first term on the RHS is the reflection loss, whereas the second is the attenuation loss. From (4.73) and (4.74), we write

$$\mathrm{IL} = 10\,\log\frac{P_o}{P_i} = 10\,\log\left(1 - |S_{11}|^2\right) + 10\,\log\frac{|S_{12}|^2}{1 - |S_{11}|^2} = 10\,\log|S_{12}|^2 = 20\,\log S_{12} \tag{4.76}$$

It is therefore sufficient to measure the transmission coefficient from port 1 to port 2.

4.3.3.4 Return loss

When the load is mismatched, not all of the available power from the generator is delivered to the load. This "loss" is called return loss (RL) and is defined (in dB) so that a matched load ($\Gamma_L = 0$) has a return loss of ∞ dB (no reflected power), while a total reflection ($|\Gamma_L| = 1$) has a return loss of 0 dB (all incident power is reflected). Note that return loss is a nonnegative number for reflection from a passive network.

$$\mathrm{RL} = 10\,\log\frac{P_i}{P_r} = 20\,\log\frac{1}{|S_{11}|} \tag{4.77}$$

The return loss is written in this fashion to guarantee that it comes out as a positive number. The term should be used with care because of its inverse meaning, that is, the better the matching of a load, the higher the return loss and vice versa.

4.4 Other measurements

The *S*-parameters can be used to measure other quantities. We have mentioned the detection of resonance. The method can be used to measure the resonant frequency rather than detecting resonance since at resonance the magnitude of the reflection coefficient is maximum.

In addition to measurement of the resonant frequency of resonators or resonating elements, one can measure the quality factor of resonators since these are related to power loss in the resonator.

Many more properties can and are being analyzed through the use of *S*-parameters, and some of these will be discussed in Chapter 8 when we discuss the network analyzer.

Through the use of these properties and others, the *S*-parameters have become the primary measurement in microwaves and are the primary process of measurement in network analyzers. This is not surprising since the incidence, reflection, and transmission are the fundamental processes by which waves interact with media and as such are the most general way of analysis.

4.4.1 Frequency measurements

Traditionally, one did not measure frequency directly since measurement of microwave frequencies by direct methods (such as through the use of counters or even through comparison methods such as beating) was not possible until relatively recently. Rather, one measured the wavelength. This was done by setting up standing waves and measuring the physical distance between two minima of the standing wave patterns. In transmission lines, the common method was to use a slotted transmission line terminated in a short or open and move a probe in the slot with the aid of a micrometer. The minima were detected with a receiver or power meter, and the distance shown by the micrometer was then converted into frequency using the relation between frequency, wavelength, and phase velocity that was known for the particular slotted line. In waveguides, the common method was to use a wavemeter—a cylindrical cavity resonator, dimensions of which could be changed through a micrometer and the change calibrated into wavelengths or frequency. The wavemeter also required a power detector to detect resonance. These methods were very awkward primarily because they tended to interfere with the operation of the devices and because they were slow, manual, and prone to errors. These methods have been, for the most part, relegated to history except in the very high-frequency ranges including in the optical domains.

The common method of measurement of frequency at low frequencies is through the use of counters. In this type of instrument, one counts the number of pulses (or peaks of sinusoidal signals) for a determined period of time and then the frequency is simply the count number divided by time. The main problem with this method is that at very high frequency, the gating time must be very short and the components must be able to respond at the frequencies of measurement. The common counter method can be extended to higher frequencies by first

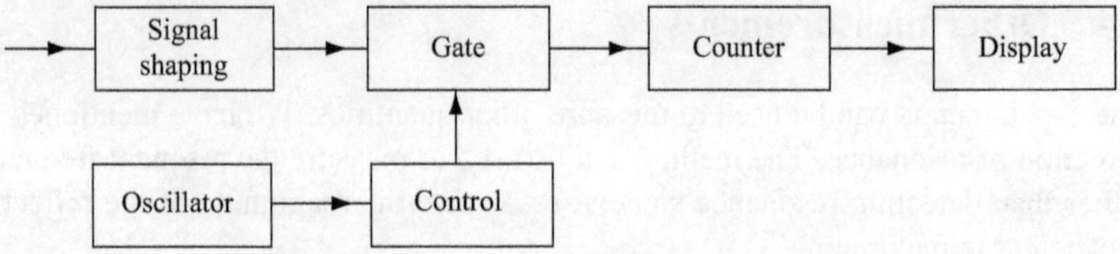

Figure 4.7 The basic frequency counter

dividing the income frequency by a given factor using prescalers and only then measuring the lower frequency after division. This approach increases the range but reduces accuracy and is not generally suitable for very accurate frequency measurements. A schematic circuit is shown in Figure 4.7. The microwave field is sampled, and a frequency divider (prescaler) is used in any combination required for the appropriate output. Digital frequency counters are used extensively because they are simple and inexpensive. The only limitation on this straightforward method is the frequency response of digital circuitry. Counters of this type exist with an upper limit of about 3 GHz, while standard universal counters measure frequency to about 500 MHz. Although perhaps not high enough for many microwave measurements, these are simple, accurate, and relatively inexpensive instruments. Typically the power needed for sampling is insignificant, and the sampling mechanism does not load the microwave source.

Although frequency counters well into the GHz region are available commercially, these are usually designed for measurements in circuits. Wave quantities can be measured with counters after conversion to electronic signals using a receiving antenna followed by amplification. However, an antenna followed by a receiver/amplifier is essentially a spectrum analyzer. Hence, often, the frequency of waves will be measured using spectrum analyzers. These are sensitive, wide-band receivers with facility for scanning over a frequency range and analyzing the signals including frequency, power, etc. Spectrum analyzers with ranges well into the GHz range exist and are probably the best choice for frequency measurements.

A simple and accurate method of measurement is to heterodyne the unknown frequency with an exactly known frequency and measure the difference between the two. This is shown schematically in Figure 4.8. The known frequency is f_a, the measured frequency is f_x. Their difference, $f_{\text{if}} = f_x - f_a$, is obtained after mixing f_x and f_a. The resulting frequency, called the intermediate frequency, is much lower than f_x and can be measured accurately by a counter. The measured frequency is then $f_a + f_{\text{counter}}$ and since both of these are known accurately, the measured frequency can be very accurate. Frequency measurement based on this process is shown in Figure 4.9. The instrument will usually require some front-end amplification and signal conditioning and filtering after mixing. Additional amplification is almost always needed before detection and counting. Unfortunately, this very simple process is seldom satisfactory. At microwave frequencies, the generation of arbitrary values of f_a is difficult especially if a wide range is needed. In more practical instruments, this is usually done by generating a fixed frequency f_0 using

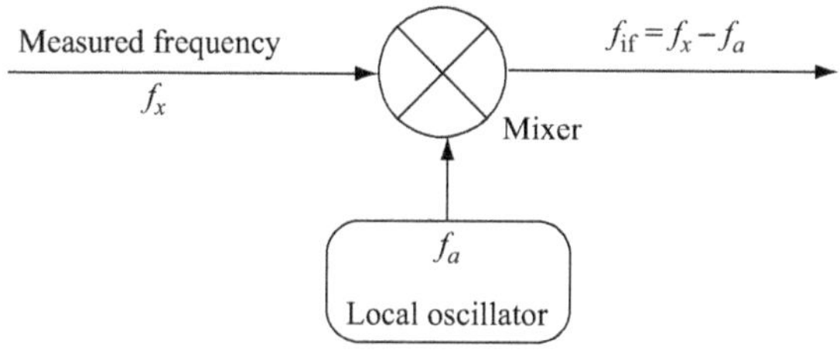

Figure 4.8 The principle of the heterodyning an unknown frequency with a known locally generated frequency to down-convert to a lower intermediate frequency

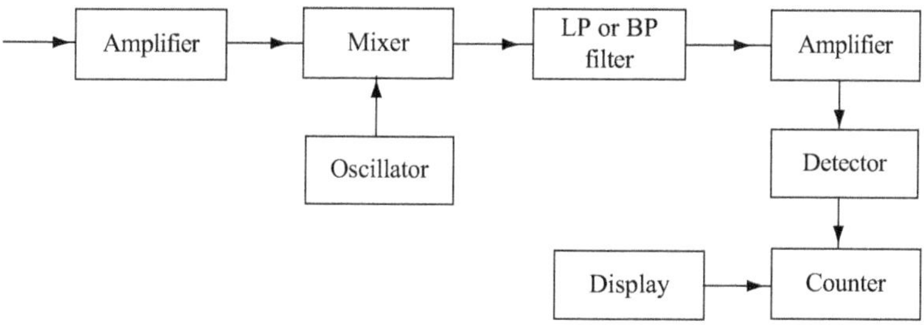

Figure 4.9 Basic heterodyne frequency meter

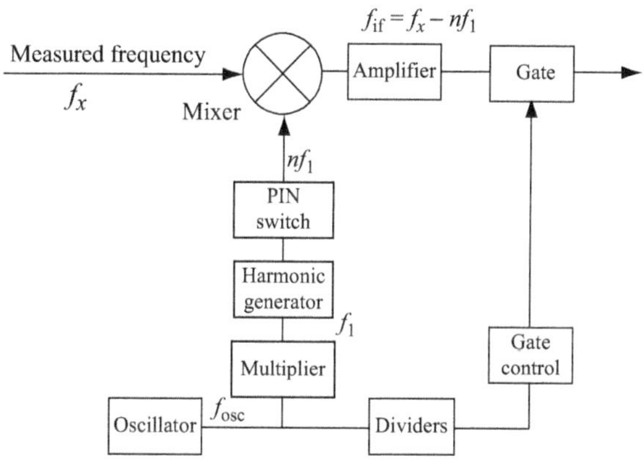

Figure 4.10 The main elements of a microwave heterodyne frequency measurement

an oven stabilized crystal generator or a cavity resonator (depending on the range of frequencies that need to be measured). This fixed frequency is then fed to a harmonic generator that generates N higher harmonics. A tuning cavity can then be used to select an appropriate harmonic $f_a = nf_0$ so that the IF frequency f_{if} is within the range of the counter, usually in the range of 100–500 MHz. This method is shown in Figure 4.10. In practical measuring equipment, either the input signal is

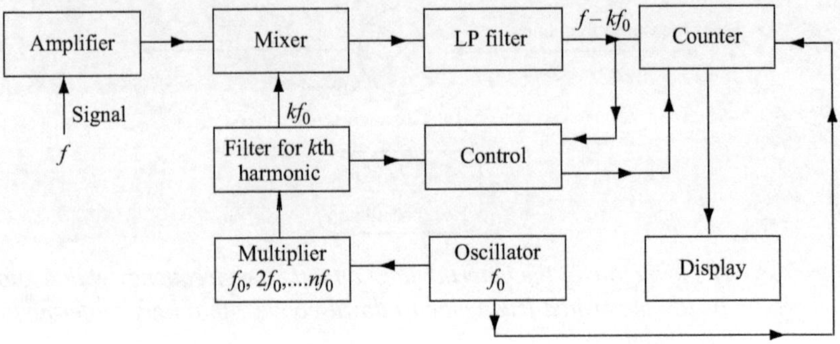

Figure 4.11 A practical heterodyne frequency meter

scanned at fixed frequency intervals, or the output from the harmonic generator is filtered for a given range, to provide a frequency difference in the range of at most a few hundred MHz which can then be counted directly. Often, the functions of harmonic frequency generation, selective filtering, and local oscillator are controlled by a microprocessor. A simple schematic of this type of frequency meter is shown in Figure 4.11, which shows additional functions necessary for practical measurements. These include again amplification and conditioning of the input signal, typically obtained with an antenna, filtering, and control mechanisms for the harmonics. Often the functions of harmonic frequency generation, selective filtering, local oscillator, and the counting process and display are controlled by a microprocessor.

4.4.2 Wavemeters

Mechanical wavemeters were in the past the main method of measuring frequency primarily because of the limitations of other instruments such as counters at high frequencies and because they were simple and inexpensive. Their accuracy is, however, lower than that of electronic frequency meters and, because they operated "off the signal," they tended to load the input. However, wavemeters are still in use either in the old mechanical configurations or in newer electronic implementations. In the upper reaches of the microwave region and in the optical domains, wavemeters are commonly used because direct measurement of frequency is not practical. We discuss them here even though in the lower range of the microwave domain, including in the range used in this work, the preference is for direct measurement using counters. Wavemeters are based on direct measurement of the length of a cavity resonator and from that one can calculate the frequency since cavity resonators are either quarter wavelength or half wavelength long (or any integer multiple of quarter wavelength). There are three types of wavemeters: transmission, reaction, and absorption wavemeters. Among these, the transmission wavemeter is most often used, primarily because it is connected in line with

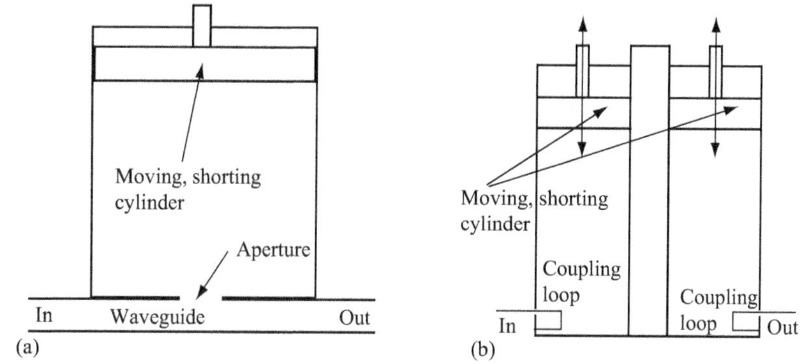

Figure 4.12 (a) Cylindrical wavemeter (reaction type) for measurements in waveguides and (b) coaxial wavemeter (transmission type) for use with transmission lines

waveguides or transmission lines, simplifying connections, and does not load the waveguide other than at resonance.

In a transmission wavemeter, power flows through the cavity, which at all frequencies except resonance acts as a short circuit. At resonance, there is a certain amount of energy absorbed by the cavity. This can be read as a dip in the output power. The output power is then used as an indication of resonance. The resonant frequency of the cavity is adjusted mechanically and the frequency read directly off the adjusting screw.

Two types of wavemeters are shown in Figure 4.12. One is a cylindrical cavity, the other a coaxial cavity. Coaxial wavemeters are more often used, primarily because they have a higher bandwidth (larger frequency band between consecutive modes). The cavity is normally a quarter wavelength cavity with the plunger being adjusted. Half wavelength cavities are also used with one of the cylindrical walls being adjusted. Coupling to the cavity is by loop probes that excite the magnetic field inside the cavity. In a transmission mode, the coaxial line is terminated in the cavity and another line couples energy out [as in Figure 4.12(a)]. Reaction type waveguides also exist, in which the transmission line is not interrupted, but the cavity is coupled to the line by exposing the inner conductor in the cavity. In effect, the inner conductor serves as a coupling loop to couple energy into the cavity. This is shown schematically in Figure 4.13. Absorption type wavemeters also exist, but are rarely used because they require more energy from the measuring environment. In all cases, the measurement is completed by adjusting the cavity to resonance at which point a dip in the signal is observed due to resonance.

Waveguide wavemeters are made of a cylindrical waveguide with an adjustable shorting wall, normally operating in the TE_{111} mode. Coupling to waveguides is by an aperture. In all cases, the frequency range of a wavemeter is that between two consecutive modes and the frequency is read directly off the adjusting screw. Figure 4.12(b) shows a waveguide wavemeter of this type.

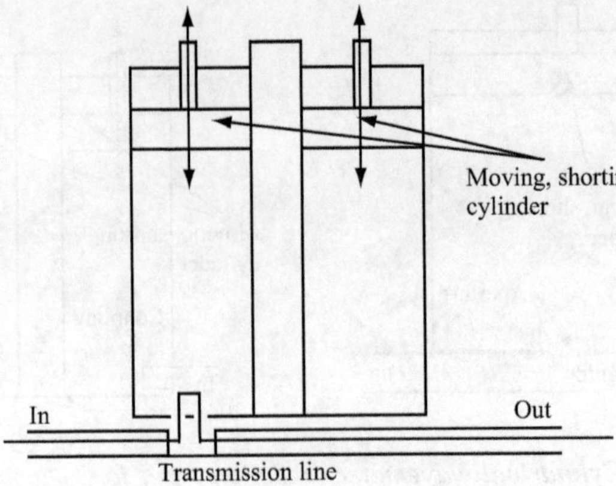

Figure 4.13 A reflection type wavemeter

4.4.3 Power measurements

The measurement of power at microwave frequencies relies on two broad methods, depending on the power level, accuracy needed, and response time. One method, particularly suited for measurement of low power, is based on rectification by a diode. The second broad method is by measuring the heating effects of microwaves and is more suitable for higher power levels. Both methods and any derivatives thereof are based on the Poynting theorem. The latter can be written as an instantaneous quantity

$$p(t) = \oint_S (\mathbf{E}(t) \times \mathbf{H}(t)) \cdot d\mathbf{s} \quad \text{(W)} \tag{4.78}$$

This gives the instantaneous power in a volume enclosed by the surface S. In most cases, when one talks about power, the time-averaged power is implied:

$$P_{\text{av}} = \frac{1}{2} \operatorname{Re}\left\{ \oint_S (\mathbf{E} \times \mathbf{H}^*) \cdot d\mathbf{s} \right\} \quad \text{(W)} \tag{4.79}$$

here $\mathbf{E}$ and $\mathbf{H}$ are phasors and * indicates the conjugate. These definitions are equivalent to the circuit definitions of power, either instantaneous:

$$p(t) = V(t)I(t) \quad \text{(W)} \tag{4.80}$$

or time averaged:

$$P_{\text{av}} = \frac{1}{2} \operatorname{Re} VI^* \quad \text{(W)} \tag{4.81}$$

here again, V and I are phasors.

In circuits and transmission lines, one can easily measure current and voltage whereas in waveguides or in space, the measurement of fields is more practical.

In many cases however, power or power density can be measured directly through properties of rectifying diodes or indirectly through the heating effects of power. As with any measurement, one must ensure that the method and sensors used to implement it are appropriate for the frequency or band of frequencies of the source.

4.5 Power sensors and detectors

4.5.1 Diode power sensors

A diode rectifies the signal and hence produces either a DC or a rectified AC signal proportional to the power in the signal since the diode is a square-law detector. The common diodes used are Schottky barrier diodes. These are useful down to about −70 dBm (about 100 pW) and up to about 50 GHz. The detection of power is based on the fact that the output of the diode, that is, its current is proportional to the square of the voltage across the diode. Given the *V–I* characteristic of a zero-bias Schottky diode in Figure 4.14, the current in the load for small signal detection (around the origin) can be written as

$$I = aV^2 \tag{4.82}$$

where a is a constant characteristic of the diode and V is the voltage across the junction. Hence, the name square law and its use as power detectors. For a sinusoidal current, this is written as

$$I = a(V \cos \omega t)^2 \tag{4.83}$$

Of course, diodes are never ideal and a more appropriate model for a real diode is more complex. In general, the *V–I* relation can be written as

$$I = a_0 + a_1 V + a_2 V^2 + a_3 V^3 + \cdots \tag{4.84}$$

or

$$I = a_0 + a_1 V \cos \omega t + a_2 (V \cos \omega t)^2 + a_3 (V \cos \omega t)^3 + \cdots \tag{4.85}$$

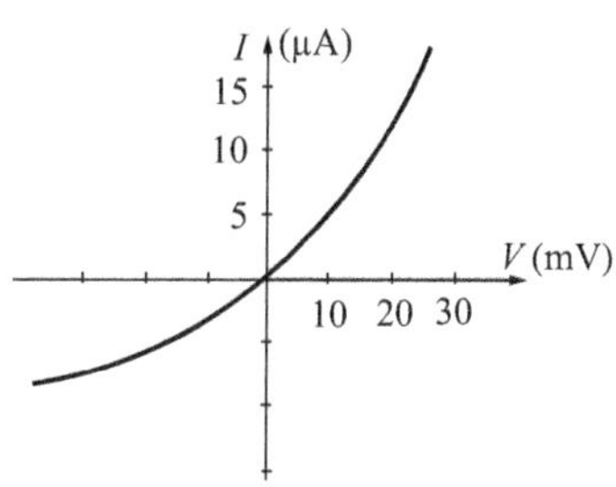

Figure 4.14 V–I characteristic of an ideal diode

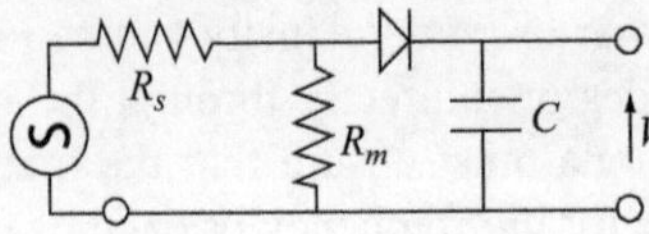

Figure 4.15 Basic power measuring circuit

The term a_0 relates to the value of the current at $I=0$, $V=0$ (origin). In an ideal diode, this is zero and in very good diodes it is negligible. The linear term is significant only at low currents, whereas the third and higher order terms are relevant at high currents. For a diode to be used as an accurate power sensor, these issues must be taken into account. First, a diode with very low leakage current is used. Then, its nonlinearities must be properly evaluated and compensated. In addition, the range of operation must be such that these undesirable effects are kept to a minimum. In general, diodes cannot be used at very low power levels or at very high power levels. The range is roughly from −70 to +20 dBm although this depends on the frequency range. Diodes can be used for broadband detection or for narrowband detection.

The power sensor itself is very simple in principle and shown in Figure 4.15. In practice, in addition to the matching network indicated by the resistance R_m and the bypass circuit (the capacitor C), there are often biasing circuits, temperature compensation circuits, and others to ensure accurate measurement of power.

4.5.2 Thermistors, bolometers, and thermocouples

An approach that is very different from that of the diode detector is the use of devices that can detect and quantify the change in temperature caused by the absorption of microwave power. This includes bolometers, thermistors, and thermocouples of various types. A bolometer is a device that changes its resistance with temperature. In the past, this was based on the positive temperature coefficient of resistance of metals (very thin wires), but the modern form uses a thermistor. Thermistors are semiconductor devices with either a negative or positive temperature coefficient of resistance, with preference to negative temperature coefficients of resistance devices. Thermistors for this purpose are very small—about 0.3–0.4 mm in diameter. This ensures a measurable change in temperature for minute input power. Because thermistors have an exponential resistance versus temperature curve (Figure 4.16), it is crucial that the operating temperature of the thermistor remains constant for all operational environmental temperature and power levels. This can be achieved either by compensation or by using power substitution methods (Figure 4.17). For example, the thermistor can be fed from a DC source to raise its temperature to a fixed level above the background temperature but monitoring its resistance and keeping this temperature constant. At this point, the power supplied to the thermistor is $V_{\text{th}}{}^{*}I_0$. R is at a value R_0 and R_{th} at a value R_{t0}. When measuring external power, say P_e, the thermistor's temperature will rise, and its resistance drops to a value lower than R_{t0}. The current in the circuit increases to a value, say I_1, and the voltage across the thermistor changes to V_{th1}.

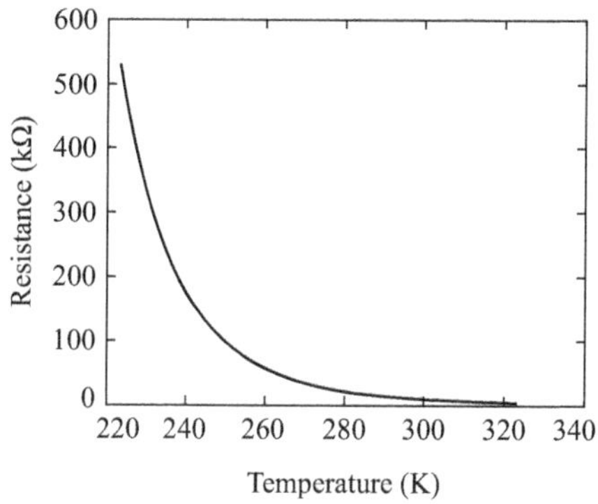

Figure 4.16 Thermistor resistance versus temperature curve

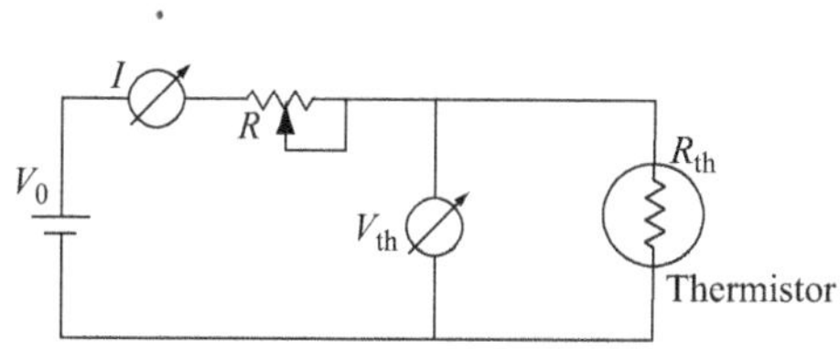

Figure 4.17 Power substitution principle. The power needed to restore the thermistor's temperature equals the external power absorbed by the thermistor

The power in the thermistor is $V_{th1}*I_1$. Now the resistance of resistor R is increased until the resistance of the thermistor is restored to R_{t0}. The change in power equals $V_{th}*I_0-V_{th1}*I_1$. This power is the reduction in power needed to restore the resistance of the thermistor (hence the original temperature), and it equals the absorbed microwave power. In practice, a more sophisticated circuit will be needed but the principle is the same.

Bridge circuits are often used for this purpose with or without sensors that sense the background temperature. A basic measuring circuit of this type is shown in Figure 4.18(a). The indication can be measured directly on the meter, or the bridge may be rebalanced and the *DC* power required to rebalance the bridge is measured. Normally the latter method is used. In Figure 4.18(b), two identical sensors are used with the upper sensor on the left branch isolated from the input power flux and hence serves to compensate for ambient temperature. In the absence of external power on Z_{S1}, the bridge is balanced ($Z_1 = Z_2$ and $Z_{S1} = Z_{S2}$ for all ambient conditions), and the voltage measured is zero.

Thermistor-based sensors can sense down below 1 μW but require particular attention to proper calibration. Obviously, not all incoming power is absorbed in the thermistor, and the absorption efficiency of thermistors is less than 100%. These quantities must be carefully measured so that the output of the thermistor measurement can be compensated for losses.

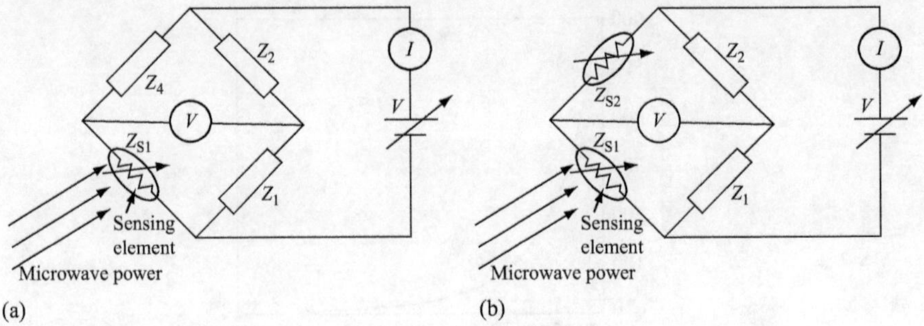

Figure 4.18 A thermoresistive power measurement configuration: (a) uncompensated bridge and (b) compensation for ambient temperature

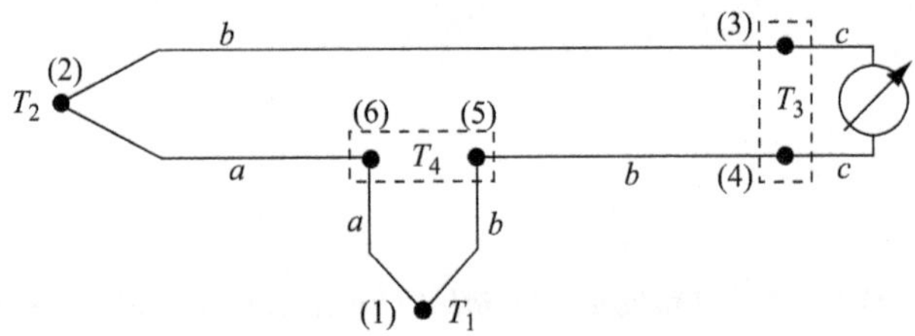

Figure 4.19 Thermocouple measurement. T_1 is the cold junction (reference) temperature

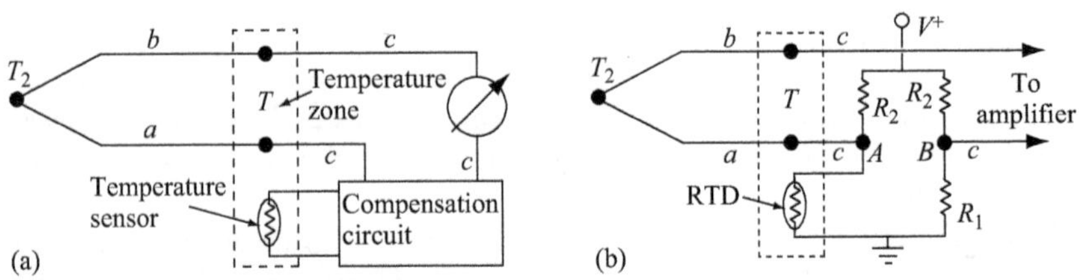

Figure 4.20 Compensation of cold junction: (a) compensation circuit based on measurement of the reference temperature and (b) implementation of the compensation

Unlike thermistors, thermocouples are passive devices that do not require external power to operate. These are junctions between two different materials—either metals or metal–semiconductor or two semiconductors that produce a DC voltage proportional to the temperature. To do so, the thermocouple sensor is made of two junctions, one held at a constant temperature, the second measuring the temperature due to microwave power (Figure 4.19). The reference junction may be replaced with a compensation circuit by measuring the reference temperature [Figure 4.20(a)]. The reference voltage, equivalent to a fixed temperature (usually

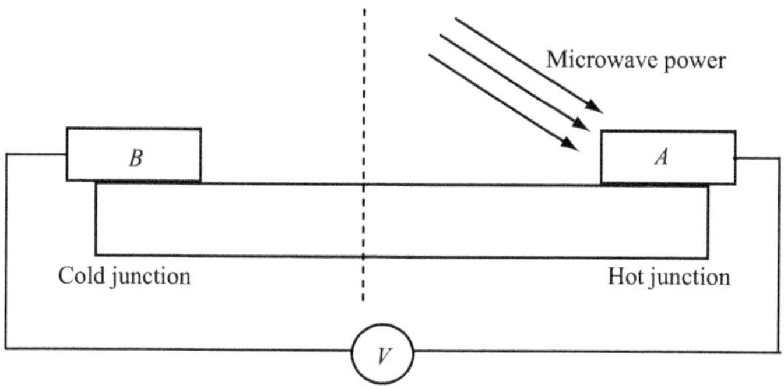

Figure 4.21 The thermocouple in a measuring circuit. The cold junction is isolated from the input power

0 °C), is inserted instead in series with the hot junction to replace the cold junction [Figure 4.20(b)]. Although the output of the thermocouple is not linear, its characteristics are available in exact relations for all practical combinations of materials. Therefore, the measurement is very accurate. Any two different metals may be used for junctions, but modern thermocouples used for microwave power measurements are semiconductors for a number of reasons including the ease of production and integration, the much larger thermoelectric outputs of semiconductor thermocouples and their smaller size and hence faster response. In practice, it is not possible to absorb power directly into the junction because the junction itself is very small and because, in many cases, one uses thermopiles made of a number of junction connected in series electrically. Instead, an absorber made of a conductor with low heat capacity (such as gold) and coated black is used, and the thermocouple or thermopile measures the temperature of the absorber. Thermocouples are often used in pairs for compensation (one thermocouple is exposed to the power source, the second is shielded from the source) and can be easily integrated with matching resistors and bypass capacitors. The basic configuration is shown in Figure 4.21. This compensation allows for direct elimination of common mode effects such as ambient temperature. Initially, both junctions are at the same temperature, and the output is zero. As microwave power is absorbed in junction *A* (hot junction) while junction *B* is shielded from the measured power, the temperature difference generates a potential difference, proportional to the power absorbed. The thermocouple marked as *B* serves as a cold junction, and in this configuration, a cold junction compensation is not applicable. Calibration of the device is normally done using simple *DC* or low-frequency power substitution. In addition to the basic circuit, a variety of methods for signal conditioning, attenuation, and amplification are sometimes used to extend measurement ranges. The output of thermocouple and thermopiles is measured directly. Sensitivity is between 100 and 200 μV/mW and unlike diodes can operate at much higher power levels of a few hundred mW.

In general, instruments using these devices for power measurements (such as network analyzers) will use diodes for low power ranges and switch to thermistors or thermocouples for higher power ranges.

It should also be clear that thermocouples and thermistors react slower than diodes and so the measurement time must be longer. All of them must be properly calibrated and compensated for losses.

It should be mentioned in passing that microwave power can also be measured by calorimetric methods. Obviously these are not suitable for instrumentation, but the standards of power measurements are based on calorimetric methods, and, when high powers must be measured directly, calorimetric methods are often used.

4.5.3 Measurement of power density

Sometimes, the total power is of less interest than the power density. This is the case, for example, when a small sample is illuminated by an antenna and only a small fraction of the total power is linked to the sample. Similarly, in radiation exposure measurements, only the power density is normally specified. The main difference between measurement of power and power density is that in power density measurements, only a small fraction of the power is sampled. One obvious method of measuring flux is to use a small loop. The loop operates as an antenna and has a relatively narrow bandwidth. Dipole antennas of various forms can be used for this purpose. The power density can also be measured using thermistors or, more often, semiconductor thermocouples. Perhaps, the most common method for power flux measurement is the use of an array of thermocouples on a substrate. The individual thermocouples are fabricated on a substrate with alternating hot and cold junctions, at intervals smaller than a quarter wavelength such that the net result is a thermopile with a number of thermocouples connected in series. The main advantage of this device is that it is nonresonant (unlike the small loop) and therefore broadband. For applications at very high frequencies, it is not normally possible to use loops for measurement but thermopiles, fabricated by integrated circuit means, can always be made small enough to be below the required quarter wavelength intervals. The use of power substitution for measurement simplifies calibration and measurement. A thermopile array is shown schematically in Figure 4.22. The array is built with alternating hot and cold junctions on a substrate, with individual elements insulated from each other and the cold junctions shielded from the hot junctions (see Figure 4.21). A sensor of this type is calibrated in power per unit area and, provided the sensor is small enough, the reading, which is an average value, can be accurate.

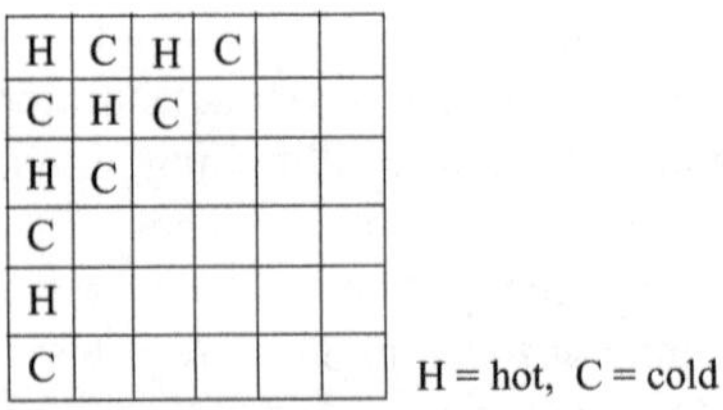

Figure 4.22 A thermopile array for measurement of power density. Junctions alternate and cold junctions are isolated from the microwave source

4.6 Measurement of Q-factor of resonators

We discussed the behavior of resonant circuits in Chapters 2 and 3, including the definition of the Q-factor. Since the sensitivity of measurements in cavity resonators in the microwave range depends directly on the quality factor of the cavity, the measurement of the Q-factor is an important issue. In the present work, we are particularly interested in open stripline resonators. In general, open resonators have lower Q-factors than the equivalent closed cavity and hence knowledge of the Q-factor is even more important than in closed cavities.

There are two fundamental methods of measuring the Q-factor of a cavity. The first measures power transmitted through the cavity and is shown in Figure 4.23(a). This requires that an input and an output port be available, and the cavity is treated as a two-port network. The second method is a reflection method shown in Figure 4.23(b). In this case, the cavity is a "load" on the transmission line connecting to its port, and it is clearly a one-port network. In the first case, the power-transmitted peaks at resonance, whereas in the second, the reflected power dips at resonance.

One can define the Q-factor of the resonator either with a load [such as the measuring instrument in Figure 4.23(a)] or without the load. These are called the loaded and unloaded Q-factors, respectively and were defined in (2.208) and (2.213). In the configuration in Figure 4.23(a), one measures the loaded Q-factor, whereas in Figure 4.23(b) one measures the unloaded Q-factor. Since the loaded Q-factor can be obtained from the unloaded Q-factor and vice versa, either method can be used.

We show here both methods starting with the transmission method, simply because that is the method used in this work. For this purpose, some of the discussion in Chapter 2 is repeated here as it pertains to the measurement of the Q-factor. The loaded Q-factor is measured directly, whereas the unloaded Q-factor is calculated from the loaded Q-factor.

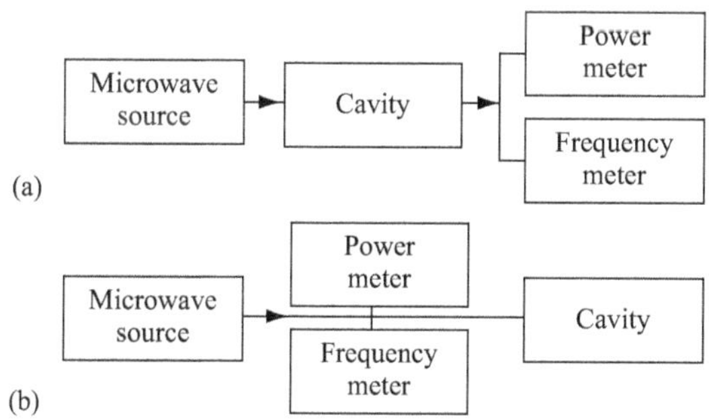

Figure 4.23 Two basic methods used to measure quality factor: (a) power transmitted through the cavity shows a peak at resonance and (b) power measured at the cavity port shows a dip at resonance

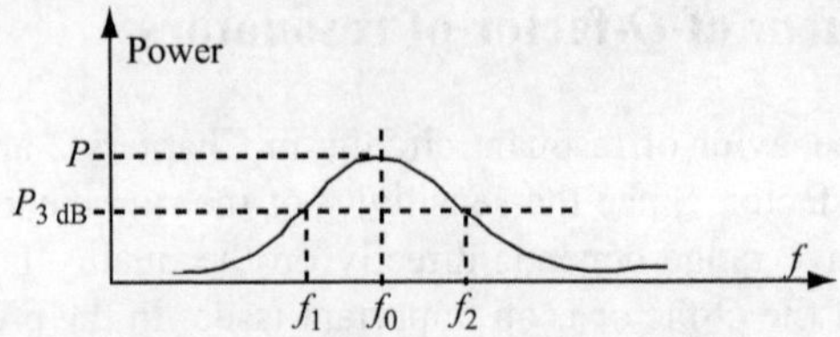

Figure 4.24 3-dB points and corresponding frequencies used to determine the quality factor of a cavity resonator

Following the discussion in Chapter 2, it is clear that there are a number of methods by which one can measure the Q-factor. The definition of the Q-factor is the starting point—one must measure the energy loss per period and the peak energy stored in the cavity.

$$Q = 2\pi \frac{\text{peak stored energy}}{\text{energy dissipated/per period}} \tag{4.86}$$

These quantities are difficult to measure, so alternative methods are used.

The loaded Q-factor can be most easily measured from the resonant frequency and the bandwidth of the resonator using (2.208):

$$Q_L = \frac{\omega_0}{\text{BW}} = \frac{\omega_0}{\omega_2 - \omega_1} = \frac{f_0}{f_2 - f_1} \tag{4.87}$$

In this relation, f_2 and f_1 are the upper and lower half-power frequencies. Their difference is the bandwidth of the resonator (see Figure 4.24). In high-Q resonators, this is particularly easy to do especially using a network analyzer. If the Q-factor is low, the frequency response is much flatter and determination of the half-power frequencies is more difficult to measure leading to larger errors. Nevertheless, this method can be very accurate for all but the poorest quality resonators, especially when an instrument like a network analyzer is used.

The unloaded Q-factor can be calculated from the following:

$$\frac{1}{Q_L} = \frac{1}{Q_0} + \frac{1}{Q_e} \tag{4.88}$$

where Q_L is the loaded Q-factor, Q_e is the external Q-factor (due to loading), and Q_0 the unloaded Q-factor.

We assume here a two-port measurement in which both the input and output impedances are Z_0 (as would be the case when the resonator would be connected to a network analyzer or in line with a transmission line of characteristic impedance Z_0).

Since we are interested in the unloaded Q-factor, we write

$$\frac{1}{Q_L} = \frac{1}{Q_0}\left(1 + \frac{Q_0}{Q_e}\right) = \frac{1}{Q_0}(1+g) \quad \rightarrow \quad Q_0 = (1+g)Q_L \tag{4.89}$$

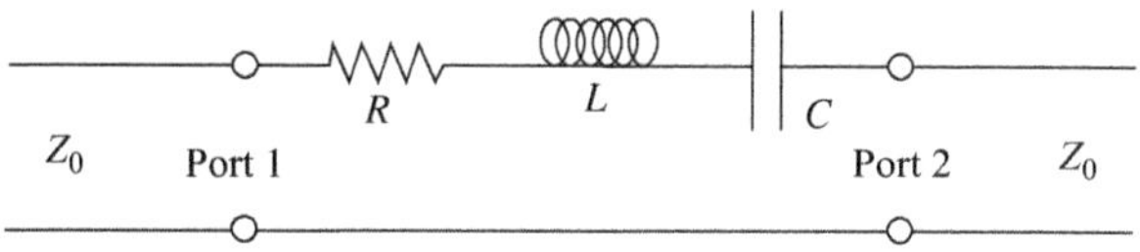

Figure 4.25 Two-port network. The network is a series resonator and is connected in series with a transmission line

where g is called a coupling factor. This relation may be used to calculate the unloaded Q-factor Q_0 from the loaded Q-factor since the latter is easier to obtain using the resonant frequency and bandwidth as shown in (4.87).

4.6.1 Q-Factors for series resonance

The external Q-factor, Q_e for series resonance was calculated in Chapter 2 as [see (2.211)]:

$$Q_e = \frac{\omega_0 L}{R_L} \tag{4.90}$$

where R_L is the load resistance external to the resonator. This gives

$$Q_e = \frac{\omega_0 L}{R_L} = \frac{\omega_0 L}{2Z_0} \tag{4.91}$$

here we note that the series load impedance is $2Z_0$ since the resonator is connected in series (see Figure 4.25).

We can now write the coupling factor as

$$g = \frac{Q_0}{Q_e} = \frac{\omega_0 L/R_L}{\omega_0 L/2Z_0} = \frac{2Z_0}{R_L} \tag{4.92}$$

Referring now to Figure 4.25, at resonance, the impedance of the resonator equals R and is minimal and hence the transmission is maximal. Off-resonance, the impedance of the resonator increases, the reflection increases, and hence the insertion loss increases. We can now calculate the S-parameters of the network starting with the $ABCD$-parameters that are easiest to evaluate. From their definition in (4.33), the $ABCD$-parameters at resonance are

$$A = 1, \quad B = R, \quad C = 0, \quad D = 1 \tag{4.93}$$

By using the transformation to S-parameters in (4.47)–(4.50), we have

$$S_{11} = \frac{A + B/Z_0 - CZ_0 - D}{A + B/Z_0 + CZ_0 + D} = \frac{R/Z_0}{2 + R/Z_0} = \frac{R}{2Z_0 + R} \tag{4.94}$$

$$S_{21} = \frac{2}{A + B/Z_0 + CZ_0 + D} = \frac{2}{1 + R/Z_0 + 1} = \frac{2Z_0}{2Z_0 + R} \tag{4.95}$$

$$S_{12} = \frac{2(AD - BC)}{A + B/Z_0 + CZ_0 + D} = \frac{2(1-0)}{1 + R/Z_0 + 1} = \frac{2}{2 + R/Z_0} = \frac{2Z_0}{2Z_0 + R} \tag{4.96}$$

$$S_{22} = \frac{-A + B/Z_0 - CZ_0 + D}{A + B/Z_0 + CZ_0 + D} S_{22} = \frac{R/Z_0}{2 + R/Z_0} = \frac{R}{2Z_0 + R} \tag{4.97}$$

Now we note that

$$\frac{g}{1+g} = \frac{2Z_0/R}{1 + (2Z_0/R)} = \frac{2Z_0}{2Z_0 + R} \tag{4.98}$$

or

$$\frac{g}{1+g} = S_{21} \tag{4.99}$$

That is, at resonance we can write

$$g = \frac{S_{21}}{1 - S_{21}} \tag{4.100}$$

or

$$Q_0 = (1+g)Q_L = \left(1 + \frac{S_{21}}{1 - S_{21}}\right)Q_L = \left(\frac{1}{1 - S_{21}}\right)Q_L \tag{4.101}$$

In other words, we can calculate the unloaded Q-factor by simple evaluating the S-parameters; in this case, we only need the S_{21}-parameter at resonance and then use (4.89) and (4.87) to evaluate the unloaded Q-factor.

If, for whatever reason, the external quality factor is needed, it can be found from (4.100) and (4.89):

$$g = \frac{S_{21}}{1 - S_{21}} = \frac{Q_0}{Q_e} \rightarrow Q_e = \left(\frac{1 - S_{21}}{S_{21}}\right)Q_0 = \left(\frac{1 - S_{21}}{S_{21}}\right)\left(\frac{1}{1 - S_{21}}\right)Q_L = \frac{Q_L}{S_{21}} \tag{4.102}$$

That is, the same measurement provides Q_e as well.

4.6.2 Q-Factors for parallel resonance

A very similar process can be followed to obtain the unloaded Q-factor for parallel resonators, but now the coefficients are different. From Chapter 2, we write for the external and unloaded quality factors:

$$Q_e = \frac{R_L}{\omega_0 L}, \quad Q_0 = \frac{R}{\omega_0 L} \tag{4.103}$$

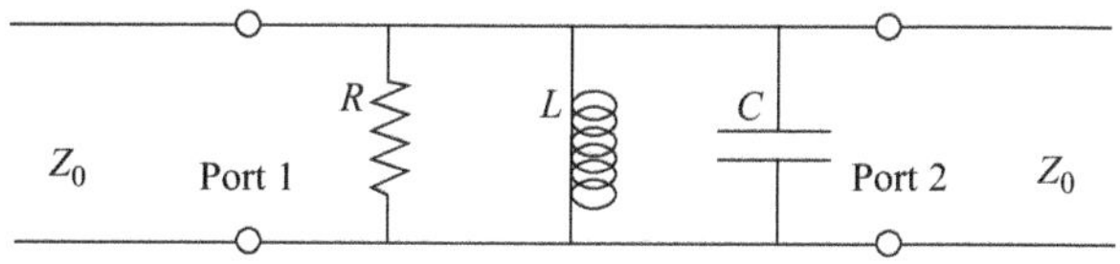

Figure 4.26 Two-port network. The network is a parallel resonator and is connected in parallel with a transmission line

Now, since the load is $Z_0/2$ (see Figure 4.26), we have

$$Q_e = \frac{Z_0}{2\omega_0 L} \tag{4.104}$$

From (4.89), we write

$$g = \frac{Q_0}{Q_e} = \frac{R/\omega_0 L}{Z_0/2\omega_0 L} = \frac{2R}{Z_0} \tag{4.105}$$

and

$$\frac{g}{1+g} = \frac{2R/Z_0}{1+(2R/Z_0)} = \frac{2R}{Z_0+2R} \tag{4.106}$$

Writing the *ABCD*-parameters from Figure 4.26,

$$A = 1, \quad B = 0, \quad C = \frac{1}{R}, \quad D = 1 \tag{4.107}$$

The *S*-parameters are

$$S_{12} = \frac{2(AD-BC)}{A+B/Z_0+CZ_0+D} = \frac{2(1-0)}{1+Z_0/R+1} = \frac{2}{2+Z_0/R} = \frac{2R}{2R+Z_0} \tag{4.108}$$

$$S_{11} = \frac{A+B/Z_0-CZ_0-D}{A+B/Z_0+CZ_0+D} = \frac{-Z_0/R}{2+Z_0/R} = \frac{-Z_0}{2R+Z_0} \tag{4.109}$$

$$S_{22} = \frac{-A+B/Z_0-CZ_0+D}{A+B/Z_0+CZ_0+D} S_{22} = \frac{-R/Z_0}{2+R/Z_0} = \frac{-Z_0}{2R+Z_0} \tag{4.110}$$

$$S_{21} = \frac{2}{A+B/Z_0+CZ_0+D} = \frac{2}{2+Z_0/R} = \frac{2R}{2R+Z_0} \tag{4.111}$$

Therefore,

$$\frac{g}{1+g} = S_{21} \quad \text{and} \quad g = \frac{S_{21}}{1-S_{21}} \tag{4.112}$$

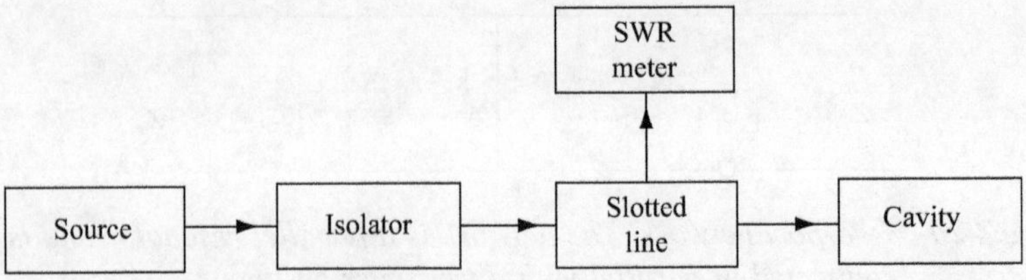

Figure 4.27 Standing wave method of measuring half bandwidth frequencies for Q-factor measurements

Hence,

$$Q_0 = \left(\frac{1}{1 - S_{21}}\right) Q_L \tag{4.113}$$

Q_e is also available from (4.102).

Note that this has the same form as (4.101), but the *S*-parameters are different.

Perhaps, a more intuitive approach to the measurement of the *Q*-factor is afforded by the use of the reflection method in Figure 4.23(b). The *Q*-factor is measured from the properties of the cavity as viewed from the input, that is, the power reflected from the cavity can be measured through a standing wave measurement as shown in Figure 4.27. Here the cavity is the load on the transmission line. Since the cavity presents a pure resistance at resonance, the SWR at resonance is the ratio between the cavity impedance and the line impedance:

$$\text{SWR} = \frac{Z_c}{Z_0} \tag{4.114}$$

if the cavity is overcoupled (if the cavity impedance Z_c is higher than Z_0) or

$$\text{SWR} = \frac{Z_0}{Z_c} \tag{4.115}$$

if the cavity is undercoupled (i.e., if the cavity impedance Z_c is lower than Z_0).

The measurement starts by tuning the cavity to resonance. The standing wavemeter reading represents power at resonance. Detuning the cavity and measuring the frequency at which the standing wavemeter reads 3 dB below that at resonance on each side of the resonant frequency gives the two 3 dB points. These frequencies are f_1 and f_2. The *Q*-factor is then the ratio between f at resonance and $f_2 - f_1$ as in (4.87). In this measurement, one obtains the unloaded *Q*-factor, or Q_0. If the loaded cavity *Q*-factor is required, this is easily calculated from the unloaded *Q*-factor and the SWR as

$$Q_L = \frac{Q_0}{1 + \text{SWR}} \tag{4.116}$$

and, based on the definition above, SWR is always greater than 1. This gives a loaded cavity Q smaller than Q_0.

4.7 Measurement of impedance

At microwave frequencies, either in transmission lines, waveguides, or in space, impedance is fundamental to operation and to many measurements. The most common method of measurement of impedance is from the reflection coefficient or, equivalently from the SWR. The reflection coefficient at the load on a line is

$$\Gamma_L = \frac{Z_L - Z_0}{Z_L + Z_0} = |\Gamma_L| e^{j\theta_\Gamma} \tag{4.117}$$

where Z_L is a load impedance to be measured and Z_0 either the characteristic impedance of a line or the wave impedance in space. Clearly, one has to measure both the magnitude of the reflection coefficient and its phase θ_Γ since impedance must be considered complex for general measurements.

The impedance becomes

$$Z_L = Z_0 \frac{(1 + \Gamma_L)}{(1 - \Gamma_L)} = Z_0 \frac{(1 + |\Gamma_L| e^{j\theta_\Gamma})}{(1 - |\Gamma_L| e^{j\theta_\Gamma})} \tag{4.118}$$

When measuring the impedance of a device, it is connected as a one-port network. Since under these conditions, the reflection coefficient is the S_{11}-parameter, the impedance is evaluated from the S-parameters as well:

$$Z_L = Z_0 \frac{(1 + S_{11})}{(1 - S_{11})} \tag{4.119}$$

Note, however, that S_{11} equals the reflection coefficient only if S_{21} is zero, otherwise S_{11} is the total reflection at port 1 and includes, in addition to the reflection coefficient, terms transmitted back into port 1 from port 2. S_{11} is always equal to the reflection coefficient in any one-port device. The same approach can be used to measure the input impedance of a loaded transmission line. The impedance measured is that of a one-port network.

4.8 Measurement of permittivity and loss tangent

The measurement of the dielectric constant and the loss tangent of materials is important to the characterization of materials and particularly so in the microwave range in which these quantities must be characterized as a function of frequency. In many instances, the permittivity is known at low frequencies but not at microwave frequencies, or it may not be known at the frequency of interest. It is therefore of paramount importance to be able to accurately measure the dielectric constant and the loss tangent and hence obtain the complex permittivity of the dielectric. Often too, as is the case in this work, one does not have a single dielectric but rather a mixture of dielectrics and the permittivity of the mixture must be evaluated.

There are a number of methods that can be used for the measurements, and there are various standards devoted to this issue. The method of choice depends on the type of dielectric and the instrumentation available. As was the case with other measurements, the emphasis here will be on those methods that can be performed with modern instruments such as the network analyzer.

In general, permittivity must be considered to be complex:

$$\varepsilon = \varepsilon' - j\varepsilon'' \tag{4.120}$$

ε' is usually referred to as the dielectric constant of the material, whereas the ratio of $\varepsilon''/\varepsilon'$ is referred to as the loss tangent

$$\tan\delta = \frac{\varepsilon''}{\varepsilon'} \tag{4.121}$$

The imaginary part of the permittivity is dependent on conductivity and frequency and is often written as

$$\varepsilon'' = \frac{\sigma}{\omega} \tag{4.122}$$

The complex permittivity is then

$$\varepsilon = \varepsilon'\left(1 - j\frac{\sigma}{\omega\varepsilon'}\right) \tag{4.123}$$

and the loss tangent is

$$\tan\delta = \frac{\sigma}{\omega\varepsilon'} \tag{4.124}$$

The loss tangent is explicitly dependent on frequency but so is ε', and therefore both must be measured at the frequency of interest or very near to it.

Depending on the dielectric to be measured and the frequency range, there are different methods and different test fixtures that may be used. These range from relatively simple capacitance and AC resistance measurements at low frequency from which the complex permittivity is deduced to samples in waveguides or coaxial test rigs to resonant methods. The type of test and the test fixture to be used also depends on the state of the dielectric. Solid dielectrics can often (but not always) be machined to specific shapes, whereas liquids require different test fixtures. In some cases, provisions must be made for high- or low-temperature testing and in still others the test sample may be too small for a particular method of testing. Thus, many methods have been developed, and it is not possible to discuss all of them here. However, the methods we discuss are characteristic to measurements in the microwave range and are relevant to the work reported in this text. It should also be mentioned that many of the methods of testing can also provide the complex permeability of the dielectric. However, since most dielectrics are non-magnetic and since the measurement of complex permeability is much rarer, it is not discussed here.

4.9 Waveguide method of measurement

One of the more common methods of measurement of permittivity (and permeability) is the use of a shorted transmission line of some sort with a sample of the medium to be measured placed in the line. Either a transmission line (usually a coaxial line) or a waveguide may be used depending on various parameters such as frequency, available fixtures, and the medium to be measured. The measurement is similar in both types except, of course, that coaxial lines support TEM at any frequency, whereas waveguides supper TE (transverse electric modes) or TM (transverse magnetic modes) above a minimum cutoff frequency. Thus, the frequency becomes important. Additional constraints may apply. For example, because cutoff frequencies in waveguides are related to dimensions of the waveguide, they are not practical at the lower frequencies in the microwave and submicrowave regions. At low frequencies, the waveguides are large and the samples needed for testing are also large. Coaxial lines propagate waves down to "zero" frequency so the issue of size is not a problem, but the sample must be in the form of a tube to fit between the inner and outer conductors of the coaxial structure. For some dielectrics, this may be difficult to produce. Coaxial lines are also best when broadband measurements are needed.

There are two fundamental methods in either waveguides or transmission line. One method requires that the waveguide or transmission line be shorted at one end and the sample to be tested placed at the short (see Figure 4.28). The reflection coefficient is measured, from which both the dielectric constant and the loss tangent may be calculated. The sample is of thickness t. The launcher is a probe that generates the required mode of propagation in the waveguide. The reference plane is where measurements are made. In practice, another probe measures the reflected wave at that point (the S_{11}-parameter). The method is often called a reflection method since only the S_{11}-parameter is measured. The second method is shown in Figure 4.29. In this, the sample is placed in the waveguide or transmission line, and both the reflection and transmission coefficients are measured hence the name reflection–transmission method. In practice, the measurements are done with a vector network analyzer; hence, either the S_{11}-parameter is measured for the reflection method or both the S_{11}- and S_{21}-parameters are measured for the reflection–transmission method. We will describe here the reflection–transmission method since the reflection method is quite similar.

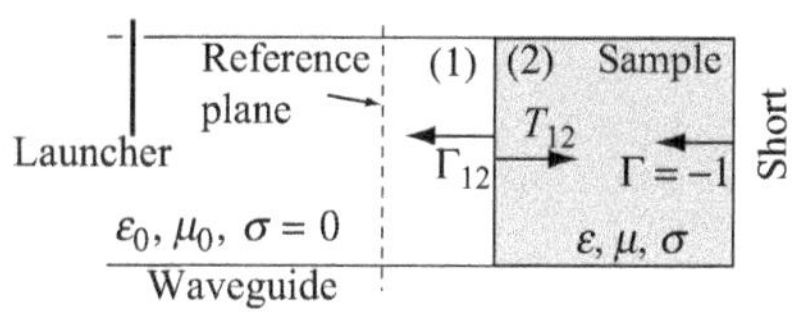

Figure 4.28 Measurement of permittivity of a sample in a shorted waveguide

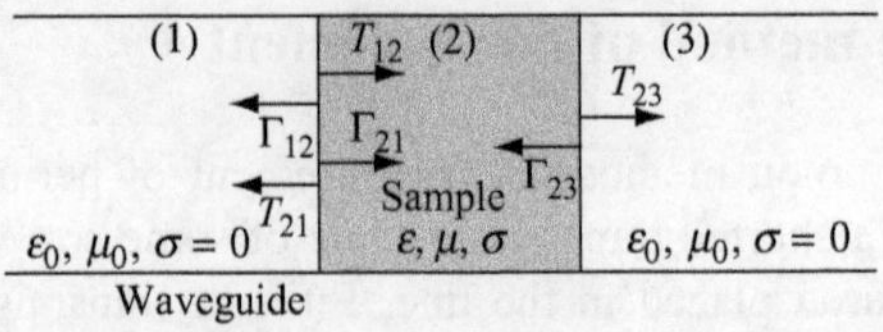

Figure 4.29 Measurement of permittivity of a sample in a waveguide

Given Figure 4.29, the reflection coefficient off the sample, that is, the S_{11} coefficient is due to multiple internal reflections within the sample. The wave is first reflected with a reflection coefficient Γ_{12}. Part of it is transmitted with a transmission coefficient T_{12} and that propagates to the opposite surface with attenuation and change in phase. At that surface, part of the wave is transmitted into medium 3 with a transmission coefficient T_{23}, and part reflected back with a reflection coefficient Γ_{23} propagates back to the left surface (with, again attenuation and phase change). Part of that is transmitted across the interface with a transmission coefficient T_{21}, and part of it reflected back into the sample with a reflection coefficient Γ_{21}. This component adds to the first reflection Γ_{12} to form part of the S_{11} term. This continues indefinitely. Using the notation shown, the reflection coefficient can be written as

$$S_{11} = \Gamma_{12} + T_{12}e^{-j\beta l}\Gamma_{23}e^{-j\beta l}T_{21} + T_{12}e^{-j\beta l}\Gamma_{23}e^{-j\beta l}\Gamma_{21}e^{-j\beta l}\Gamma_{23}e^{-j\beta l}T_{21} + \cdots. \tag{4.125}$$

Similarly, we can write S_{21} as the sum of all transmitted components after multiple internal reflections

$$S_{21} = T_{12}e^{-j\beta l}T_{23} + T_{12}e^{-j\beta l}\Gamma_{23}e^{-j\beta l}\Gamma_{21}e^{-j\beta l}T_{23} + T_{12}e^{-j\beta l}\Gamma_{23}e^{-j\beta l}\Gamma_{21}e^{-j\beta l} \times\Gamma_{23}e^{-j\beta l}T_{23} + \cdots. \tag{4.126}$$

Now, denoting

$$\begin{aligned} &\Gamma_{12} = \Gamma, \quad \Gamma_{21} = -\Gamma, \quad \Gamma_{23} = -\Gamma, \quad T_{12} = 1 + \Gamma, \\ &T_{21} = 1 - \Gamma, \quad T_{23} = 1 - \Gamma \end{aligned} \tag{4.127}$$

and

$$e^{-\gamma l} = T \tag{4.128}$$

where l is the length of the sample and γ the propagation constant in the sample. We have

$$\begin{aligned} S_{11} &= \Gamma + (1+\Gamma)T(-\Gamma)T(1-\Gamma) + (1+\Gamma)T(-\Gamma)T(-\Gamma)T(1-\Gamma) + \cdots \\ &= \Gamma + \frac{(1+\Gamma)T(-\Gamma)T(1-\Gamma)}{1-\Gamma^2T^2} = \frac{\Gamma(1-T^2)}{1-\Gamma^2T^2} \end{aligned} \tag{4.129}$$

and

$$S_{21} = (1+\Gamma)T(1-\Gamma) + (1+\Gamma)T(-\Gamma)T(-\Gamma)T(1-\Gamma) + (1+\Gamma)T(-\Gamma)T(-\Gamma)T(1-\Gamma)(-\Gamma)T(1-\Gamma) + \cdots = (1+\Gamma)T(1-\Gamma)\frac{1}{1-\Gamma^2T^2} = \frac{T(1-\Gamma^2)}{1-\Gamma^2T^2} \tag{4.130}$$

In these relations, Γ is the reflection coefficient at the interface and T the transmission coefficient. Assuming now that the S_{11}- and S_{21}-parameters are measured (magnitude and phase), the process continues through calculations in one of a number of methods. Perhaps the best known is the Nicholson–Ross–Weir (NRW) method. It requires the calculation of the reflection coefficient Γ followed by the transmission coefficient T. From these, one can then calculate the complex permittivity and the complex permeability. In the NRW method, the reflection coefficient is written in the following form:

$$\Gamma = X \pm \sqrt{X^2 - 1} \tag{4.131}$$

The sign is selected so that $|\Gamma| \leq 1$. The roots are

$$X = \frac{S_{11}^2 - S_{21}^2 + 1}{2S_{11}} \tag{4.132}$$

Substituting back into (4.131):

$$\Gamma = \frac{S_{11}^2 - S_{21}^2 + 1}{2S_{11}} \pm \sqrt{\left(\frac{S_{11}^2 - S_{21}^2 + 1}{2S_{11}}\right)^2 - 1} \tag{4.133}$$

Hence, the reflection coefficient is described directly from the S-parameters. From (4.129) and (4.130), the transmission coefficient is

$$T = \frac{S_{11} + S_{21} - \Gamma}{1 - (S_{11} + S_{21})\Gamma} \tag{4.134}$$

T is given in (4.128) in terms of the propagation constant and Γ is given in (4.133). The permeability and permittivity are determined from Γ and T as follows. The complex permittivity is written as

$$\varepsilon_r = \varepsilon r' + j\varepsilon r'' = \frac{\lambda_0^2}{\mu_r}\left(\frac{1}{{\lambda_c}^2} - \left[\frac{1}{2\pi l}\ln\left(\frac{1}{T}\right)\right]^2\right) \tag{4.135}$$

where λ_c is the cutoff wavelength of the waveguide ($\lambda_c = \infty$ for coaxial lines but is finite and well defined for waveguides). From this, one calculates the relative complex permittivity.

If the relative permeability is not 1 (i.e., for magnetic materials), then the relative permeability is calculated first as

$$\mu_r = \frac{1-\Gamma}{\Lambda(1-\Gamma)\sqrt{(1/\lambda_0^2)-(1/\lambda_c^2)}} \tag{4.136}$$

where λ_0 is the free-space wavelength, and Λ is found from

$$\frac{1}{\Lambda^2} = -\left(\frac{1}{2\pi l}\ln\left(\frac{1}{T}\right)\right)^2 \tag{4.137}$$

Because the S-parameters are complex, the reflection and transmission coefficients are also complex. Therefore, the term $\ln(1/T)$ is also complex. Although the real part presents no problems, the imaginary part is not unique and depends on the length of the sample. That means that (4.135) has multiple solutions. This can be resolved in a number of ways. The most obvious is to solve for all possible solutions with an imaginary part equal to $j(\theta + 2\pi n)$ where $n = 0, \pm 1, \pm 2, \ldots$ where $n = \text{int}(l/\lambda_g)$ and λ_g the wavelength in the sample and select the solution that one expects since in many cases the permittivity is known approximately. Another method is to make the sample thin enough (l small) to guarantee that $n = 0$ and hence get the correct solution. This is usually accompanied by loss of accuracy. There are also methods of exactly calculating n from the group delay on the line, a quantity that can be measured directly by a network analyzer. Since the group delay can also be calculated, one can then select the correct permittivity solution by equating the measured and calculated group delays. There are additional issues that often need to be addressed such as compensation for losses in the waveguide or transmission line walls, but these are secondary issues and discussing them here would only complicate the discussion without adding to understanding.

The method described here is the choice method of measurement and, in fact, is recommended in standards for permittivity measurements. It can be done entirely with vector network analyzers. Although it can be done at any frequency and over a range of frequencies, it is important that the frequency is not a resonant frequency. The method fails at resonance, and the results obtained at resonance should not be used.

Finally, it is obvious that the process can be applied to the reflection method in Figure 4.28 with the obvious changes in the reflection coefficient and the fact that only the S_{11}-parameter is measured. In fact, we can use (4.125)–(4.128) directly by setting $\Gamma_{23} = -1$ and $T_{23} = 0$. Then the process described in (4.131)–(4.137) produces the permittivity and permeability as in the previous method.

4.10 Cavity perturbation method

This method is particularly useful when the permittivity of a small sample of the dielectric is to be tested at a specific frequency. The method consists of placing the sample in a resonant cavity at the location where the electric-field intensity is maximum and the magnetic-field intensity is zero if the complex permittivity is measured (complex permeability is measured by placing the sample at the location of maximum magnetic-field intensity and zero electric-field intensity). The method requires a priori knowledge of the field distribution and, most importantly, the sample must be small enough so that the change the field distributions of the empty cavity is minimal (see Section 3.5.2 on the theory of cavity perturbation). Under these conditions, one can assume that the fields remain the same as for the empty cavity, but the resonant frequency of the cavity changes due to the dielectric constant of the test sample and the quality factor of the cavity changes due to the loss tangent of the sample.

Given an empty cavity resonating at a frequency ω_0, it will resonate at a frequency ω after the introduction of the dielectric sample. Similarly, if the loaded Q-factor of the empty cavity is Q_0, the loaded Q-factor changes to Q_s after the introduction of the sample. The cavity perturbation relations discussed in Chapter 3 [see (3.81) and (3.84)] provide the necessary relations. From (3.83) and (3.84):

$$\frac{\omega - \omega_0}{\omega_0} \approx -(\varepsilon' - \varepsilon_0)\frac{\int_{V_s} \mathbf{E}_i \cdot \mathbf{E}_0^* dv}{2\int_V \varepsilon_0 |E_0|^2 dv} \approx -(\varepsilon_r' - 1)\frac{\int_{V_s} |E_0|^2 dv}{2\int_V |E_0|^2 dv} \tag{4.138}$$

where V_s is the volume of the sample and V the volume of the cavity. Note that the fields in the sample and in the cavity are assumed to be the same in the second form in (4.138). The negative sign simply indicates that the resonant frequency goes down as the sample is introduced into the cavity ($\omega < \omega_0$). For a small sample, the equation can be simplified to

$$\frac{\Delta\omega}{\omega_0} = \frac{\Delta f}{f_0} \approx -(\varepsilon_r' - 1)\frac{|E_0|^2 V_s}{2\int_V |E_0|^2 dv} \tag{4.139}$$

Similarly, the quality factor is given as [see (3.108)]:

$$\frac{1}{Q} \approx \varepsilon_r'' \frac{|E_0|^2 V_s}{\int_V |E_0|^2 dv} \tag{4.140}$$

where

$$\frac{1}{Q} = \frac{1}{Q_s} - \frac{1}{Q_0} \tag{4.141}$$

By measuring the resonant frequency of the perturbed cavity and the resonant frequency of the empty cavity, one obtains the change in resonant frequency $\Delta\omega$. However, the measurement of the field E_0 in the cavity is next to impossible since that would require introduction of probes which then would themselves change the resonant frequency. For successful use of this method, one relies on cavities for which the field can be eliminated from the calculation. As an example, consider a cavity of dimensions a, b, d in the TE_{103} mode shown in Figure 4.30. The cavity only has nonzero electric-field intensity E_y:

$$E_y = \frac{\lambda}{2a}\sin\left(\frac{\pi x}{a}\right)\sin\left(\frac{3\pi z}{d}\right) \tag{4.142}$$

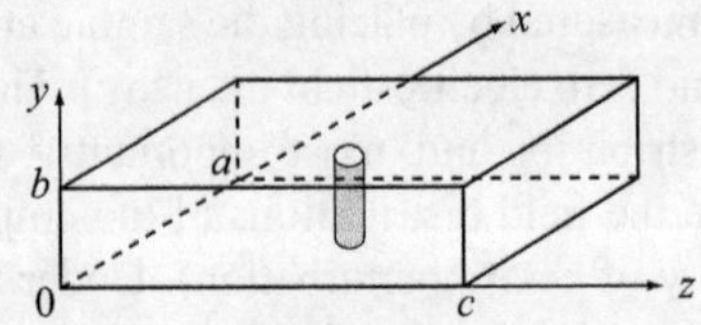

Figure 4.30 Small sample at the center of a rectangular cavity resonator

By placing the sample at $x = a/2$, $z = d/2$, the two sine terms are eliminated:

$$E_y = \frac{\lambda}{2a} \tag{4.143}$$

That is, the electric-field intensity is maximum at this point.

Therefore, the nominator in (4.139) and (4.140) becomes

$$|E_0|^2 V_s = \left(\frac{\lambda}{2a}\right)^2 V_s \tag{4.144}$$

The integral in the denominator is now evaluated:

$$\begin{aligned}\int_V |E_0|^2 dv &= \int_{x=0}^{a}\int_{y=0}^{b}\int_{z=0}^{d} \left(\frac{\lambda}{2a}\sin\left(\frac{\pi x}{a}\right)\sin\left(\frac{3\pi z}{d}\right)\right)^2 dxdydz \\ &= \left(\frac{\lambda}{2a}\right)^2 \frac{abd}{4}\end{aligned} \tag{4.145}$$

Substituting these into (4.139) and (4.140), we get

$$\frac{\Delta f}{f_0} \approx -2(\varepsilon_r' - 1)\frac{V_s}{V} \tag{4.146}$$

and

$$\frac{1}{Q} \approx 4\varepsilon_r''\frac{V_s}{V} \tag{4.147}$$

Thus

$$\varepsilon_r' = 1 + 0.5\left(\frac{V}{V_s}\right)\frac{f_0 - f}{f} \tag{4.148}$$

and

$$\varepsilon_r'' = \left(\frac{V}{4V_s}\right)\left(\frac{1}{Q_s} - \frac{1}{Q_0}\right) \tag{4.149}$$

If the empty cavity has a high Q (low losses), the second term in the brackets can be neglected and only the Q-factor for the cavity with the sample need be measured.

It should be noted again that, by necessity, both Q_0 and Q_s are loaded Q-factors since in practice, only loaded Q-factors can be measured, whereas unloaded Q-factors are calculated from the loaded Q-factors if needed. By using a rectangular cavity, resonating at a specific mode, and placing the sample at a particular location, the measurement is reduced to that of measuring the resonant frequencies with and without the sample and that of measuring the quality factors with and without the sample. The volume of the cavity is known ($V = abd$) and that of the sample is given.

In practice, there are certain precautions that must be followed. The sample must be small but not too small, or the changes in frequency and Q-factor will be small leading to errors in measurement. It must be machined with smooth surfaces and accurately placed in the cavity.

Also, it should be obvious that other types of cavities may be used. Of these, the most common is a circular cavity resonating at the TM_{010} mode.

Of all the methods available for measurement of dielectric properties, the resonant methods are the most sensitive and, often the more accurate, especially when only small samples are available.

4.11 Other methods

There are, of course, other methods of measurement of permittivity of materials. We already mentioned the capacitive method, which is particularly useful at lower frequencies. The permittivity can also be measured in free space when other methods are not practical such as at high temperatures or in hostile environments. The measurement is shown in Figure 4.31. The sample is usually large and the antenna directive, so that ideally all power is either transmitted through or reflected from the sample. The method proceeds by evaluating the S-parameters with the sample holder in place but no sample and then, again, evaluate the S-parameters of the sample with the sample holder. By a technique called deembedding, the effect of the sample holder is removed so that one is left with the S-parameters of the sample itself. After that, the process is the same as for the reflection–transmission method described above. Equations (4.135)–(4.137) are used here as well since it is assumed that the antennas receive all reflections. The limitations of the method are primarily due to multiple reflections from conducting surfaces including the antennas and diffraction at edges of the sample, especially when the sample cannot

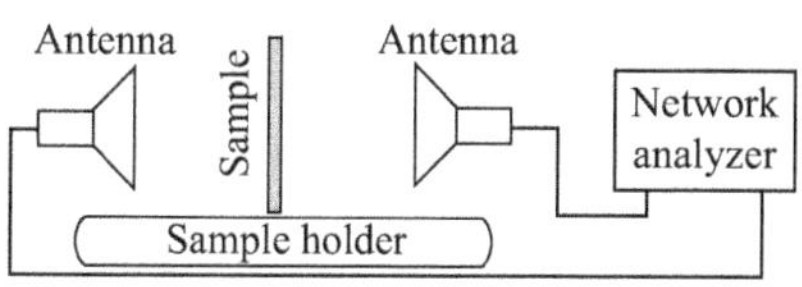

Figure 4.31 Free-space measurement of permittivity

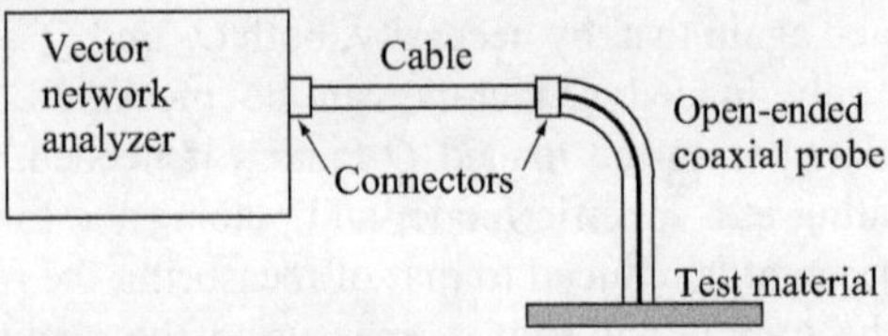

Figure 4.32 Open-ended coaxial probe used to measure the complex permittivity of a material

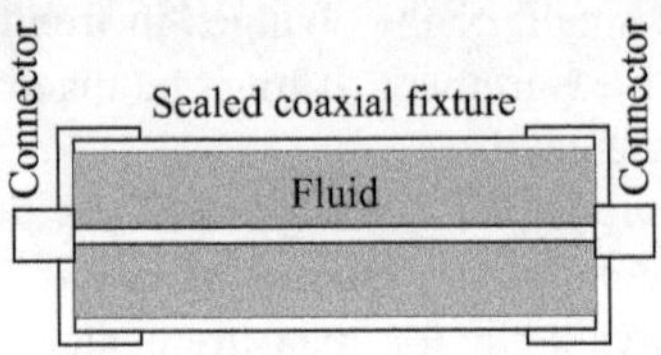

Figure 4.33 Coaxial fixture for measurement of the permittivity of liquids. The liquid fills the entire space between the connectors

be made large enough or if the antennas are not sufficiently directive or placed too far apart.

Another useful method is based on an open-ended coaxial probe shown in Figure 4.32. The probe is placed against the dielectric to be measured and the reflection coefficient at the interface between the probe's end and the material is found. If properly calibrated, the phase differences and any minor mismatching due to the cables and connected would have been taken into account, and the vector network analyzer produces the correct *S*-parameters. The material properties are calculated from that. This method is particularly useful for dielectrics that cannot be machined such as biological tissue or in nondestructive testing of existing structures where in situ measurements are essential. Other applications are in liquids and in environmental tests. The method typically produces the complex permittivity of the test sample.

The methods for measurement of permittivity are in most cases designed for measurement of permittivity of solid dielectrics, primarily because the methods require sample preparation and that often means machining of solids. Liquids can also be measured under some restrictions. For example, the reflection method in Figure 4.28 can be used for liquids by simply holding the waveguide or coaxial line vertically so that the liquid rests on top of the short and fills the structure to a height l. The transmission–reflection method can be used for liquids by using an enclosed section of transmission line or waveguide and filling it partially or completely with the liquid. Resonant methods are also useful in this regard, especially for lossless or very low loss liquids since the perturbation method can be used as well for whole-cavity material perturbation. The change in resonant frequency in this case is entirely due to the change in material properties and both the permittivity and the loss tangent may be evaluated. Figure 4.33 shows schematically an arrangement for evaluation of the dielectric constant of a liquid using the transmission–reflection method in a coaxial sample holder. The liquid fills the space between the inner and outer conductors so that the length of the sample equals the length of the coaxial fixture. As long as the lines are matched, the fact that the liquid fills the section does not pose a problem. By filling the sample holder entirely, the holder can be held vertically or horizontally. Obviously the same configuration can be used with the reflection method by shorting the output (lower connector). In this case, the holder must be vertical. The transmission line section is connected to a network analyzer using matched coaxial cables, and the measurement proceeds by evaluating the *S*-parameters.

Bibliography

The literature on microwave measurements is rich and extensive. Early measurements relied on direct effects on fields, but modern methods of measurements rely on the evaluation of *S*-parameters whenever possible and, because of that, many measurements are done with network analyzers, sometimes supplemented by spectrum analyzers. These will be discussed in some detail in Chapter 8 of this work.

The theory behind the evaluation of *S*-parameters and their relations to other types of parameters can be found in any book on microwaves. The following is a short list. References [7–10] discuss the newer *X*-parameters.

[1] D. M. Pozar, "Microwave Engineering," 2nd edition, John Wiley & Sons, Inc., New York, NY, 1998.

[2] P. R. Karmel, G. D. Colef and R. L. Camisa, "Introduction to Electromagnetic and Microwave Engineering," John Wiley and Sons, Inc., New York, NY, 1998.

[3] R. E. Collin, "Foundations for Microwave Engineering," 2nd edition, McGraw-Hill, New York, NY, 1992.

[4] A. Das and S. K. Das, "Microwave Engineering," McGraw-Hill, Boston, 2008.

[5] R. J. Collier and A. D. Skinner, "Microwave Measurements," 3rd edition, IET, London, 2007.

[6] J. P. Dunsmore, "Handbook of Microwave Component Measurements: With Advanced VNA Techniques," John Wiley, New York, 2012.

[7] D. E. Root, "X-Parameters*: The Emerging Paradigm for Interoperable Characterization, Modeling, and Design of Nonlinear Microwave and RF Components and Systems," Agilent Technologies, 2010.

[8] D. E. Root, J. Horn, L. Betts, C. Gillease and J. Verspecht, "X-Parameters: The New Paradigm for Measurement, Modeling, and Design of Nonlinear RF and Microwave Components," Microwave Engineering Europe, December 2008, pp. 16–21.

[9] D. E. Root, J. Xu, J. Horn, M. Iwamoto and G. Simpson, "Device Modeling with NVNAs and X-Parameters," 2010 IEEE MTT-S INMMiC Conference, Gotenborg, Sweden, April 26, 2010.

[10] D. E. Root, J. Verspecht, J. Horn and M. Marcu, "X-Parameters: Characterization, Modeling, and Design of Nonlinear RF and Microwave Components," Cambridge University Press, Cambridge, 2013.

The subject of measurement of permittivity, permeability, and loss tangent has received considerable attention because of their importance in design of microwave system and the behavior of fields in materials. The basic methods and variations can be found in published articles and reports and have even found their way into standards of measurements. The following are some examples:

[11] ASTM Standard D5568-01 Standard method for measuring relative complex permittivity and relative magnetic permeability of solid materials at microwave frequencies.

[12] ASTM Standard D2520-01 Standard test methods for complex permittivity (dielectric constant) of solid electrical insulating materials at microwave frequencies and temperatures to 1650 °C.

[13] ASTM Standard D150-98 Standard test method for AC loss characteristics and permittivity (dielectric constant) of solid electrical insulation.

[14] D. McGinnis, "Electromagnetic Properties of ECCOSORB MF-190, FERMILAB-PBAR-NOTE-576," 1997.

[15] D. Sun, "Measurement of Complex Permittivity and Permeability of Microwave Absorber ECCOSORB MF 190," FERMILAB PBAR NOTE 576, 1997.

It should be noted that almost all methods of measurement of material properties use a variation of the methods in [11–17], and all rely on rather complex computations using the raw measurements. These computations, which are essential in the extraction of the complex permittivity and complex permeability, have been developed by Weir [16] and Nicholson and Ross [17] and are referred to as NRW method. One can find these algorithms in, for example, [11,12] but also programed as routines within network analyzers. Needless to say, these are not the only methods of extraction and many variations exist.

[16] W. Weir, "Automatic measurement of complex dielectric constant and permeability at microwave frequencies," Proceedings of the IEEE, Vol. 62, No. 1, January 1974, pp. 33–36.

[17] A. M. Nicholson and G. F. Ross, "Measurement of the intrinsic properties of materials by time-domain technique," IEEE Transactions on Instrumentation and Measurements, Vol. IM-19, No. 4, 1970, pp. 377–382.

Modern microwave measurements make extensive use of network analyzers, spectrum analyzers, impedance analyzers, and other modern instruments. These instruments simplify measurements and provide very accurate results. But they also "hide" the details of the measurement from the user. It is sometimes useful to consult some of the older, classical sources on microwave measurements as these provide the details necessary for understanding of the theory and parameters involved. One of the classic books on the subject is Ginzton's book [18]. Notable, because of the extent of material, is the three-volume book edited by Sucher and

Fox [19]. Lavergheta [20] discusses a wide range of methods from a practical point of view with extensive discussion on equipment. Other sources are [21–26]. Many of these books are out of print but are available either as PDF files or as reprints.

[18] E. L. Ginzton, "Microwave Measurements," McGraw Hill Book Company, Inc., New York, NY, 1957.
[19] M. Sucher and J. Fox, Eds., "Handbook of Microwave Measurements," Polytechnic Press, New York, NY, 1963 (Vol. I, II, II).
[20] T. Laverghetta, "Handbook of Microwave Testing," Artech House, Boston, 1981.
[21] A. E. Bailey, "Microwave Measurements," IEE Electrical Measurement Series, Peter Peregrinus Ltd., London, 1988.
[22] H. E. Thomas, "Handbook of Microwave Techniques and Equipment," Prentice-Hall, Inc., Englewood Cliffs, NJ, 1972.
[23] P. I. Somlo and J. D. Hunter, "Microwave Impedance Measurements," IEE Electrical Measurement Series, Peter Peregrinus Ltd., London, 1985.
[24] F. L. Warner, "Microwave Attenuation Measurements," IEE Electrical Measurement Series, Peter Peregrinus Ltd., London, 1977.
[25] J. A. Lane, "Microwave Power Measurement," IEE Electrical Measurement Series, Peter Peregrinus Ltd., London, 1972.
[26] F. E. Gardiol, "Introduction to Microwaves," Artech House, Norwood, MA, 1984.
[27] M. Golio and J. Golio, "RF and Microwave Circuits, Measurements, and Modeling," CRC Press, Boca Raton, FL, 2007.
[28] V. Teppati, A. Ferrero and M. Sayed, (Eds.), "Modern RF and Microwave Measurement Techniques," Cambridge University Press, Cambridge, 2013.
[29] G. H. Bryant, "Principles of Microwave Measurements," IET, London, 1993.
[30] P. Yip, "High-Frequency Circuit Design and Measurements," Chapman and Hall, London, 1990.
[31] N. B. Carvalho and D. Schreurs, "Microwave and Wireless Measurement Techniques," Cambridge University Press, Cambridge, 2013.

Additional sources specific to material properties testing are listed in [32–37].

[32] R. N. Clarke (Ed.), "A Guide to the Characterisation of Dielectric Materials at RF and Microwave Frequencies," Published by The Institute of Measurement & Control (UK) & NPL, London, 2003.
[33] J. Baker-Jarvis, M. D. Janezic, R. F. Riddle, *et al.*, "Measuring the Permittivity and Permeability of Lossy Materials: Solids, Liquids, Metals, Building Materials, and Negative-Index Materials," NIST Technical Note 15362005.
[34] "Test methods for complex permittivity (Dielectric Constant) of solid electrical insulating materials at microwave frequencies and temperatures to 1650°C," ASTM Standard D2520, American Society for Testing and Materials.

[35] M. Janezic and J. Baker-Jarvis, "Full-wave analysis of a split-cylinder resonator for nondestructive permittivity measurements," IEEE Transactions on Microwave Theory and Techniques, Vol. 47, No. 10, October 1999, pp. 2014–2020.

[36] J. Krupka, A. P. Gregory, O. C. Rochard, R. N. Clarke, B. Riddle and J. Baker-Jarvis, "Uncertainty of complex permittivity measurement by split-post dielectric resonator techniques," Journal of the European Ceramic Society, Vol. 10, 2001, pp. 2673–2676.

[37] "Basics of Measuring the Dielectric Properties of Materials". Agilent Application Note. 5989-2589EN, April 28, 2005.

Chapter 5
Design of sensors for rubber thickness and fabric-coating monitoring

5.1 Introduction

This chapter discusses the design of resonator sensors for two applications of interest. One is a wide latex-coated fabric moving on a production line with the sensors monitoring the coating thickness or, alternatively, the moisture content in the fabric for the purpose of controlling the amount of latex on the fabric. The second is a rubber sheet moving on a calender (a cylindrical drum) in which the interest is the thickness of the sheet. Selection of physical parameters, dimensions and operational parameters, and the simulations necessary are discussed. Alternative designs including multiple sensors and moving sensors for full coverage of the fabric are weighed, and an appropriate design is reached. The details of design are given in full with alternatives and justification so that the reader has full accounting of what the design involves and what to expect from the final product. The most common alternative measurement methods of low-density dielectrics such as fabrics, paper, and thin rubber are either nuclear (by measuring absorption of gamma or beta particles from a radioactive source) or through transmission and/or reflection of electromagnetic waves. In some cases, the methods are more primitive than that; a sample of the final product is cut and weighed to ascertain that the coating is within the required limits. Because this must be done after the production process, it is extremely wasteful and whole production runs may need to be scrapped due to insufficient or overcoating. The purpose of the designs described here is to eliminate this uncertainty and monitor the production in real time to offer feedback for continuous correction of the coating thickness.

As is often the case in microwave systems, design is a mix of science and art. The design of stripline resonators is no exception but, fortunately, some of the uncertainties can be eliminated by simulation. The use of simulation tools is emphasized at every step of the process.

Although the discussion here focuses on the rubber and tire industry needs, the sensors and the design process apply to other industries including the production of paper, fabrics, plastics, wood veneers, food and the like, and in monitoring of various processes. Some details of the production process will be given for the purpose of understanding the sensor and its use, but these are not really part of the design. Both sensors are designed as open cavity resonators to allow free movement of the fabric or rubber at production speeds and are built and installed without modifications to the production lines.

We start with the sensor design, followed by simulation and concluding with discussion of the applications in which the sensors are used. The following chapters will deal with evaluation of the sensor's performance, implementation, and testing.

5.2 Sensor design for fabric coatings

To understand the design of the sensor, it is useful to first look at the required parameters from the point of view of the sensed quantity since that, to a large extent, dictates what can be done and what sensing parameters are critical.

In the first problem to be addressed, a nylon or polyester fabric is dipped in a latex solution to coat it with a certain amount of latex. The fabric in question is then dried and, in this case, used for production of tires. The quantity that the sensor is intended to sense is the amount of latex retained on the fabric after drying. Clearly, too little latex will inhibit the function of the fabric and may cause delamination in the tire, whereas too much latex, in addition to the cost, weakens the layered structure of the tire. The latex solution is about 94% water, 5% solids, and small amounts of additives. After drying, the solids remain bonded to the fabric. Although the fabric material has a relative permittivity of about 3.13 for nylon and 2.92 for polyester (the permittivity is frequency and temperature dependent), the fabric is thin and sparse so that when wet, the dominant quantity is water, whereas when dry, the dominant quantity is the deposited solids. The interest is the coating of solids retained after drying but, by monitoring the thickness (or, alternatively the weight per unit area) of the wet fabric, one can then control the coating thickness, provided proper calibration of the sensor can be done. The fabric is wide—the width can vary but typically is around 180 cm, and its thickness is under 1 mm (before coating). The coating thickness depends on the application, but it needs to be controlled within strict tolerances and must be uniform across the width and length of the fabric. Typically, the amount of latex is controlled by removing excess material after dipping using blowers (see Figure 5.1). The purpose of the sensor is to control these blowers to obtain the proper amount of material on the fabric, prior to drying. In Figure 5.1, the sensor is placed after the vacuum dewebbers, the purpose of which is to remove excess latex off the web. The dewebbers are an arrangement of fans that move air through the fabric to remove excess dip material by a suction-like action. In a complete system, the signal from the sensor is used to control the dewebbers to maintain a set coating. A more complete description of the dip system and its components is available in Section 5.4 in conjunction with online testing of the sensor.

Because the fabric is thin, it is important that the sensor selected be sensitive to variations in permittivity at the location of the fabric. Similarly, given that the coating is produced in a continuous length at relatively high speeds, it is important that the sensor does not interfere in any way with the motion of the fabric or with its coating and must allow sufficient space for splices between sections of the fabric to

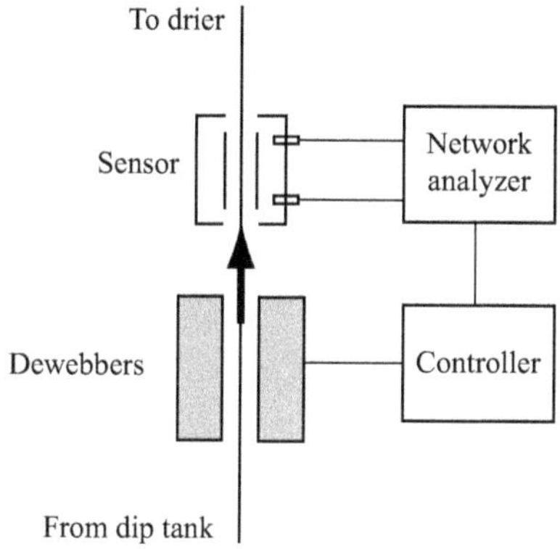

Figure 5.1 The relation of the sensor to the dip system

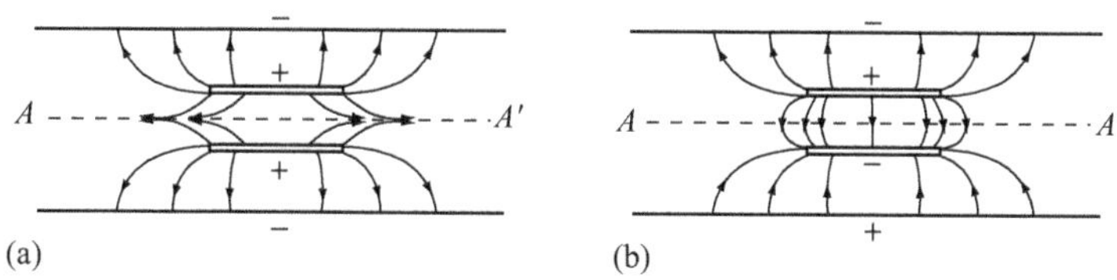

Figure 5.2 Electric-field distribution in broadside coupled transmission lines: (a) even and (b) odd modes

pass through. Clearly, the most obvious solution, one that uses capacitance to sense permittivity variations is not applicable—the variations in capacitance due to variations in retained solution are too small. On the other hand, an open cavity resonator will not only allow for free motion of the fabric but can be made sensitive enough to variations in permittivity of the fabric. Of the possible resonant structures, the most promising is the broadside coupled stripline resonator discussed in Section 3.4. As was discussed in Chapter 3, the resonator is open and the distance between the striplines can be significant. However, the main attractive feature of this type of resonator is in its electric-field distribution within the space between the striplines. Figure 5.2 shows the field distributions for the odd and even modes of propagation in coupled transmission lines. These are very similar at resonance except, of course, that the resonator is finite in length and hence the fields tend to be highest at the edges of the center plates. The odd-mode fields are vertical across the centerline of the cavity. This clearly means that the sensor will be less sensitive to position of the fabric or changes in its permittivity since around the centerline the field is more or less uniform. On the other hand, in the even mode, in which the fields are horizontal around the centerline, the field gradient is high around the centerline and therefore the resonant frequency will be highly sensitive to changes in the conditions around the centerline. A sensor made as a broadside coupled stripline resonator can satisfy both the physical requirements of a nonintrusive, open sensor, and the sensitivity requirements. Further, although the odd mode is not sensitive to the fabric properties, it can be used for compensation purposes since both modes are roughly equally sensitive to bulk properties in the space between the stripline, such as temperature and humidity.

As with any design, various decisions will affect the performance of the sensor. In the case of the cavity resonator for the fabric sensor, there are three primary considerations to be weighed:

1. the resonant frequency, f_0
2. distance between striplines
3. area covered by the sensor

Since the change in frequency is the measure used to infer changes in permittivity (i.e., changes in coating thickness), the higher the base frequency, the larger the changes in frequency for a given change in coating thickness. However, the resonant frequency relates to the physical dimensions of the cavity, meaning that increasing the frequency necessarily reduces both the distance between the striplines and the area covered by the sensor. The distance between the striplines is dictated primarily by the need to clear splices in the fabric, and it is relatively easy to decide upon. A distance of about 10–15 cm should be sufficient for most applications. In applications such as paper production or a continuous distribution such as grain on a belt, the distance may be reduced to the minimum necessary to clear the product. This distance affects the Q-factor of the cavity as well as possible external interference such as due to proximity of objects or personnel. Smaller distances produce higher Q-factor sensors with less interference from the outside and higher sensitivity. The downside of this is that maintaining the fabric at the centerline becomes more critical and motion of the fabric must be smoother. The area covered by the sensor is also important in that the larger the area the more of the width of the fabric is monitored but, at the same time, its properties are averaged over a larger area. That is, a higher frequency sensor will be more sensitive to localized variations in fabric properties, but a smaller section of the fabric is covered.

There are additional, secondary considerations that must be taken into account. These include weight of the sensor, measurement of the resonant frequency, rigidity of the cavity, shielding effects, and others.

Obviously, there is no single solution to this tradeoff between the various considerations and in the end, one has to decide on a set of parameters that optimizes the application at hand. Assuming that the most important decision is the resonant frequency (because it defines sensitivity), we start with that. In an open cavity of the type shown in Figure 5.2, it is very difficult to calculate the exact resonant frequency, but we can perform some simple calculation to come up with an acceptable design. To do so, we start with the required sensitivity. In the application at hand, the separation between the striplines is set at 12 cm. For purposes of initial design, we assume the fabric is 1-mm thick, and the sensor covers an area S (to be defined). The properties of the fabric and latex mixture are as follows:

Relative permittivity of water: 78.
Relative permittivity of solids in the dip mixture: 2.5.
Relative permittivity of the fabric: 2.5.
The mixture is approximately 20% solids and 80% water.
After dipping in the solution, the solution occupies 35% of the fabric volume.
Required sensitivity: 1% in solids retention (after drying).
These values are not exact, nor do they need to be as long as the final design allows for a sufficient margin in sensitivity.

For resonance, once the dimensions of the resonator have been set, the important quantity is the permittivity within the resonator. If the space in the resonator contains a uniformly distributed medium, then the permittivity is that of the medium. In the cases related to this work, that is never the case since the fabric or the rubber sheet only occupies a small portion of the space, whereas the rest is air. The air itself contains water in the form of moisture. The fabric being evaluated is made of many constituents, the most important being water and the base material (nylon, polyester, or aramid). Thus, one can only talk of an effective permittivity, something that is difficult to come by even when the mixture of materials is uniform within the cavity and certainly when the media are distinctly separated. The effect of, say, the fabric depends where the fabric is within the resonator. We place it at the center because the even-mode fields are highest at that point and hence the effect of the fabric on the resonant frequency is larger than if it were off-center. All that means that calculation of an effective permittivity is difficult and fraught with errors. For proper calculation, one has to take into account many parameters including relative volumes, permittivities, location of the various media, any polarization that may exist, shapes of particles if individual particles can be identified, not to mention environmental quantities such as temperature and pressure. There are many methods of calculation of effective permittivity for various conditions and varying levels of complexity. These are known by various names including mixture formulas, homogenization formulas, and more. However, we will take a simple approach here since the purpose of calculating the effective permittivity is to obtain an estimate rather than an exact value. To do so, we will assume that the materials within the fabric are homogeneously distributed (or mixed) and also assume the same for the air and fabric within the resonator. This approximation provides a starting point and a simple way of estimating the sensitivity of the sensor.

There are a few mixing formulae for the relative permittivity homogeneous mixtures of dielectrics. One of the simplest and commonly used is

$$\sqrt{\varepsilon_{\mathrm{eff}}} = \sum_{i=1}^{N} v_i \sqrt{\varepsilon_i} \tag{5.1}$$

To this, one can add an even simpler (and less accurate) approximation as

$$\varepsilon_{\text{eff}} = \sum_{i=1}^{N} v_i \varepsilon_i \tag{5.2}$$

In these equations, ε_{eff} is the effective relative permittivity, v_i is the volume fraction of the ith dielectric constituent, ε_i its relative permittivity, and N the number of constituents in the mixture.

By using these formulae to calculate the effective relative permittivity of 20% solids, 80% water in the dip mixture, we can write using (5.1):

$$\sqrt{\varepsilon_r} = \sqrt{78} \times 0.8 + \sqrt{2.5} \times 0.2 = 7.38 \quad \rightarrow \quad \varepsilon_r = 54.49 \tag{5.3}$$

To see how the amount of dip affects the dielectric constant and therefore the resonant frequency, we assume a "solid" material, 1 mm thick with 35% solution (relative permittivity equal to 62.9), whereas the remaining 65% of the total volume has relative permittivity of 2.5. Using again the same mixture formula, the permittivity of the material is now

$$\sqrt{\varepsilon_r} = \sqrt{54.49} \times 0.35 + \sqrt{2.5} \times 0.65 = 3.611 \quad \rightarrow \quad \varepsilon_r = 13.04 \tag{5.4}$$

A 5% change in solution pickup by the fabric should result in 1% change in solids retention. Suppose now the fabric picks up 40% (1% increase in solids retention). The relative permittivity of the wet fabric now becomes

$$\sqrt{\varepsilon_r} = \sqrt{54.49} \times 0.4 + \sqrt{2.5} \times 0.6 = 3.9 \quad \rightarrow \quad \varepsilon_r = 15.22 \tag{5.5}$$

The uses of the other mixture formulas produce different values, and these are shown in Table 5.1.

In either case, the change in permittivity is very significant and should be easily detectable. But the question is how to decide on the resonant frequency of the cavity. Since analytical calculations using, for example, the perturbation method are rather difficult to perform, we opt for numerical calculations. The process is as follows:

1. Start with an initial design by deciding on a set of dimensions for the sensor. If a preference for any of the parameters is available, use that and adjust the other parameters accordingly. For example, one may have a preference or a restriction on the resonant frequency or on the area covered by the sensor. In the case

Table 5.1 Effective permittivity in the resonator for two mixing formulas

	ε_r of the dip mixture	**ε_r of the fabric at 35% dip pickup**	**ε_r of the fabric at 40% dip pickup**
Formula (5.1)	55.49	13.04	15.22
Formula (5.2)	62.9	23.64	26.66

described here, because the fabric is wide, the sensor will necessarily be smaller than the width so as to avoid averaging over the whole width of the fabric. An initial design of the sensor covering about 1/4 of the width is reasonable. Then, one can use either multiple sensors to cover the width of the fabric (with appropriate overlap) or incorporate a single moving sensor to "scan" the width of the fabric. Each method has its advantages, and we will discuss both.

2. Calculate the resonant frequency with the fabric present and without it (using the relative permittivities above) and calculate the shift in resonant frequency caused by the change in retained solids. From these results, one can calculate the expected sensitivity of the sensor based on the smallest reliable measurement of frequency shift than can be obtained.
3. If the sensitivity is not sufficiently high, reduce the dimensions of the sensor to increase the resonant frequency and hence the shift in resonant frequency due to changes in solid pickup until an acceptable resolution is obtained. It is important to recognize that the calculations are approximate because of the approximate nature of the properties used and hence a margin must be built into the sensitivity calculation.
4. Verify the results experimentally. This of course means building a prototype of the sensor for the purpose. At this point, the measured sensitivity can be compared with the calculation, and perhaps readjusted higher or lower by repeating steps (3) and perhaps (4).
5. Improve the sensor based on the experimental and numerical results. This may entail adjusting the resonant frequency to a convenient value, adding shielding to the sensor to reduce outside influences improving rigidity and the like. Additional calculations may be necessary, but the numerical process is fast and efficient and should not present a significant increase in time or cost.

An initial design is shown in Figure 5.3. It consists of two ground planes, each 50-cm long and 35-cm wide with the two striplines 30-cm long and 6-cm wide, separated 12 cm apart and centered between the two ground planes. The height of the resonator is 28 cm with the two striplines at 7 cm from their respective ground planes. These dimensions, especially the widths, are somewhat arbitrary. The length of the striplines corresponds to a theoretical resonant frequency of 500 MHz for a $\lambda/2$ resonator:

$$f_0 = \frac{3 \times 10^8}{0.3 \times 2} = 500 \text{ MHz} \tag{5.6}$$

This somewhat arbitrary frequency is a compromise between the need for as high a frequency as possible to increase sensitivity and practical issues of coverage as well as the need to use lower frequencies so that the electronic components can be easily incorporated. The actual resonant frequency is expected to be somewhat different because of the influence of the ground planes and the other dimensions selected for the design. The dimensions of the ground plane were selected to cover a reasonable portion of the fabric width (in this case about one quarter). The width of the ground planes was selected sufficiently wide to contain the field of the stripline to avoid

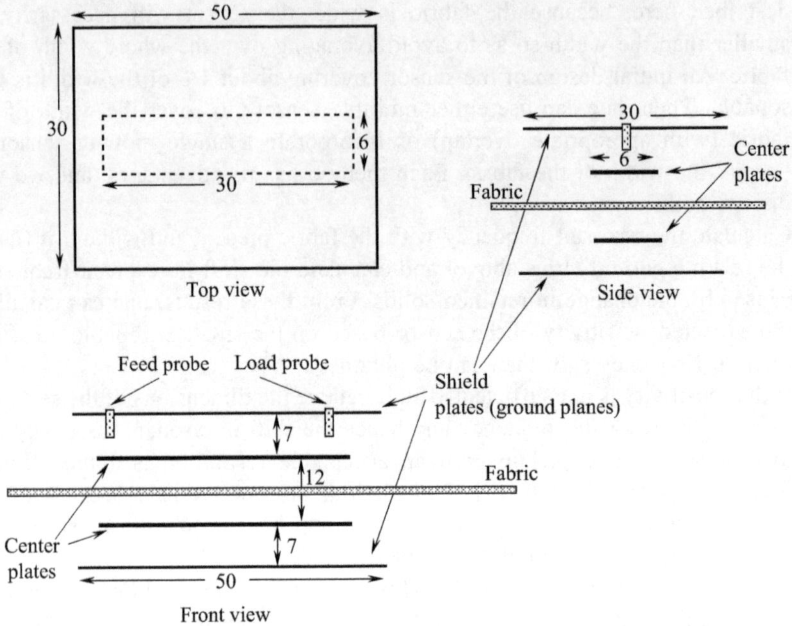

Figure 5.3 Prototype broadside coupled stripline resonator. The resonator is placed with the longer dimension perpendicular to the direction of motion of the fabric

interference with external objects. Ideally, these dimensions should be much larger but, because these dimensions are used in a simulation, they were deemed sufficiently large. In addition, because it was anticipated that partial shielding of the resonator will be incorporated, the issues of interference from external bodies were not considered important at this stage.

As can be seen in Figure 5.3, the resonator is fed through a probe, whereas the second probe is used as a load to sense the resonant frequency and Q-factor. In the simulation, only the exciting probe need be modeled since resonance can be detected from the magnitudes of the fields, but in experimental verification the sensing or load probe is used by a network analyzer to measure resonance.

The fabric moves on the centerline of the resonator, and both the even and odd modes are calculated. The purpose of the initial simulations is simply to establish the expected sensitivity of the resonator and to provide data for any modifications that may be needed prior to implementation and testing. Because the initial design is so very simple, it is relatively easy to build it so that measurements can be performed as additional verification. This is not always possible but whenever it is, it becomes useful and provides confidence in the design.

The simulation was done with a general purpose finite-difference time-domain (FDTD) program, described in Appendix C, over a span of 1 GHz. Figure 5.4

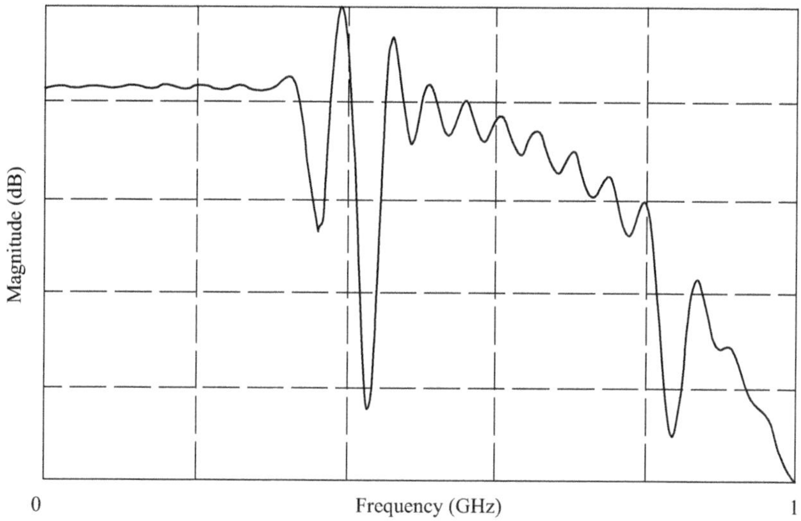

Figure 5.4 Computed resonant frequencies of the empty cavity. The first is an even mode at 380 MHz followed by an odd-mode resonance at 422 MHz. A second higher order even mode occurs at 832 MHz

shows the FDTD simulation showing the first even and odd modes as expected at 380 and 422 MHz, respectively, and a higher order (even) mode at 832 MHz. Note that the odd resonant frequency is higher than the even resonant frequency.

The two probes shown in Figure 5.3 are used as a source and load probe for the network analyzer to monitor the *S*-parameters in a two-port configuration. The excitation is through the gap between the probe and upper plate. To detect resonance, only the S_{11}-parameter is needed since at resonance the S_{11}-parameter exhibits a sharp drop. Figure 5.5 shows a screen snapshot showing the even and odd resonant frequencies of the unloaded cavity in Figure 5.3 showing an even mode at 384.715 MHz and an odd mode at 428.243 MHz. The scan is between 300 and 450 MHz so that only the first order resonances are visible. The difference between the simulation and experiment is mainly due to small differences between the nominal dimensions and the actual dimensions of the cavity but also due to environmental conditions (moisture) between the plates. Both are lower than those expected from a $\lambda/4$ resonator as was explained above.

To estimate the sensitivity of the resonator, the configuration described above, consisting of a 1-mm-thick fabric with various amounts of solutions, can be used to compute the resonant frequency. This is shown in Table 5.2 for 30%, 35%, and 40% solution in the fabric. The table also shows the effective permittivity of the fabric based on the calculations above.

Thus, a 5% change in solution pickup, which is a 1% change in solids pickup, causes a shift in resonant frequency of 0.7 MHz. Assuming a minimum detectable shift on the network analyzer of 10 kHz, the system should be able to

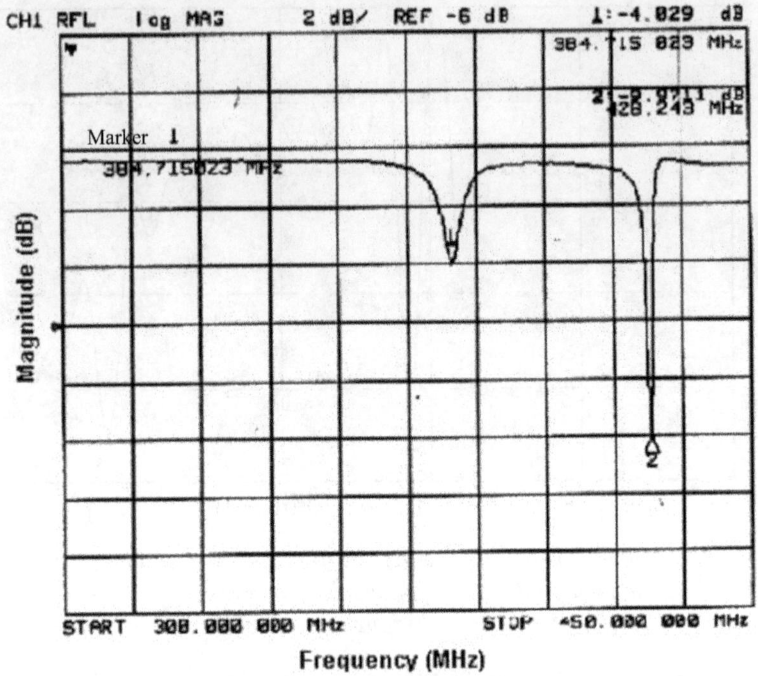

Figure 5.5 Screenshot of a network analyzer for the empty resonator showing the even and odd modes at 384.715 and 428.243 MHz, respectively

Table 5.2 Sensitivity-to-solid coating: solids 20% ($\varepsilon_r = 2.5$), water 80% ($\varepsilon_r = 78$), fabric ($\varepsilon_r = 2.5$)

% Solution	ε_{eff}	Resonant frequency even mode (MHz)	Resonant frequency odd mode (MHz)
30	20.62	374.4	424.3
35	23.64	373.7	424.3
40	26.66	373.0	424.3

detect a change of 1%*10/700 kHz = 0.014% in solids pickup. The network analyzer can in fact detect shifts as low as 1 kHz, but the figure given here should be a good guideline. The high sensitivity indicated here is mainly due to the large permittivity of water. In other words, the system does not detect the change in amount of solids but rather the change in amount of water. For this reason, the resonator can also be used to sense the drying process of the fabric.

It is also useful to note that the odd-mode resonant frequency is not affected by the change in permittivity of the fabric. This is consistent with the analysis of the fields in Section 2.12.

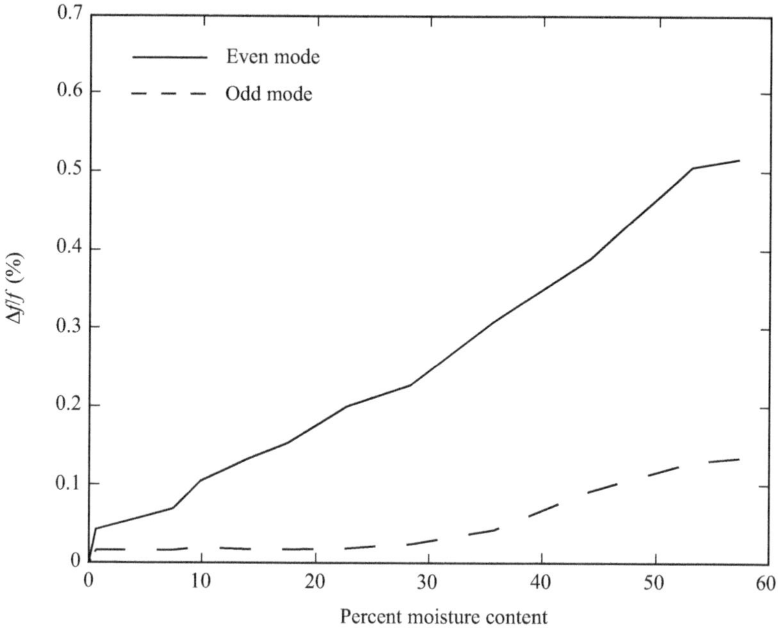

Figure 5.6 Experiment showing behavior of the even and odd resonant frequency with moisture content in a fabric

Experimental verification of sensitivity was also carried out although under different conditions. A thin fabric was used with a very low permittivity to test the sensitivity to moisture (water). The dip solution was not used for these tests primarily because of the difficulty in handling it and because it dries quickly. Nevertheless, the response with respect to moisture content is similar to that in the simulation. The results can be seen in Figure 5.6. The change in frequency is shown with respect to the moisture content in the fabric. In this experiment, the moisture content was measured by weight. The fabric was weighed first, then water was sprayed on it until its weight corresponded to the volume fraction of 100% moisture, that is, water fills the equivalent volume of the fabric. The latter was calculated from the thickness of the fabric and its area. This of course is not very accurate, but the figure shows two important results. First, the change in the resonant frequency of the even mode is linear with moisture content (within the accuracy afforded by the sample preparation method described above). Second, and equally important is that the odd-mode resonant frequency is much less affected by the moisture content as one would expect. Nevertheless, there is a slight effect, especially at higher moisture levels. This indicates both that the odd-mode frequency can be used for compensation of environmental conditions and that this compensation is not perfect. More will be said about this in Chapter 7 when we discuss this issue in the context of calibration.

5.2.1 Sensor modifications and optimization

The sensor in Figure 5.3, although certainly sufficiently sensitive for the purpose, has a number of shortcomings that need to be addressed before it can be used in an industrial environment. One obvious issue is the rigidity of the structure. The ground planes are relatively large and hence prone to vibrations. They also need to be attached to an external structure while allowing the space between the center plates open. The center plates themselves must be attached to and electrically insulated from the ground planes. A second problem is the fact that the center plates are exposed and hence subject to two separate issues. One is mechanical. Splices in the fabric can easily snag in the protruding center conductors and possibly damage the sensor. The second is electrical and manifests itself in the fact that the resonant frequency can be influenced by structures and bodies outside the sensor. Because of the wide-open structure, any substance, member, structure, or even personnel may affect the resonant frequency.

Next comes the question of coverage of the fabric. The sensors are only 50-cm wide (the center plates are perpendicular to the direction of motion of the fabric) and hence cannot cover the whole width of the fabric. Multiple sensors in various configurations or moving sensors can be employed, each method having its advantages and disadvantages.

A regime of simulations and tests is necessary to establish the final configuration of the sensors and to modify the sensors to satisfy the requirements of the system. The simulations also establish limits on working parameters that need to be followed for the sensor to perform as designed. These issues are described next.

5.2.1.1 The simulation regime

Integral to the design of the sensors is a simulation regime that attempts to reduce the design time and anticipate modifications and the optimization of the sensors and the sensing environments. All simulations results given here and elsewhere in this work were done using an FDTD program available commercially. The fundamentals of the method are described in some detail in Appendix C, but some details of the setup for simulation are given here, prior to presentation of optimization results leading to an acceptable design. The importance of the simulation process cannot be overstated—it is an absolutely necessary step that allows the designer to obtain results without the need for extensive experimentation and to weed out those changes that are either not necessary or that have little impact on the design. Simulation is much faster and less expensive in comparison with experimentation and should always precede prototyping.

The simulation starts by defining the geometry of the sensor. This includes the folded ground planes and the center plates as shown in the center of Figure 5.7. This is done through the human–machine interface (HMI) of the software and, while the HMI for various software programs may be different, it usually involves either drawing the object or specifying its coordinates and dimensions as well as the type of materials. The ground and center planes are modeled as perfect conductors. In addition, one has to specify the driving source for the simulation. This is the

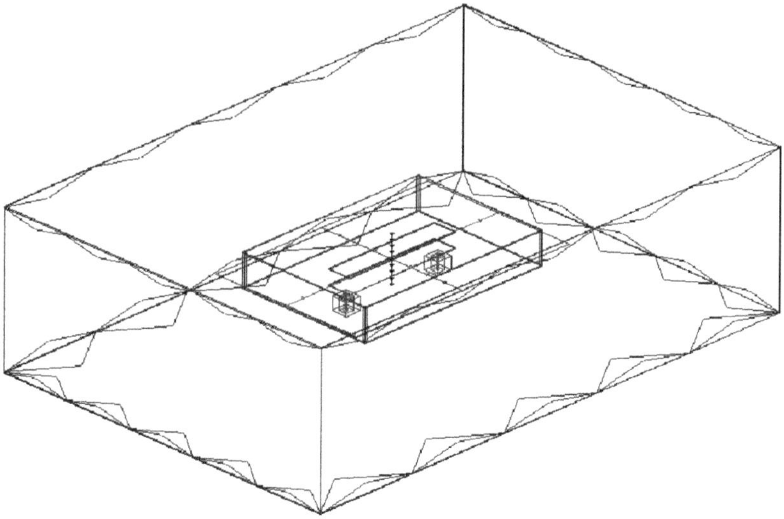

Figure 5.7 FDTD model for the resonant sensor in Figure 5.3

source (or feed) probe in Figure 5.3 or Figure 5.10. In the FDTD program used here, the probe is replaced with an input plane, shown as the rectangular structure (port) at the left side, below the lower center plane in Figure 5.7. The second, identical structure is the load port at which the fields are monitored to detect resonance (or, in other terms, to compute the S_{21}-parameter).

These structures, which look like small square-based towers, are designed as well to present the proper impedance to the transmission line connecting to them (not part of the simulation). In most cases, this would be 50 Ω and represent the connectors to the transmission lines. Because the resonator is open, one must assume that the electric and magnetic fields outside the resonator extend to infinity, something that computer models cannot simulate. To take this into account, artificial boundaries are created to enclose the geometry at some reasonable distance away (shown in Figure 5.7 with triangular markings). These boundaries are used to perform near-to-far-field transformation so that the net effect is that fields at any distance can be calculated without the need to model large volumes. An explanation of the specification of sources, boundary conditions, and near-to-far-field transformations is given in Appendix C. Figure 5.7 shows an empty sensor, but the addition of a fabric simply means defining the geometry, dimensions, location, and material properties of the fabric.

Once the material properties of the space, the fabric, and the boundaries are specified, the simulation can begin. Again, the details of the method are many and fully explained in Appendix C, but in simple terms it follows the idea that once the source is switched on, all we have to do is, follow the evolution of the fields within the cavity in time until the resonant frequency is detected on the basis of the fact that fields are highest at resonance or, more likely, that either the S_{11} is maximum

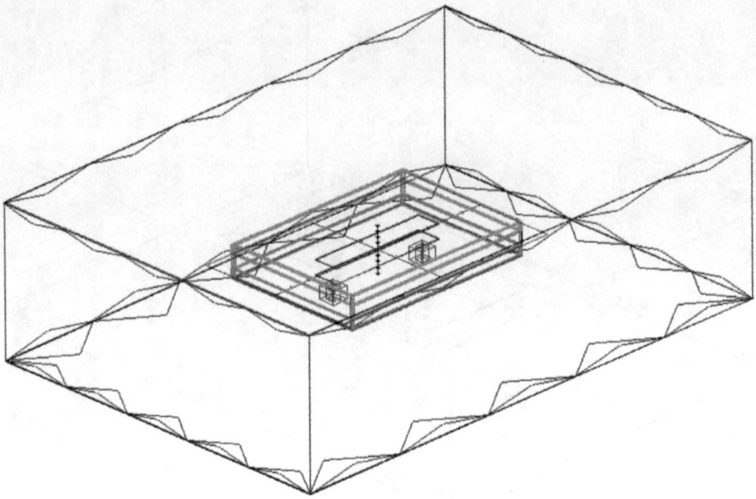

Figure 5.8 FDTD model for the resonant sensor in Figure 5.10

(maximum reflection caused by the fact that the resonator impedance is minimal) or the S_{21} is minimum (caused by the fact that minimum power is transferred from the source to the load port). This can be done by supplying a short pulse to the source port or by scanning the source port over a range of frequencies using a sinusoidal source. The latter is how a network analyzer measures resonance, and it is the method used here (see, e.g., Figure 5.4). In the simulations performed here, we are only interested in the resonant frequency and the Q-factor; hence, the field levels are of no interest. However, given the field at the source port, one can calculate the actual electric- and magnetic-field intensities, power, losses, and any other quantity derivable from the fields. Note as well that the model in Figure 5.7 corresponds to the geometry in Figure 5.3 (the ground planes are not folded). The same model with folded ground planes is shown in Figure 5.8 and corresponds to the geometry in Figure 5.10.

Clearly much more complex structures can be simulated. An example, also modeled as part of the design process, is shown schematically in Figure 5.23. In this configuration, a number of sensors, placed side by side, are attached to a large metal shield, the purpose of which is to both shield and support the individual sensors so that a larger area can be covered. Figure 5.9 shows the FDTD model for one of the sensors at the center of the shield. Note that, in this simulation, the ground planes are folded as in Figure 5.8. This sensor is then duplicated and displaced to form the overall structure (these are not shown for clarity) and, following application of sources and boundary conditions, the resonant frequency of each of the sensors can be calculated either together (all sensors driven at the same time) or, as was done here, one at a time because only one sensor is usually driven, while the others are off. A simulation of this type reveals the effect of the physical structure of neighboring sensors on the driven sensor.

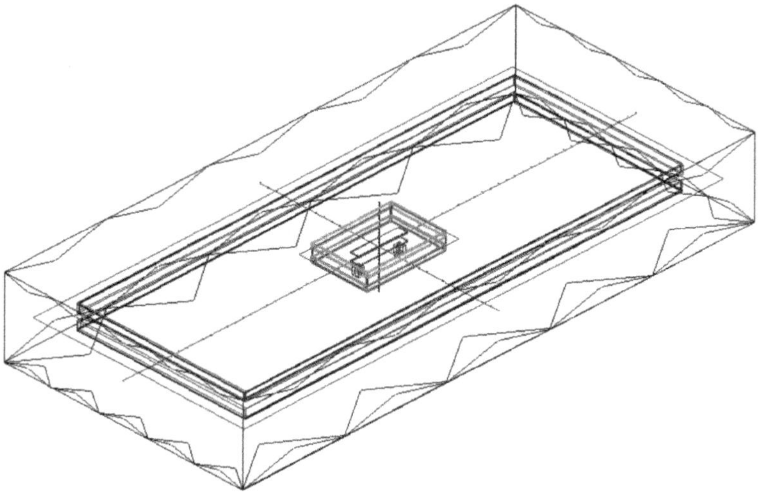

Figure 5.9 FDTD model for the simulation of a sensor placed at the center of a large external shield (see Figure 5.23)

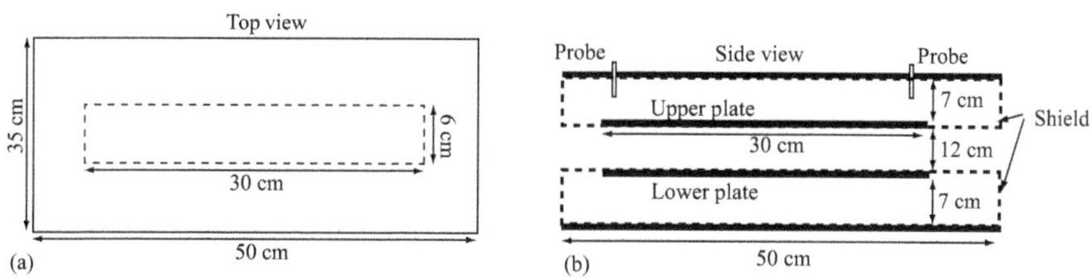

Figure 5.10 The resonator with folded-over ground planes (shown by the dotted lines) to form a partially shielded open resonator. (a) top view (b) side view

5.2.2 Shielding of the sensor

As a microwave resonator, the ideal structure is a closed cavity, a structure that maximizes the Q-factor and eliminates outside influences. Obviously, the sensor discussed here must remain open. However, the space above the upper center plate and below the lower center plate is not used, and hence that space can be shielded. This is done by folding the upper and lower plates down to the level of the center plates (or slightly past them) as shown in Figure 5.10(b). Introducing this partial shield has a number of consequences. The most obvious is the fact that it protects the center plates from possible damage from splices in the fabric. It also adds rigidity to the outer plates and to the center plates. Because the shielded sensor approximates a closed cavity, it improves the Q-factor as well and reduces sensitivity to external influences, as will be shown shortly. To see the effect of the shield, the "static" electric-field intensity with and without the shield is shown in

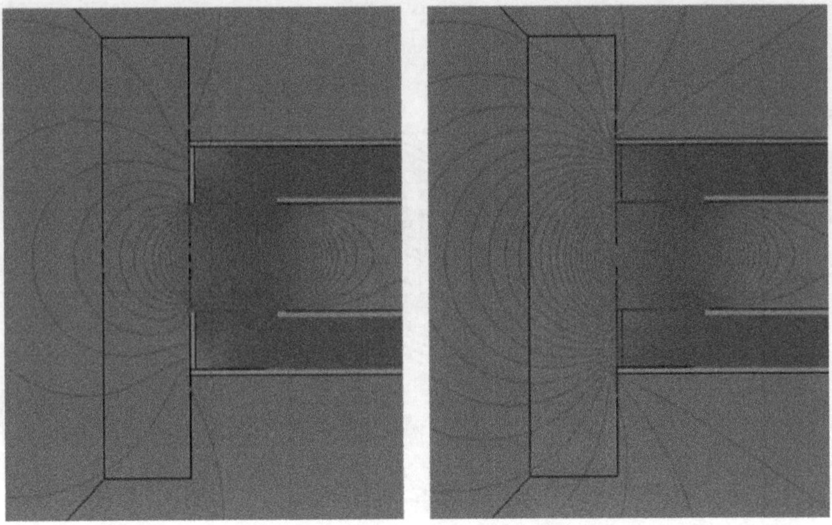

(a) (b)

Figure 5.11 The effect of the partial shield on the static electric field: (a) with the bent-over plates (partial shield) and (b) no shield

Figure 5.11. The effect of the shield can be clearly seen. The fields shown in Figure 5.11 are quasistatic, that is, they are calculated at the correct frequency but not as a resonant structure. At resonance, this effect is much more significant because the fields are larger. However, because the shield is partial, the fields still extend outside the cavity. The extent depends on the distance between the two folded-over ground planes as well as on the width and length of the cavity.

To see the effect of the folded-over plates on the resonant frequency in the presence of personnel, consider Figure 5.12. It shows the shift in resonant frequency due to a dielectric block at various distances. At distances of about 1 m and above, from the side of the shielded sensor, the effect due to the presence of personnel is negligible, whereas the unshielded sensor requires a distance roughly three times larger to produce the same effect. It is, however, interesting to note that at very close proximity (less than about 20 mm), the unshielded sensor is less sensitive to the presence of foreign bodies. As a rule, a perimeter of 1–1.5 m laterally from the sensor must be considered an exclusion zone from which personnel and any moving or temporary foreign bodies should be kept out. Fixed structures however do not pose a problem even though they may change the resonant frequency since the sensor can be calibrated in their presence. This only applies to stationary sensors. If the cavity resonator moves, then any fixed structures will also alter the reading of the sensor, and this cannot be calibrated for. Good conductors and good dielectrics in particular will only affect the resonant frequency, whereas lossy dielectrics in close proximity may affect the Q-factor of the resonator and hence its performance.

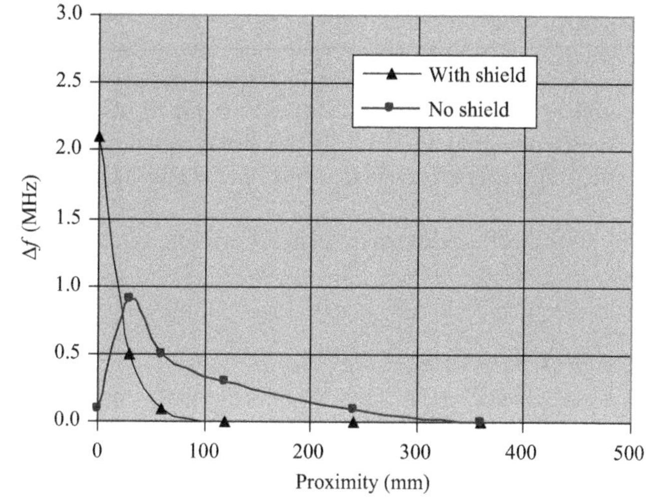

Figure 5.12 Effect of personnel in the proximity of the open side of the sensor. Distance between center plates is 120 mm

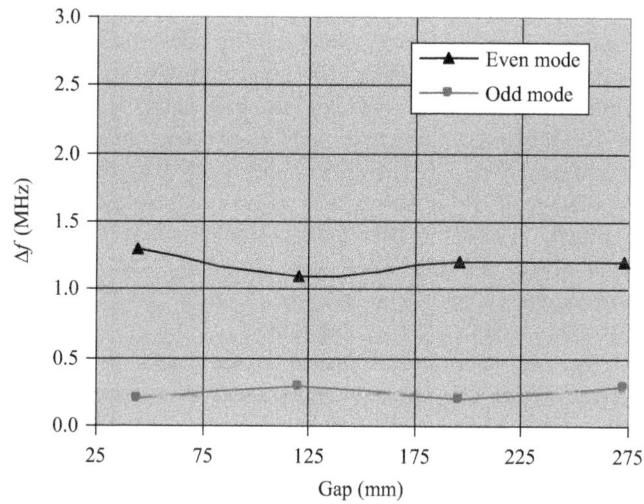

Figure 5.13 Effect of the gap between the upper and lower shields on the resonant frequency. The distance between the center plates remains fixed at 120 mm

Although the shield must be open, the gap between the upper and lower shields does not have to be the same as that between the center plates. To see the effect of the gap on the sensitivity of the sensor, the gap was varied from 40 to 275 mm (nominal distance between center plates is 120 mm) as shown in Figure 5.13.

Clearly, the effect of the gap on the resonant frequency is minimal. The shift shown is due to the fabric relative to an empty resonator. The resonant frequencies (both odd and even) depend primarily on the length of the center plates and, of course, on dielectric materials within the resonator. However, in practical applications, the gap between the shields is roughly the same as that between the center plates, and this distance is critical in terms of sensitivity and can be seen in the simulations that follow.

The upper and lower shields were made of a 6-mm aluminum plate with all seams welded.

5.2.3 Simulation and optimization

Although the initial simulation has shown that the sensor is quite sensitive and the addition of the folded shield limits the effects from external sources as well as increase the *Q*-factor of the resonator, it is important to simulate other parameters of the sensor to ensure an optimal configuration. One of the most critical issues in the sensor is the separation between the center plates. This requirement is imposed by the fabric, and the nominal 120-mm separation was the minimum distance acceptable for practical implementation. For the current application, a larger separation is desirable but not critical, whereas in other applications, smaller separation may be possible (such as in production of paper). Figure 5.14 shows the effect of the gap between the center plates on the odd- and even-mode resonant frequency. Both increase dramatically with the decrease in gap indicating that the sensitivity of the sensor increases (more or less) exponentially with the decrease in the gap. The 120-mm separation used here is a compromise between physical requirements and sensitivity. It should also be noted that the separation has an effect on the *Q*-factor of the cavity as well as on external influences since, as the separation is reduced, the gap between the two shields is also reduced, more closely approximating a closed cavity. The sensitivity in Figure 5.14 is entirely due to the separation between the center plates since the distance between the shields has no effect on the resonant frequency (see Figure 5.13).

Another influence on the performance of the sensor is the position of the exciting probe relative to the (upper) center plate. While keeping the probe on the centerline (axis) of the plate, its position can be moved laterally along the centerline. Figure 5.11 shows a simulation of the sensor with the position of the probe relative to the center of the upper center plate along its axis. There is little to be said other than, perhaps, that the even-mode sensitivity is slightly higher around 120 mm. Experiments have shown similar results at about 130 mm. In the final sensor design, the two probes (one exciting, one load) were placed symmetrically about the upper plate, separated 25 cm apart so that they are 25 mm from the edges of the plates. The results in Figure 5.15 were obtained with both probes on one side of the sensor so that the two probes are placed next to the two edges of one center plate. The probes can also be placed at opposite sides of the sensor either across from each other or diagonally opposite from each other. However, because the fields are symmetric, the results are the same regardless of how the probes are arranged.

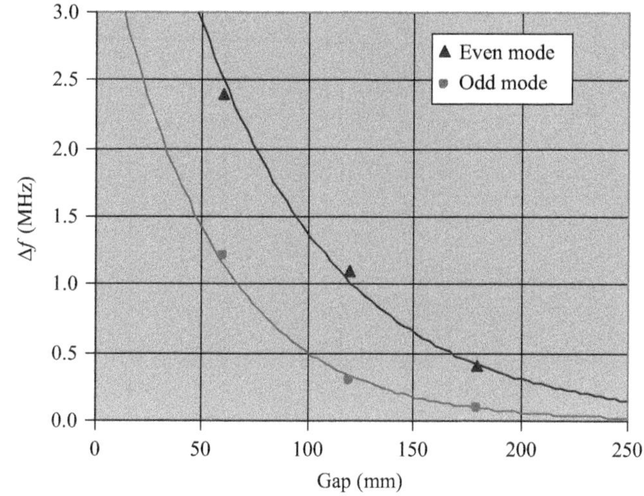

Figure 5.14 Relation between separation of the center plates and resonant frequency

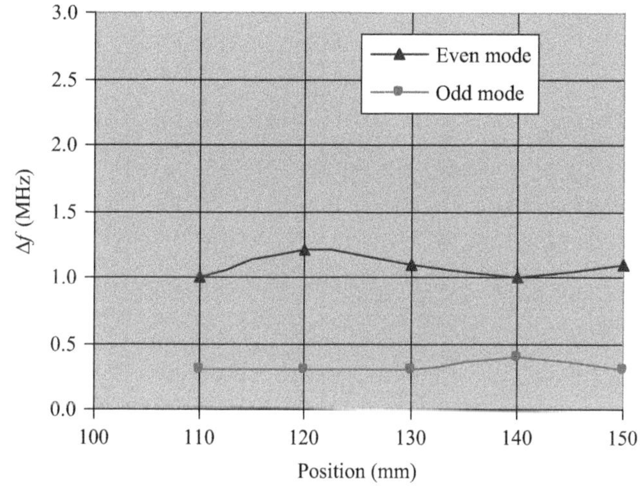

Figure 5.15 Effect of exciting probe position relative to the center of the upper center plate

Because the center plates have the largest influence on the performance of the resonator, optimization of these was in order although the simulations show that the shape, thickness, and width of the center plates have minimal or no influence on the resonant frequency. For example, Figure 5.16 shows the results of simulation of butterfly-shaped center plates. The center of the plate was narrowed, and the results plotted for a ratio of 0.2/1–5/1 between the narrow dimension and the wide

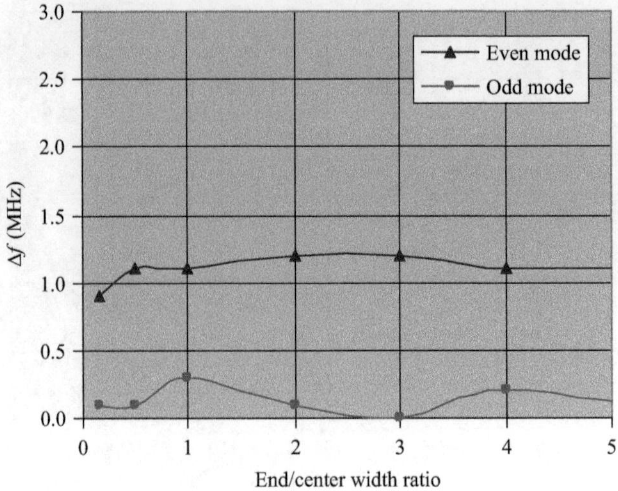

Figure 5.16 Effect of center plate shape on the resonant frequency. The center plate is butterfly-shaped

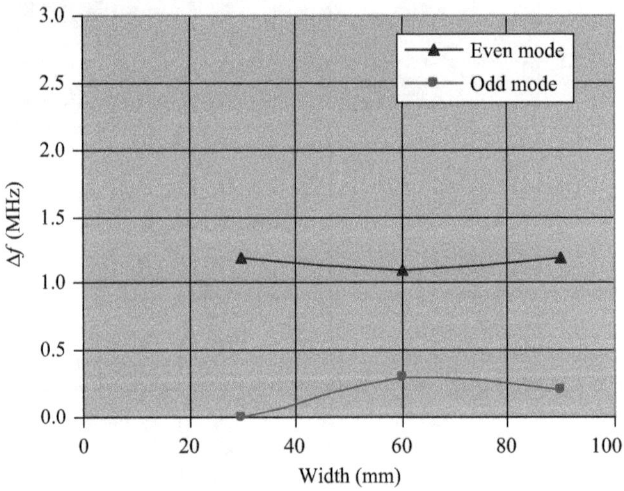

Figure 5.17 Effect of center plate width on the resonant frequency. The center plate is rectangular

dimension. As long as the center plates are reasonably wide (above about 20 mm), the response is flat. Although this result is less than interesting, it showed that the design can use virtually any shape, and in the end, a simple rectangular center plate with rounded edges was used as a practical design.

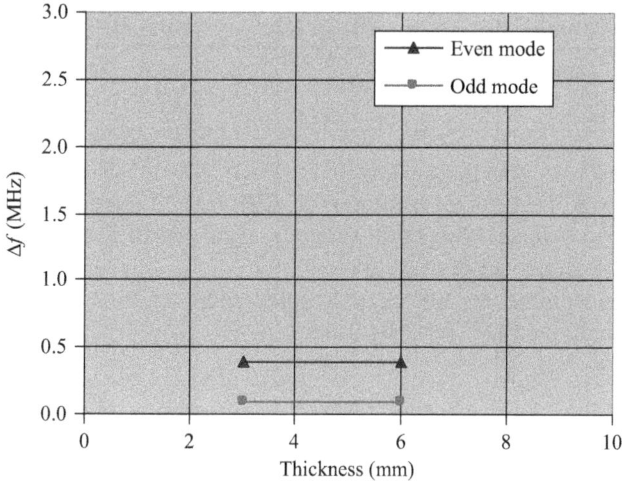

Figure 5.18 Effect of center plate thickness on the resonant frequency. The center plate is rectangular

Similarly, the width of the rectangular center conductor width is not very important as long as it is above about 20 mm as shown in Figure 5.17. However, because the width of the ground planes is fixed at 30 cm, the width must be kept well below that, or the distribution of the fields within the sensor will change and with it, its performance. The length of the center plate was kept constant at 35 cm.

One can, of course, optimize further, especially with regard to the center plates. For example, Figure 5.18 shows that the thickness of the plates has no effect on performance although this can be deduced without simulation. The plate thickness was changed to increase its stiffness, and the only effect that one could anticipate is a slight change in resonant frequency due to the fact that the thicker plate occupies a larger volume in the cavity and hence the resonant frequency can increase slightly. However, as long as the separation between the plates remains unchanged, the thickness of the plates has no effect. There are other effects that can have larger influences on the performance. Very thin plates can vibrate and hence introduce errors in the sensed resonant frequency based on the sensitivity to the distance between the plates. Similarly, any foreign material in the sensor or changes in humidity and temperature will have much larger influences on the resonant frequency. These will be discussed separately.

5.2.4 Sensitivity to motion of the plates

To see how the motion of the plates (due to vibrations, separation, etc.) influences the resonant frequency, consider the following:

Results	Even mode (MHz)	Odd mode (MHz)
Nominal	374.4	424.3
1.	374.4	424.0
2.	374.4	424.3

1. Starting with the nominal separation of 120 mm and nominal solution of 40% (as described above), the plates were now separated to 122 mm while keeping the fabric entered between the plates.
2. Keeping a nominal separation of 120 mm and nominal solution of 40%, the web fabric moved off-center by 2 mm.

These simple results show that the even mode does not change as the plates are separated but the odd mode does shift. Also, the motion of the web within the resonant cavity does not seem to affect the resonant frequency (or the effect is too small to be seen within this small change and within the 100-kHz resolution of these calculations).

This clearly indicates that if the plates are allowed to vibrate, the result will change the odd mode, and if this mode is used for temperature and/or humidity compensation, it will misrepresent the results. For this reason, the cavity must be made as rigid as possible.

5.2.5 Mechanical design

The electrical design of the sensor does not, for the most part, take into account mechanical issues although it has been shown that vibrations are apt to introduce errors in the measurements and, for example, that exposed center plates are likely to snag splices in the fabric. To minimize the effects of vibrations, the center plates and the ground planes are made of 6-mm aluminum sheet stock with all edges welded. The center planes are attached to the ground planes using solid blocks of Teflon as shown in Figure 5.19 for the upper half of the sensor (an identical block is used on the lower half of the sensor as shown in Figure 5.20). The support block is 45 mm thick, and the center plates are held down with four countersunk bolts threaded into the Teflon block. These blocks are shaped to allow space on either end for the two probes. Note also that they are 64 mm deep so that the lower face of the upper center plate is flush with the edge of the shield, again to eliminate the possibility of snagging the fabric. The bent shields are flush with the center plates, and their edges

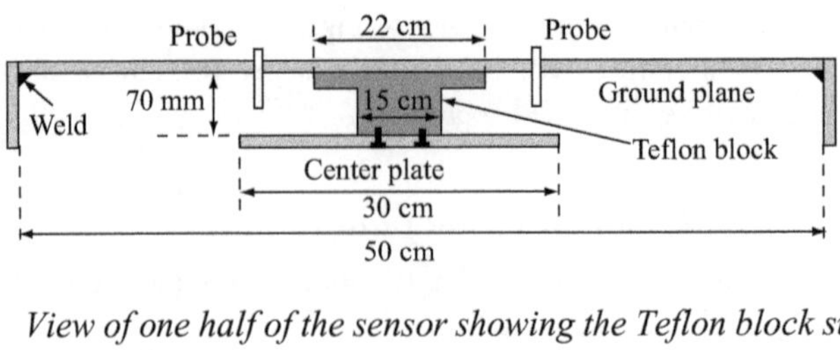

Figure 5.19 View of one half of the sensor showing the Teflon block support of the center plate

rounded as are the edges of the center plates. The final dimensions were modified to take into account the thickness of the material and are shown in Figure 5.20. The spacing between the two halves was left at 120 mm, but the other dimensions were modified to leave the inner dimensions intact. Therefore, the external dimensions of each half of the sensor are now 51.2 by 36.2 by 7.6 cm as shown.

One issue that was alluded to previously is the fact that the sensor cannot cover the entire width of the fabric. Hence, a number of possibilities were considered to rectify this deficiency. The most obvious solution is to use multiple sensors in some configuration. A possible arrangement is shown in Figure 5.22. Figure 5.21(a) shows the schematic arrangement, and Figure 5.21(b) shows the coverage in terms of the center plates. Four sensors are fixed at distances of between 10 and 15 cm from each other covering a fabric width of 127 cm. The separation is needed to minimize coupling between adjacent sensors. The four sensors represent the minimum needed for physical coverage of the fabric. Initial simulations of this configuration has shown nonuniform sensitivity to different sections of the fabric because the electric-field intensity is largest around the edges of the center plates and these only cover small sections of the fabric. The center plates themselves only

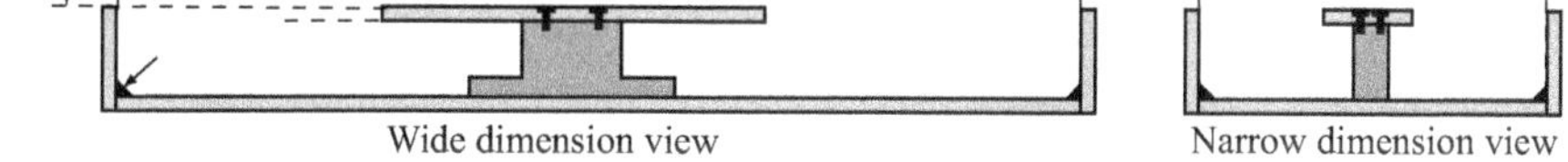

Figure 5.20 Final dimensions of the sensor with internal dimensions unchanged and separation between the two halves left at 120 mm

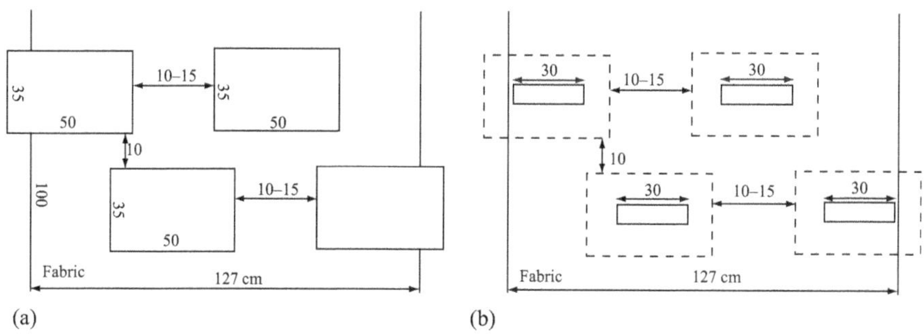

Figure 5.21 (a) Possible coverage of a 127-cm fabric with four sensors and (b) coverage in terms of center plates

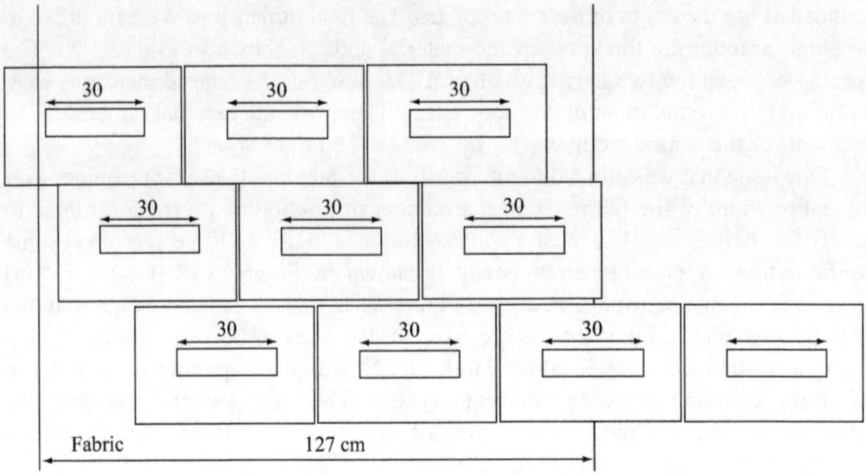

Figure 5.22 An arrangement of staggered sensors that allows better coverage of the width of the fabric

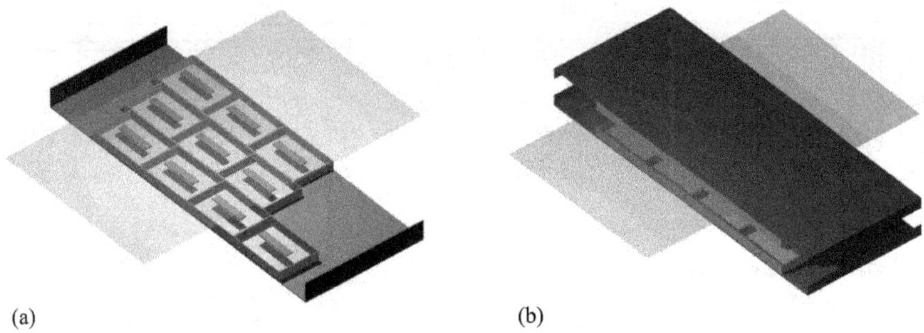

Figure 5.23 Use of multiple sensors to cover the width of the fabric: (a) one half-shell and (b) the whole assembly

cover 90 cm of the width of the fabric. Since fabrics are wider than this (180 cm is common), more sensors are needed and they need to be placed closer together to allow better coverage. To alleviate this, the configuration in Figure 5.22 may be used. This configuration consists of nine sensors across the fabric plus one off the fabric to be used as a calibration sensor but also to sense environmental effects of temperature and humidity. The sensors are arranged so that there is overlap between consecutive center plates. In practice, the nine sensors are attached to large stiff plates to hold them in place, keep the separation the same for all nine sensors, and minimize vibrations and variations between their positions [Figure 5.23(a) and (b)]. This configuration produces a more uniform sensitivity because of the staggered position of center plates.

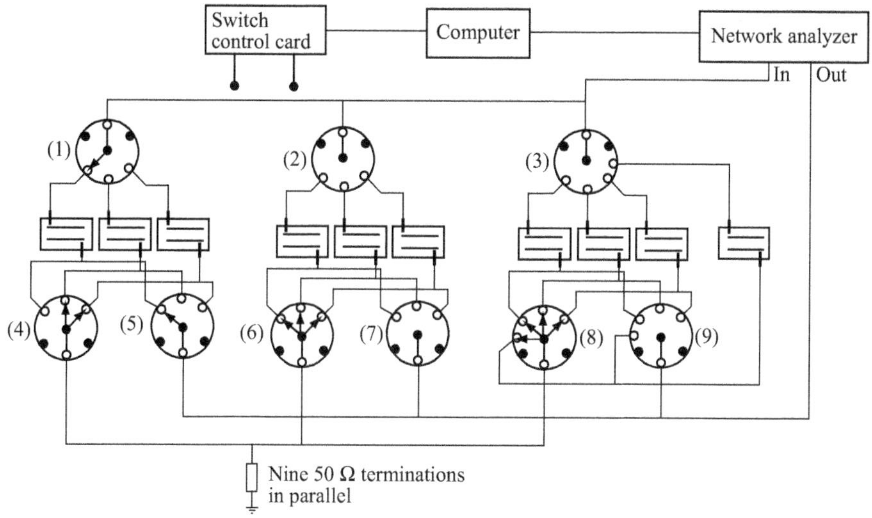

Figure 5.24 Control scheme with microwave switches

Since the resonant frequency is measured using a network analyzer, a switching control system is needed to switch in and out the nine sensors sequentially. The control system consists of nine coaxial switches, a control I/O card, and a power I/O card and is shown in Figure 5.24. The control of the switches is rather involved. In Figure 5.16, the first sensor (leftmost) is connected to the network analyzer through switches (1) and (5). At the same time, switches (2), (3), (7), and (9) are off (open). Switches (4), (6), and (8) connect the load probe of all the other sensors (sensors 2–10) in parallel to a load equal to nine 50-Ω terminators. In the next step, sensor 2 will be connected to the analyzer and all other sensor's input probe disconnected, whereas the load probes connected to the load. It is possible to do the switching with six switches by not terminating the inactive sensors removing the terminations and switches (4), (6), and (8), but then the settling time after connection is longer. Two types of switches were evaluated for this application. Both were SP4T type, one designed for operation between 10 MHz and 1 GHz, with a 1.5-dB maximum insertion loss and 250-ns switching speed and the other designed for operation up to 35 GHz and 0.15 dB maximum insertion loss. The switching was controlled by a computer, and as each sensor is switched on, its resonant frequency is measured by the network analyzer. Because of the switching in and out of the network analyzer, the switching speed possible is much lower than the maximum speed the switches are capable of. Time must be allowed for the signals to settle after connection, and sufficient time must be available for the measurement itself before the network analyzer is switched to the next sensor.

In testing the switches, it turns out that they do not perform very well in repetitive fast switching resulting in repeated failures in spite of the fact that some of the switches were rated for 10^7 cycles. Considering the cost of switches, the complexity of the switching system, and the downtime associated with switch replacement, it was decided to use a single sensor, permanently connected to the network analyzer and cover the fabric width by moving the sensor back and forth across the fabric at an appropriate rate that will allow the network analyzer to sample sections of the fabric's width. This simplifies the electrical aspects of the system but, obviously, adds a mechanical dimension that did not exist previously. The motion of the sensors was accomplished using a pair of belts, one for the upper half of the sensor and the other for the lower, driven by a pair of high-power stepper motors. By using ribbed belts and cogs and tight control over the motors, the two halves were kept in fixed positions relative to each other as they moved. The need for two belts rather than one was necessitated by the need to keep the structure open so the whole arrangement could be positioned over the fabric without the need to modify or interfere with the production line. There are, of course, additional considerations for the mechanical structure. The two halves of the sensor must be kept parallel to each other and vibration free. To do so the whole motion system is mounted on a heavy frame as will be described in Section 7.2.

5.3 Sensor design for rubber thickness sensing

The sensing of rubber sheet thickness can be sensed in a manner similar to the fabric described above. In fact, the same sensor may be used without modifications provided access is available to the rubber sheet so that the two halves of the sensor may be placed with the rubber sheet passing between them.

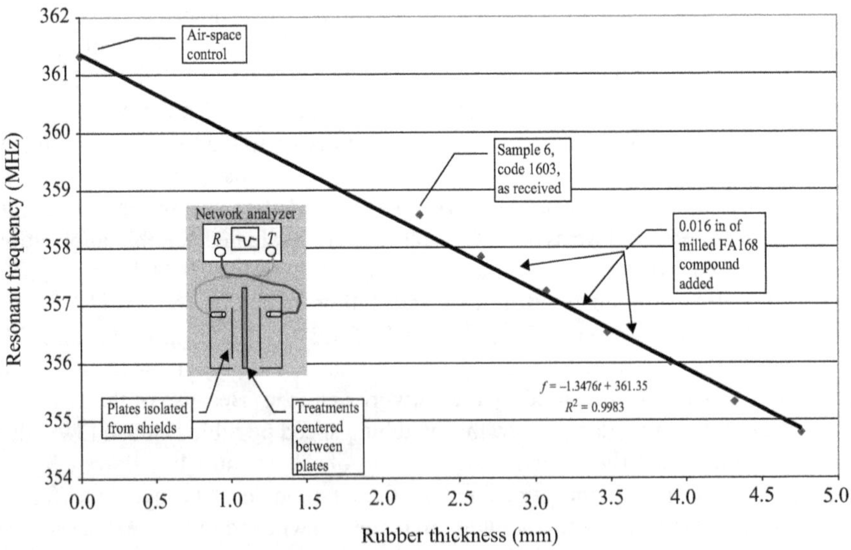

Figure 5.25 Resonant frequency as a function of rubber thickness. Source and load probes on the opposite sides of the resonator

We start with the previous sensor design by measuring the resonant frequency of the sensor with respect to rubber thickness to establish a baseline and, as well, to show the difference between the broadside coupled stripline resonator and the modified resonator that will eventually be used for rubber thickness gauging as will be described shortly. Figure 5.25 shows the resonant frequency of the cavity (see the inset) with the rubber sheet at the center of the cavity. The rubber sheet varied from 2.25 to 4.75 mm in thickness. As can be seen, the sensitivity is high and the behavior is linear. The sensitivity can be estimated from this figure or, more appropriately, from the linear least-square representation of the measured data:

$$f = -1.3476t + 361.35 \tag{5.7}$$

where t is the rubber thickness in mm and f the resonant frequency in MHz. The sensitivity is approximately 1.35 MHz/mm. Since the measurement will be done with a network analyzer and the analyzer can resolve well below 1 kHz, the sensor can sense changes in thickness of rubber of the order of 1 μm (0.742 μm/kHz). The measurements shown in Figure 5.25 were made by adding rubber sheets 0.4 mm thick to a base sheet 2.25 mm thick to obtain the various thicknesses shown. Part of the reason that the points deviate slightly from a straight line is that the thin sheets are difficult to lay flat without slight stretching.

However, in the production of continuous thin rubber sheets, the most convenient place for sensing is where the rubber moves over a rotating calender as shown schematically in Figure 5.26. The calender roll is a steel cylinder of a relatively large diameter and at least as long as the width of the rubber sheet. At that location, the rubber sheet is flush with the calender and hence sensing at this location is ideal as the position of the rubber does not vary and is perfectly flat on the roll. In addition, the surface is easily accessible and relatively clean as this is toward the end of the production line.

However, this also means that the sensor must be modified since now the rubber cannot pass between the two halves of the sensor. To design a sensor for this application, one can start with the classical stripline resonator but, because the

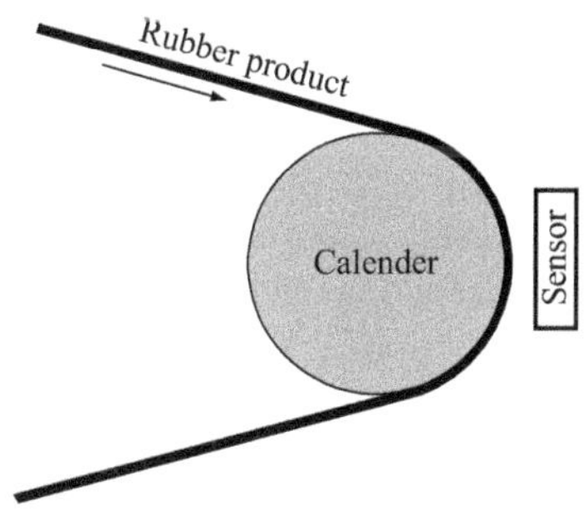

Figure 5.26 Rubber sheet moving on a calender. The sensor is most conveniently placed on the section in which the rubber makes contact with the calender

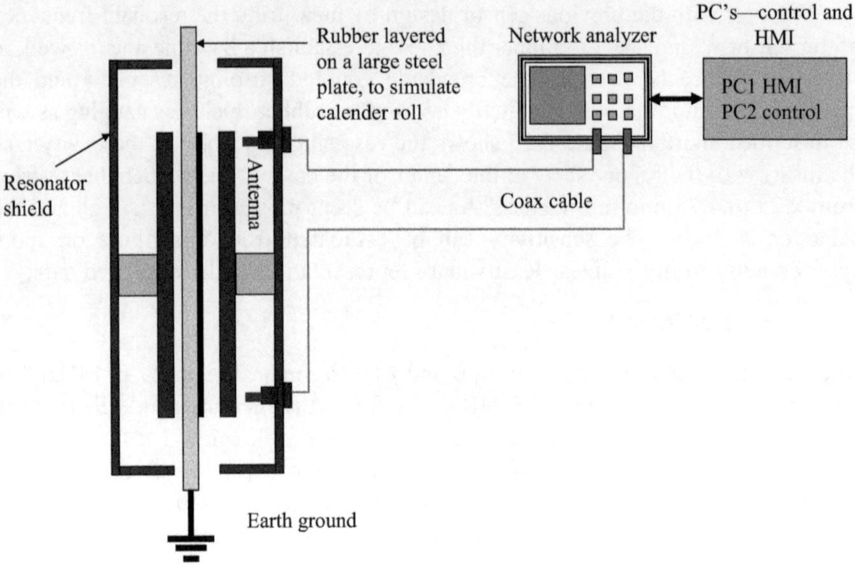

Figure 5.27 The transition from the fabric sensor to the rubber thickness sensor

calender is conducting, only one half of the sensor can be used so that the rubber is between the center plate and the calender.

The transition from the fabric sensor to that of the rubber against calender sensor starts with Figure 5.27. In this initial approach, the rubber material is still stacked on a flat steel sheet and both are placed inside the unmodified fabric sensor except for the fact that both probes are now on the side of the rubber. The reason is obvious—resonance occurs in the space between the rubber side of the steel plate and the half-shell to its right. The figure also shows the connections to the network analyzer and from it to two computers. One is the computer used to control the motion of the resonator and to record the data, whereas the other contains the HMI through which all commands are given. The reason for the separation is that in actual testing and measurements, the HMI is on a remote machine and commands may be given over the web.

The configuration in Figure 5.27 was used as an initial step toward optimization of a sensor adapted to this type of gauging. The first measurements are shown in Figure 5.28 and were intended to define an acceptable separation between the steel plate and the stripline. The figure shows two measurements, one at 35-mm separation and the second at 100-mm separation. The least-square representations of the measured resonant frequency (MHz) as a function of rubber thickness (mm) are

$$f = -0.1169t + 380.88 \text{ for 100-mm separation} \tag{5.8}$$

$$f = -1.8026t + 411.77 \text{ for 35-mm separation} \tag{5.9}$$

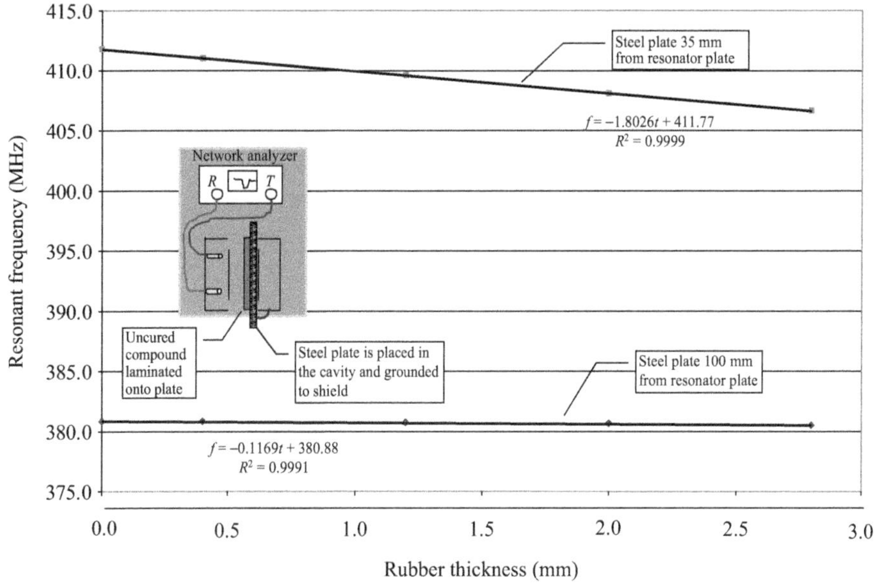

Figure 5.28 Resonant frequency versus rubber thickness for 35- and 100-mm separation between center plates. Source and load probes on the same side of the resonator

Table 5.3 Simulation of the resonator in Figure 5.27

Thickness (mm)	First resonant frequency (MHz)	Second resonant frequency (MHz)
0	436.6	863.4
2	433.2	856.9
1	434.8	860.2
0.5	435.8	861.8
0.25	436.4	862.4

The first thing to note is that in either case, the resonant frequency increases simply because the volume of the cavity decreases. Although the empty cavity in Figure 5.25 resonates at 361.35 MHz, the cavity in Figure 5.27 resonates at 380.88 MHz with the 100-mm separation and at 411.77 MHz at 35-mm separation. On the other hand, the sensitivity is much lower at the 100-mm separation but much higher at the 35-mm separation. This is because in these measurements, the rubber is against the steel plate and hence, for the same separation as in Figure 5.25, one would expect lower sensitivity. However, the sensitivity increases with the reduction in separation and at a separation of 35 mm, the sensitivity is higher. At 100-mm separation, the sensitivity is 116.9 kHz/mm or 117 Hz/μm. At 35 mm, the sensitivity is 1.8 MHz/mm or 1.8 kHz/μm. These values would allow measurements down to 8.55 μm at 100-mm separation or 0.55 μm at 35-mm separation (given a 1-kHz resolution on the network analyzer). Clearly then, the lower the separation, the better it is. This is true with the fabric sensor but much more so with the rubber thickness sensor.

Simulation results have shown the same tendency in terms of sensitivity, although the results are less linear than in the measurements. A set of simulated results is shown in Table 5.3 for the configuration in Figure 5.27 with a separation of 3.5 mm. The table shows the first and second resonant frequency. To be noted is that these are not dual frequencies since the cavity now is not a broadside-coupled cavity but rather a regular cavity that resonates at multiple modes. The sensitivity for the first resonant frequency varies between 1.2 and 1.6 MHz/mm, depending on the thickness, with an average of 1.6 MHz/mm. This is not quite identical to the results of the measurement that seem to be more linear. A possible reason may lie with the way the boundary conditions were set. In this case, because the cavity is closer to a closed cavity, the boundaries were set at the conductors and the gap was closed as well, making it into a closed cavity. This of course neglects any fields outside the cavity. The second resonant frequency may be used as well with a sensitivity twice as high. Indeed, one can use higher resonant frequencies, but, in keeping with the original idea of using the lowest frequency necessary to obtain the required sensitivity and to keep in line with the resonant frequency of the fabric sensor, we opted for the first resonant frequency. A sensitivity under 1 μm is more than sufficient for this and many other applications.

The transition from the flat steel sheet to the calender is shown in Figure 5.29. Clearly the one half of the original sensor (left side in Figure 5.27) serves no purpose so it is removed. The right side of the original sensor together with the conducting calender now forms a resonator, but because of this modification, the properties of the resonator also change as will be seen shortly. To adapt the original structure to the present configuration, the resonator is modified as follows:

1. The calender is used as one of the ground planes.
2. The center conductor and the second ground plane are curved to keep a constant distance between the ground plane and the calender. The basic configuration is shown schematically in Figure 5.30(a). The figure also shows, schematically, the connection to the network analyzer.
3. The ground plane can also be bent over to create a partially enclosed cavity as was done with the previous sensor [Figure 5.30(b)]. By doing so, the Q-factor increases and the effects of external influences decrease.
4. Because the rubber sheet is thin and tight over the calender, the gap between the center plate and the calender can be relatively small and from the values given above must be kept at a minimum to increase sensitivity.

Starting with the dimensions of the broadside coupled sensor, an appropriate sensor that implements the changes above is shown in Figure 5.31. The sensor incorporates the bent plates, and the dimensions reached at in the broadside coupled sensor, but the gap has been reduced to 35 mm (to the centerline) instead of the 60 mm in the previous case, based on the measurements in Figure 5.28 and on the simulated

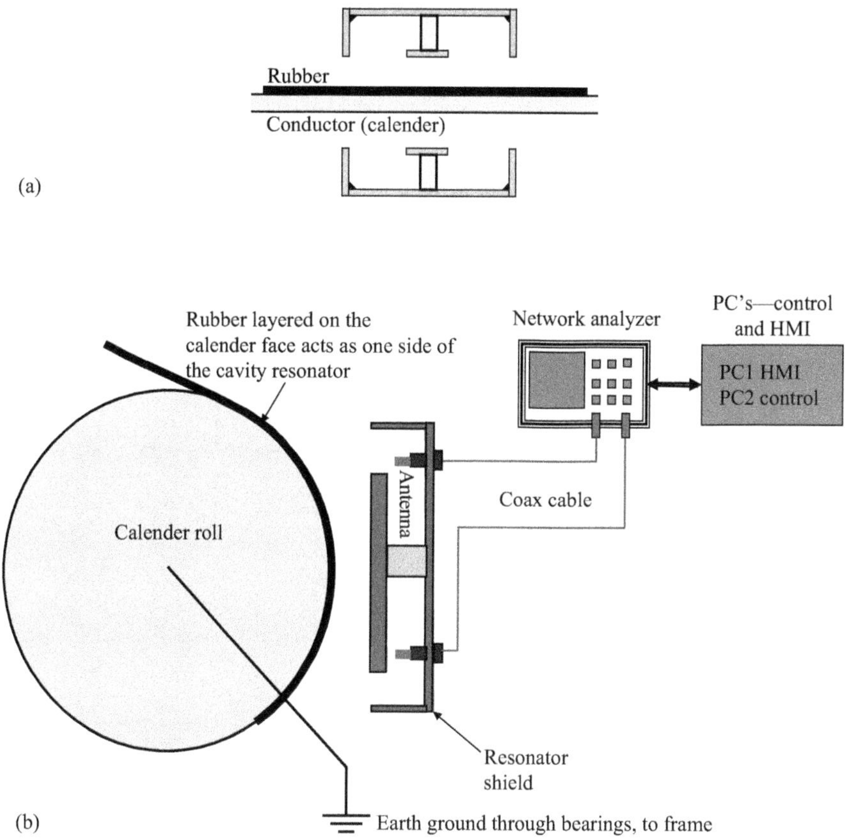

Figure 5.29 Adaptation of the stripline resonator to sense a rubber sheet on a conducting surface: (a) initial experiments and (b) the half-shell against the calender roll

values. The ground plane measures 50 by 35 cm (interior dimensions), the center conductor measures 30 by 6 cm, and both are made of 6 mm thick aluminum. Both the ground plane and the center conductor are curved to create a uniform separation between the sensor and the calender. The purpose of this shape is to ensure that the change in resonant frequency due to rubber thickness variations is as linear as possible. Since the electric-field intensity is largest at the edges of the center plane, the ports are placed near the edges of the center plate. The long dimension of the sensor is oriented axially with the calender [Figure 5.31(b)]. Figure 5.31(a) shows how the constant distance between the calender and the sensor is completed by properly curving the geometry of the sensor. The dimensions of the sensor are approximately the same as in Figure 5.20. Results of a simulation of the curved sensor are shown in Table 5.4. In comparison with Table 5.3, the resonant frequency is a little lower indicating that the volume of the resonator increased slightly. On the other hand, the resolution sensitivity increased slightly with

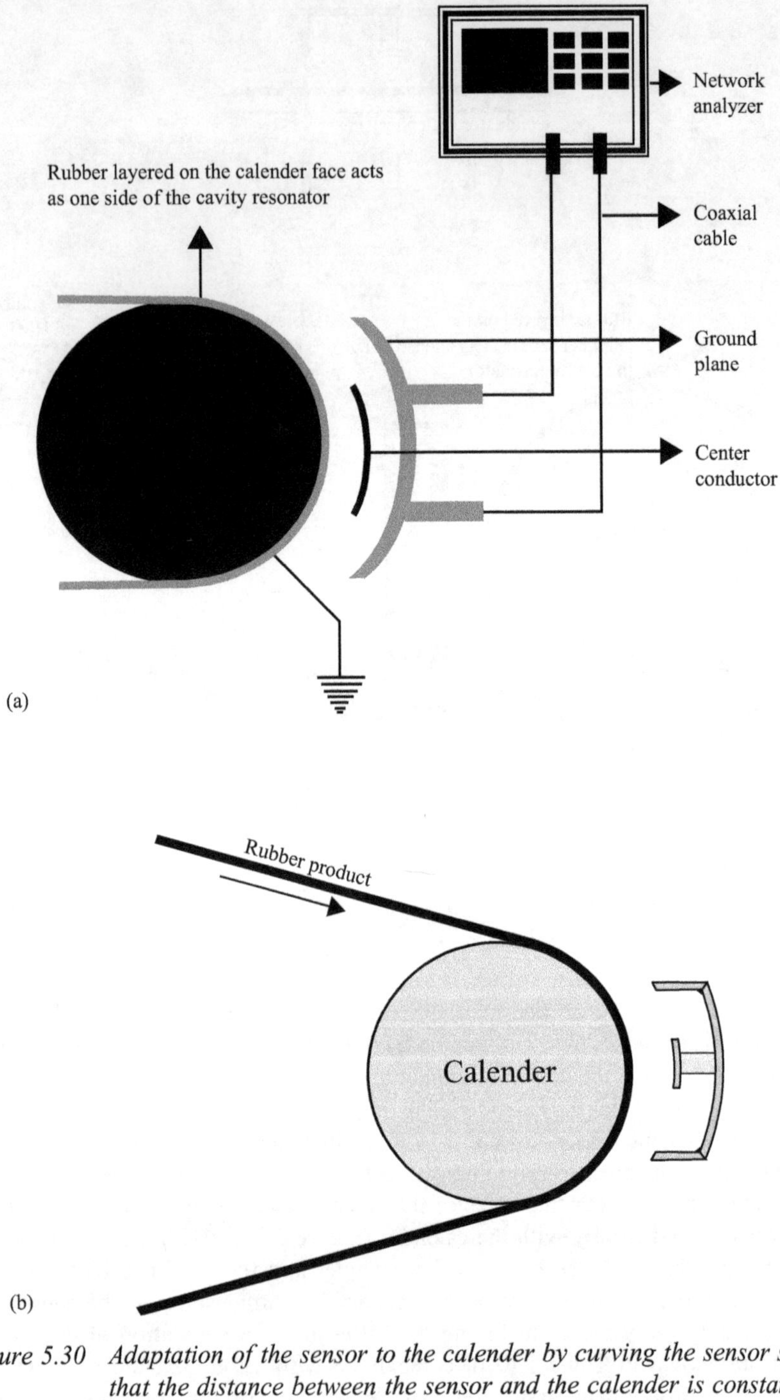

Figure 5.30 Adaptation of the sensor to the calender by curving the sensor so that the distance between the sensor and the calender is constant: (a) nonshielded prototype and (b) shielded prototype

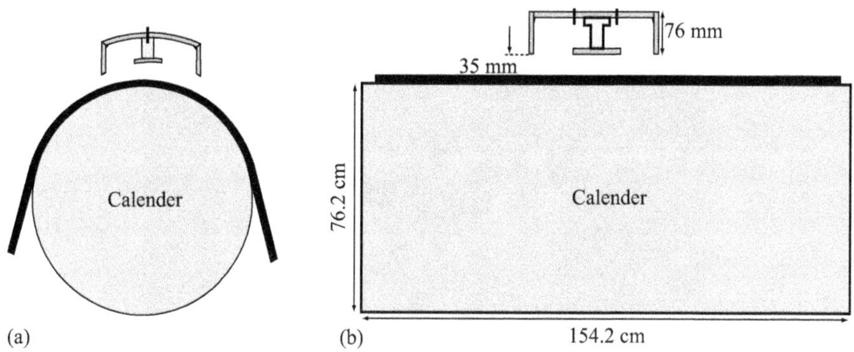

Figure 5.31 Two views of the sensor. Note the curvature in (a) and the two ports in (b). The figure is not up to scale

Table 5.4 Simulation of the resonator in Figure 5.31

Thickness (mm)	First resonant frequency (MHz)	Second resonant frequency (MHz)
0	422.2	829.6
0.5	420.5	826.3
1	419.8	824.8
1.5	418.7	822.6
2.0	417.6	820.4
2.5	416.6	818.2
3.0	415.6	816.0

average sensitivity of 2.4 MHz/mm. The linearity has also improved considerably. As in Table 5.3, the sensitivity at the second resonant frequency is higher, in this case an average of 4.4 MHz/mm.

As was the case with the fabric sensor, the width of the rubber sheet cannot be covered entirely by the sensor, and the same considerations as for the fabric sensors apply here.

It should be noted here again that the sensor as shown in Figures 5.27, 5.29, or 5.30 is not a broadside coupled resonator and hence the center plate does not seem to be necessary. In fact, the results in Figure 5.25 show as much. One could, by proper location of the probes, remove the center plate and obtain the same results. However, simulations without the stripline and with the probes in the same location show poor performance with a very low Q-factor. It was therefore decided to keep the stripline as in the original sensor. This has the advantage that the rubber thickness sensor and the fabric sensor are driven in exactly the same fashion and the measurements treated identically. The main difference between the two sensors lies in the fact that the odd-mode resonance does not exist in the rubber thickness sensor. Since the odd-mode resonance is used for compensation of environmental

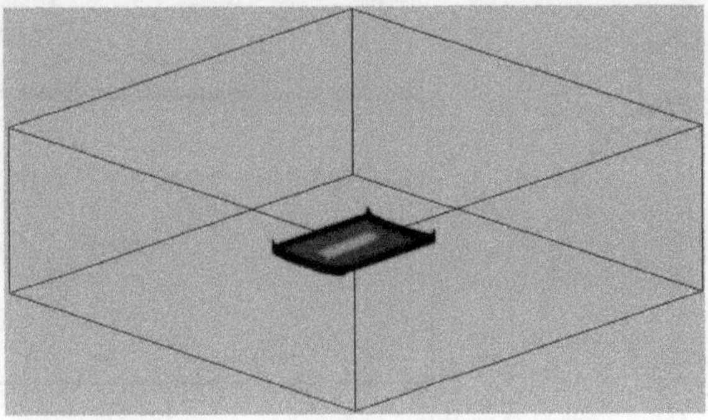

Figure 5.32 Simulator input for the curved half-shell for rubber thickness sensing

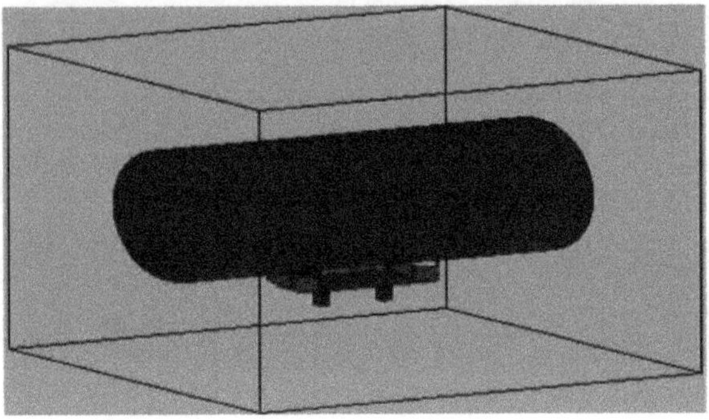

Figure 5.33 Simulator input for the rubber thickness sensor showing the calender above the curved half-shell

conditions, this also means that the compensation cannot be applied to the rubber thickness sensor.

5.3.1 Simulation and optimization

The first step in the simulation process is to establish the resonant frequency of the empty resonator (no rubber). The FDTD program is used here as well. Figures 5.32 and 5.33 show the geometry as input to the program. In Figure 5.32, the calender has been removed to show the model of the sensor. The center plate can be seen at the center, and the folds of the ground plane clearly show the curvature. Figure 5.33 shows the geometry with the calender. Note also the outer rectangular box—this delineates the space modeled. In this view, the two ports are also visible.

The resonant frequency is calculated by scanning over a range of frequencies, calculating the fields at each frequency (for an arbitrary amplitude at the input port), and sensing the field at the load port. Figure 5.34 shows a scan between 300 MHz and 1 GHz. Resonant frequencies at 422.2 MHz and at 829.6 MHz can be

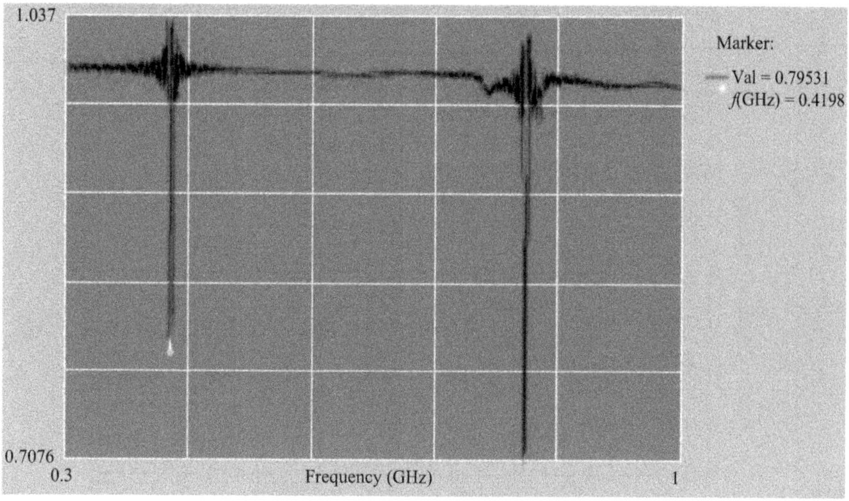

Figure 5.34 Screen capture of a simulation of the resonator in Figure 5.20

Table 5.5 Resonant frequency as a function of rubber thickness for the first and second resonant modes

Thickness (mm)	First resonant frequency (GHz)	Second resonant frequency (GHz)
0.00	0.4222	0.8296
0.05	0.4205	0.8263
1.00	0.4198	0.8248
1.50	0.4187	0.8226
2.00	0.4176	0.8204
2.50	0.4166	0.8182
3.00	0.4156	0.8160

easily seen. These are single-mode resonances since now one cannot distinguish between odd and even modes. One can select any of the two frequencies and other higher order frequencies beyond 1 GHz, but, as we have done previously, the lower frequency resonances are preferred.

The sensitivity of the sensor to rubber thickness variations is also established by the same simulation process by including a rubber sheet on the calender of varying thickness. Table 5.5 shows the results for rubber sheets between 0.05 and 3-mm thick. These are also shown in Figure 5.35. Although the change in frequency is not linear, it is approximately 2.2 kHz/μm in the first-order resonance and approximately 4.53 kHz/μm in the second-order resonance. Since the network analyzer can easily detect a change lower than 1 kHz, the sensitivity of the sensor is better than 0.5 μm in rubber thickness variation. The slope of the curves in

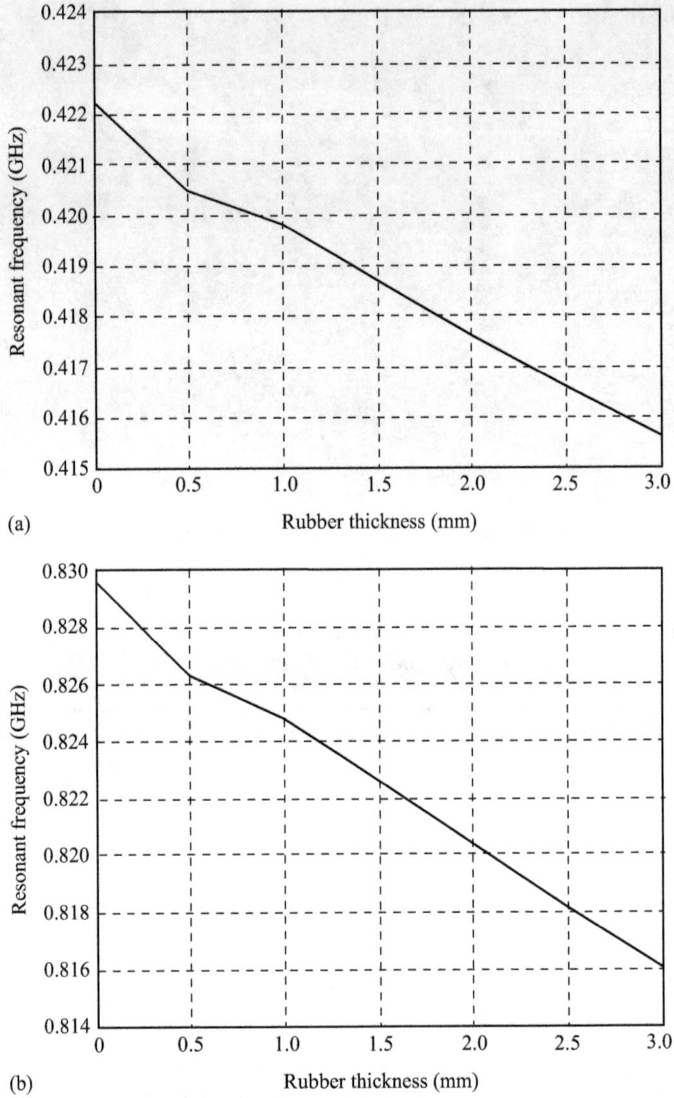

Figure 5.35 Resonant frequency as a function of rubber thickness: (a) first resonant mode and (b) second resonant mode

Figure 5.35 is constant except for a small deviation at very thin rubber sheets; hence, the sensitivity is constant for any thickness.

The optimization process follows that established for the fabric sensor but also looks at the shape of the center plate in some detail. Since we use essentially the same sensor, there is little value in repeating the simulations associated with

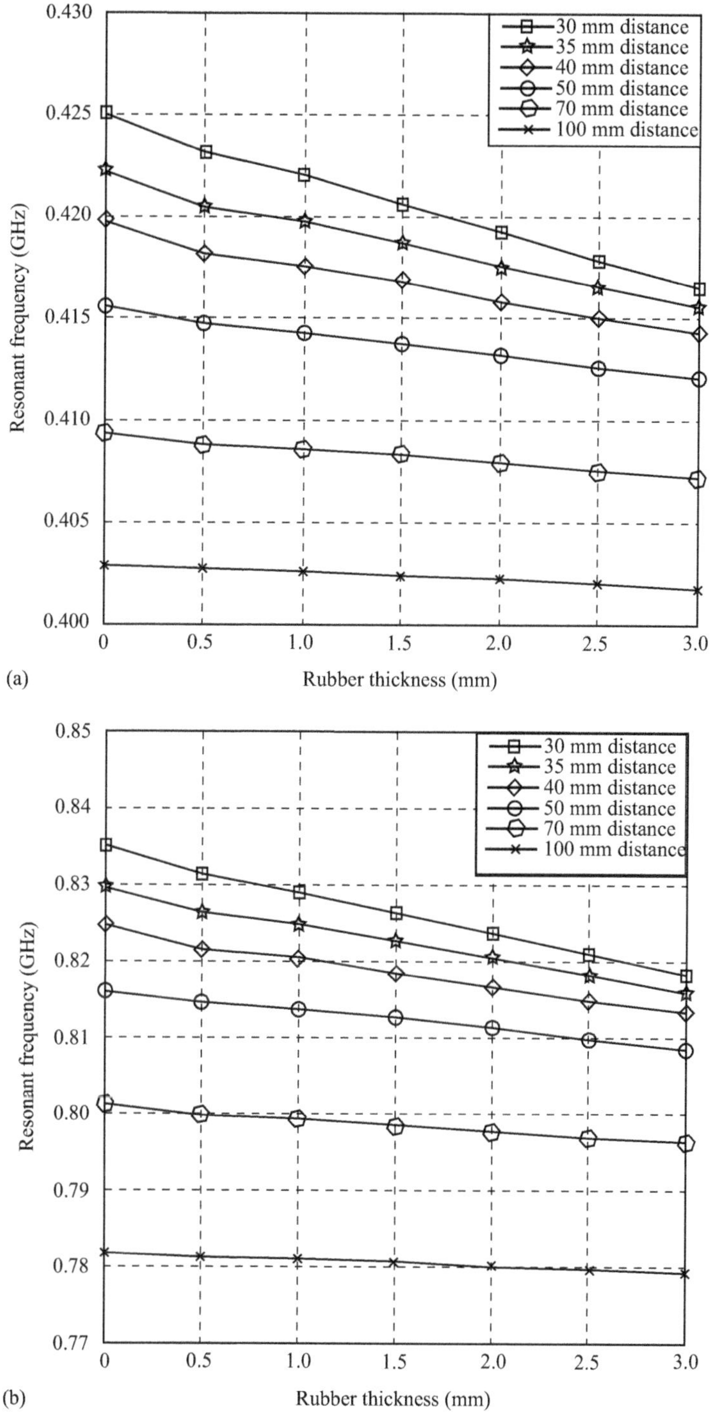

Figure 5.36 Resonant frequency as a function of distance between the center plate and the calender for varying rubber thickness: (a) first resonant mode and (b) second resonant mode

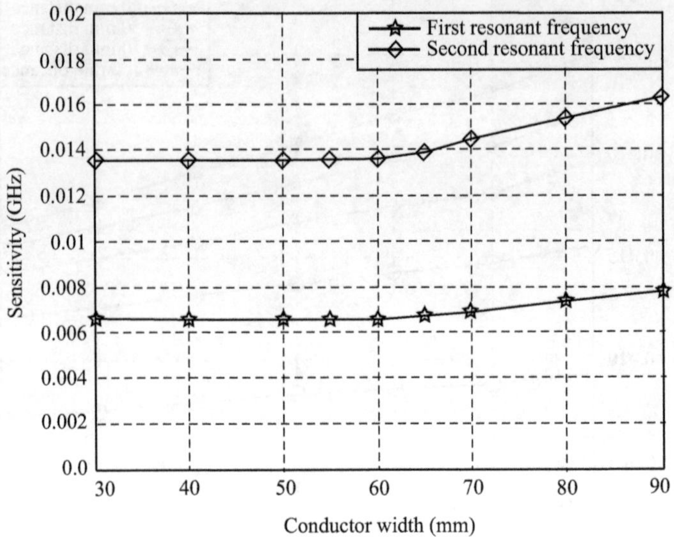

Figure 5.37 Sensitivity of the sensor as a function of center plate width for a change of 3 mm in rubber thickness. The sensitivity is 2.2 MHz/mm in the lowest resonant mode and 4.53 MHz/mm in the second resonant mode

the folded ground plane. Perhaps the most important parameter of the sensor is the distance between the center plate and the calender's surface; hence, we start with that. The distance is varied between 30 and 100 mm for rubber thicknesses between 0 and 3 mm. The results shown in Figure 5.36 confirms the expected behavior. The sensitivity decreases with increased separation. At a separation of 30 mm, the sensitivity in the first resonant mode is approximately 2.8 kHz/μm, whereas at 100 mm, the sensitivity is approximately 400 Hz/μm. This is still a respectable sensitivity but significantly lower than at 30 mm. The sensitivity in the second resonant mode is approximately twice as high as one would expect.

The center conductor of this modified stripline is, of course, an important element of the sensor, and it is important to try to optimize its dimensions and shape. However, as can be seen from the following simulations, it is, for the most part, insensitive to changes in dimensions and shape. Nevertheless, some changes can be made to improve sensitivity of the overall system. The following simulations discuss these issues.

Figure 5.37 explores the sensitivity of the sensor as a function of width of the center plate. The sensitivity is measured as the change in frequency of the resonator as the rubber thickness changes from 0 to 3 mm. For example, at a center plate width of 50 mm, the sensitivity is $6.6/3 = 2.2$ MHz/mm in the lowest resonant mode and $13.6/3 = 4.53$ MHz/mm in the second resonant mode. These increase to 2.6 and 5.43 MHz/mm at a plate width of 90 mm. As can be seen, the value selected

Table 5.6 Resonant frequency for various center conductor lengths, empty resonator

Length (mm)	First resonant frequency (GHz)	Second resonant frequency (GHz)
200.00	0.5157	0.9696
225.00	0.5108	0.9630
250.00	0.4859	0.9301
275.00	0.4296	0.8404
300.00	0.4222	0.8296
325.00	0.4135	0.8138
350.00	0.3388	0.6690
375.00	0.3351	0.6635
400.00	0.4296	0.6564

Table 5.7 Resonant frequency for various center conductor lengths, 3-mm rubber on the calender

Length (mm)	First resonant frequency (GHz)	Second resonant frequency (GHz)
200.00	0.5121	0.9627
225.00	0.5073	0.9560
250.00	0.4825	0.9227
275.00	0.4255	0.8313
300.00	0.4156	0.8160
325.00	0.4099	0.8061
350.00	0.3352	0.6615
375.00	0.3317	0.6563
400.00	0.3273	0.6494

for this (60 mm) is acceptable although a slight improvement can be had with wider plates. Increasing the width to 90 mm would lead to an improvement in sensitivity of about 18%. Increasing the width of the center plates should be undertaken carefully since the ground planes are only 350 mm wide. Too wide a center plate may require widening the ground planes and that of course would reduce the resonant frequency and the spatial resolution of the sensor.

The length of the conductor also influences the sensitivity but, in this case, the resonant frequency itself also changes (see Tables 5.6 and 5.7). Sensitivity is calculated for center plates between 200 and 400 mm keeping the folded ground plane fixed at 500 by 350 mm. The plot in Figure 5.38 shows the change in resonant frequency as the 3-mm rubber sheet is inserted into the sensor. At shorter and longer center conductors, the sensitivity is about the same (1.13 and 2.47 MHz/mm) with an increase to 2.2 and 4.53 MHz/mm at the selected length of 300 mm. This result is particular to the geometry and is influenced by the ground plane. A larger

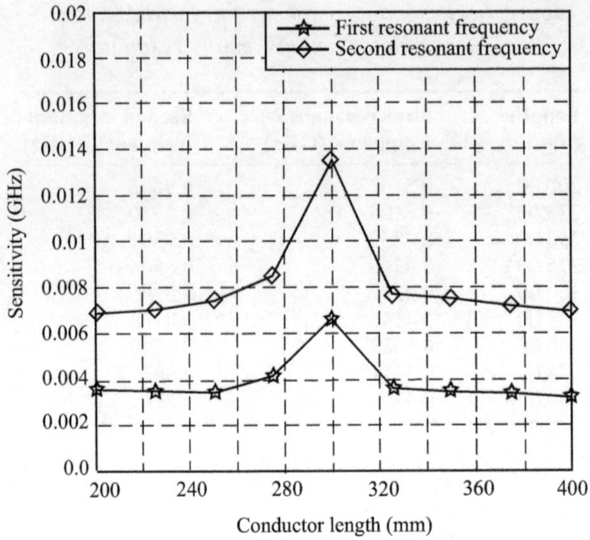

Figure 5.38 Sensitivity as a function of plate length for a change in rubber thickness of 3 mm, keeping the dimensions of the folded ground plane unchanged

ground plane would produce a maximum in sensitivity at a longer center plane, but the sensor will resonate at a lower frequency as well. Perhaps, the most useful conclusion from this is that the length of the center plate relative to the length of the resonator is the important parameter, and one cannot be optimized without optimizing the other.

In the design of the fabric sensor as well as in the present design, the distance between the ground and center planes was selected as 70 mm assuming that this dimension will have minimal influence on either the resonant frequency (the latter being defined primarily by the length of the center plate) or on sensitivity. The simulation in Figure 5.39 shows that a smaller distance would tend to reduce sensitivity, whereas a higher distance will have almost no influence on the sensitivity.

Although in most striplines, the center plates are rectangular, in resonators, because of the finite length of the center plate, its shape can be modified quite easily to obtain better sensitivity. The following are a few simple modifications to the center plate that can be easily simulated and implemented. There are of course many others.

Figure 5.40 shows the modification of the plate into a butterfly shape by widening the ends from 60 to 120 mm in increments of 10 mm while keeping the center at 60 mm. Figure 5.41 shows the sensitivity to a 3-mm change in rubber thickness. The sensitivity climbs from 2.2 and 4.53 to 4.7 and 11.53 MHz/mm for the two resonant frequencies respectively when the ends are widened to 90 mm but goes down beyond that. This effectively doubles the sensitivity of the sensor.

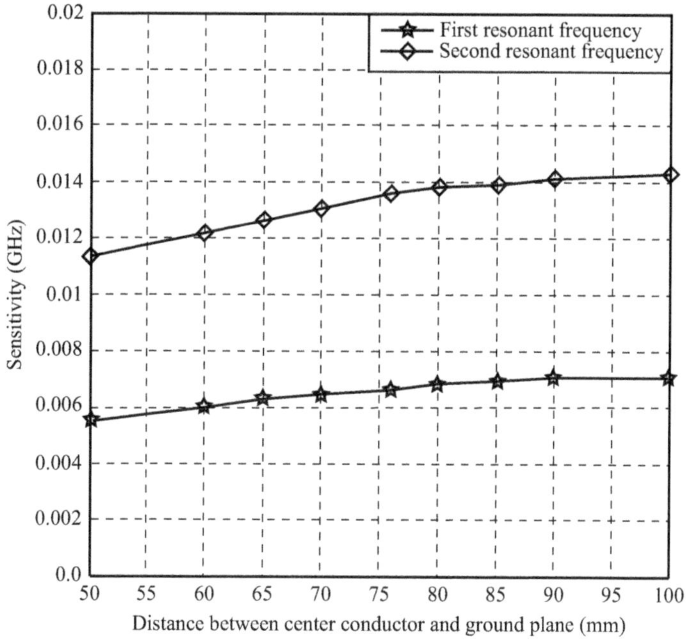

Figure 5.39 Sensitivity as a function of distance between the center plate and the ground plane for a change in rubber thickness of 3 mm, keeping the dimensions of the folded ground plane unchanged

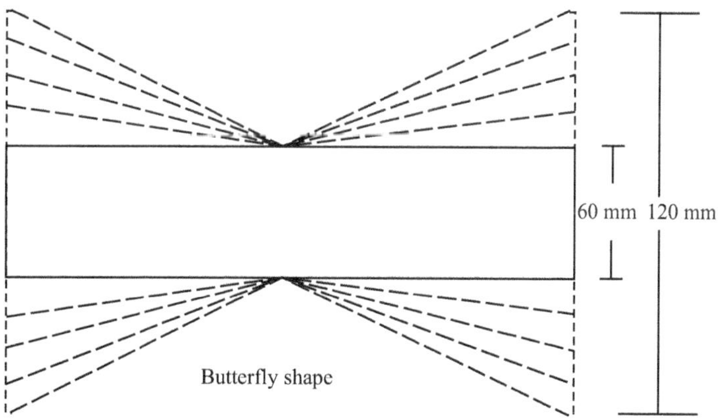

Figure 5.40 Modification of the center plate into a butterfly geometry. The plate is modified by widening the ends in increments of 10 mm

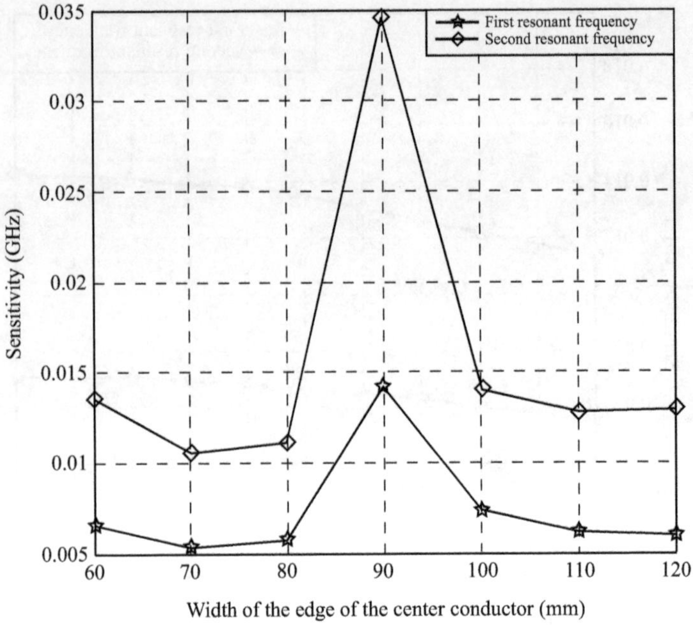

Figure 5.41 Sensitivity for the butterfly shaped center plate. Note the significant increase in sensitivity when the ends are 90 mm

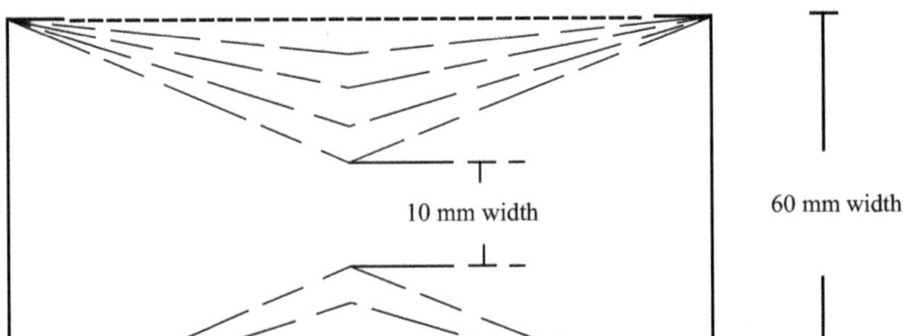

Figure 5.42 Modification of the center plate into a butterfly geometry. The plate is modified by narrowing the center in increments of 10 mm

A different way of building a butterfly-shaped plate is shown in Figure 5.42. Here, we start with the 60-mm plate and narrow it at the center, again in increments of 10 mm down to a minimum of 10 mm. However, unlike the previous case, the results are not as useful. The narrower the center, the lower the resolution as can be seen in Figure 5.43. In addition, a plate with a center that is too narrow is likely to vibrate more easily and be less sturdy.

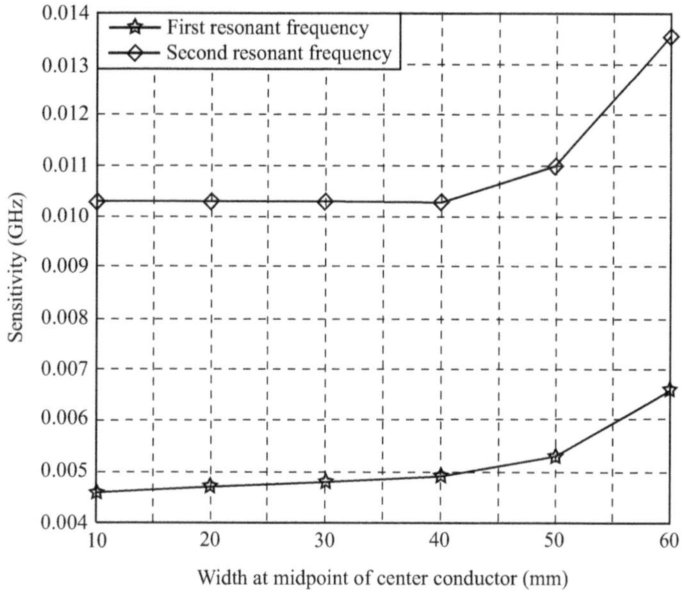

Figure 5.43 Sensitivity for the butterfly-shaped center plate in Figure 5.42. The sensitivity is lower than that of the rectangular plate

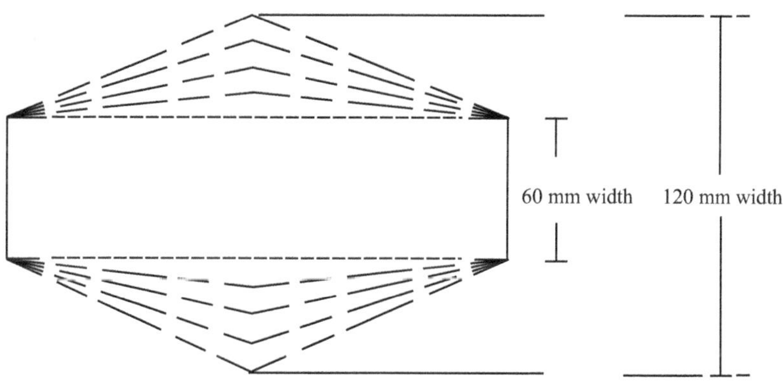

Figure 5.44 Modification of the center plate into a diamond-shaped geometry. The plate is modified by widening the center in increments of 10 mm

Another simple modification is the diamond-shaped center plate in Figure 5.44 with the accompanied results in Figure 5.45. Clearly there is no advantage in this design since the sensitivity decreases compared to the rectangular plate.

The simulation steps above clearly show that as far as sensitivity is concerned, two parameters offer the most advantages: the distance between the calender and

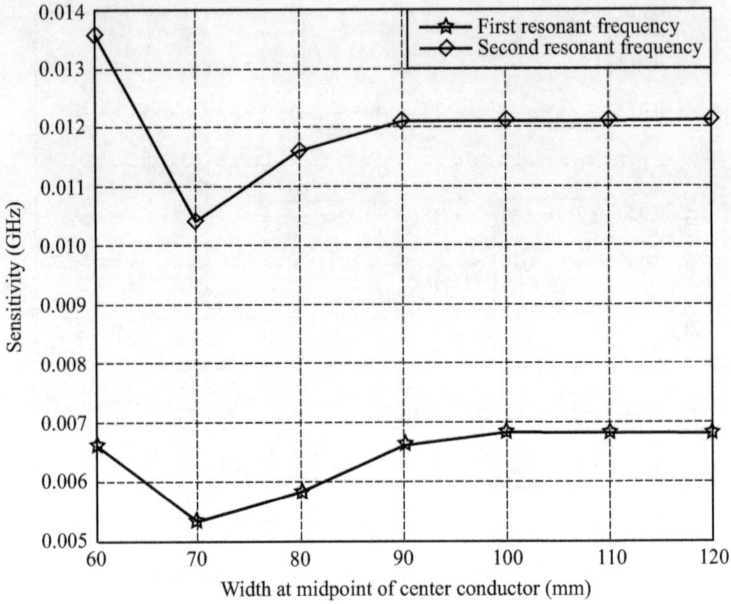

Figure 5.45 Sensitivity for the diamond-shaped center plate in Figure 5.44. The sensitivity is lower than that of the rectangular plate

the center plate and the shape of the plate with some minor improvement that can be had by changing the dimensions of the plate. Although the butterfly-shaped plate increases sensitivity by a factor of 2, it may not be very practical since the low-angle corners are likely to snag loose or torn rubber, especially if the gap between the center plate and the calender is small. In addition, the corners of the plate would need to be supported to avoid vibrations. The distance between the center plate and the calender can be reduced somewhat, but the improvement in sensitivity is minimal. Because of these reasons and to keep the design simple, the center plate was left in its rectangular shape especially since the sensitivity is very high to begin with and improvement, even by a factor of 2 is not critical to the performance of the sensor. Nevertheless, in other applications, these considerations may change and the additional sensitivity may be useful.

A final note on the simulations performed in this work. The simulations were done with commercial software based on the FDTD approach. In this approach, one "follows" the development of the electromagnetic fields in the region of interest (in this case, within the cavity) taking into account the geometry of the cavity and the material properties within the cavity. The FDTD method is not necessarily the most efficient method of simulation, but it is perhaps the easiest to understand and setup, especially in geometries with conducting boundaries. A short description of the method with additional details is given in Appendix C.

5.4 Alternative sensing strategies

The previous sections described the sensors, its design, simulations, and, of course, the reasons why the resonant sensor was selected. But, as with any design, there are always alternative methods, and sensing of moisture content or rubber thickness are no exception. Now that the needs of the sensors and the functional parameters have been understood, it is an appropriate point to discuss the alternatives that were considered and discarded in favor of the present design in spite of the fact that the sensor relies extensively on the capabilities of a network analyzer and hence is, necessarily, a relatively expensive design.

In the fabric sensor what is measured is an effective permittivity of the fabric, whereas in the rubber thickness sensor, the measurement is a dimension (thickness of the rubber sheet); to both, there are a number of possible alternatives. These include electromagnetic transmission sensors, electromagnetic reflection sensors, and, perhaps the most obvious, capacitive sensors. One can also use beta gauges based on the absorption of radiation by the fabric or rubber. These systems rely on a low-intensity source emitting on one side of the fabric and a detector on the other side. For obvious safety reasons, the gap provided for the fabric to move through must be very small, a feature that requires the device to be removed when splices are made. Beta gauges are also prone to contamination by the dip material because of the proximity of the moving fabric. Beta gauges and systems based on them are available commercially and an important motivation of the introduction of the current sensors was to avoid the use of radiation sensor because of concern for radiation and the accuracy needed. Thus, these sensors were not considered as alternatives to the present design.

5.4.1 Capacitive sensors

Capacitive sensors are the easiest to understand and simplest to implement. In essence, all one has to do is place two conducting plates, one on each side of the fabric, and measure the capacitance between the plates as shown in Figure 5.46(a). In the case of the fabric gauging, the plates can be spaced to allow the necessary clearance (120 mm was the minimum requirement in this work). The plates can either be large to cover, say, the width of the fabric or smaller to allow for better resolution across the fabric. The capacitance can be monitored with a simple capacitance meter or with a more sophisticated *RLC* meter if better accuracy is needed. The plates can also be modified by adding guard electrodes, and one can

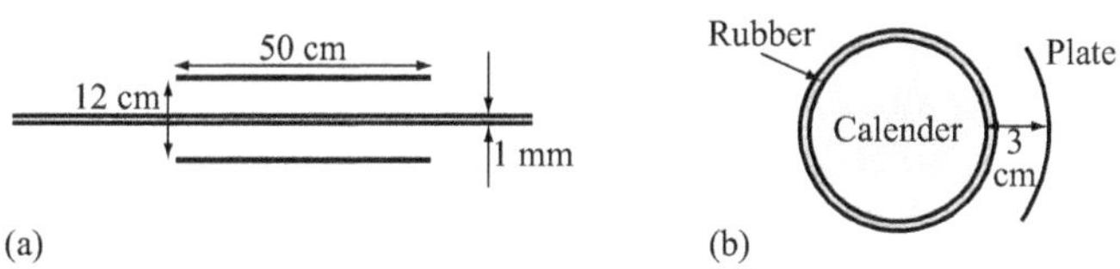

Figure 5.46 Capacitive gauging of permittivity in a dielectric: (a) fabric moisture sensing and (b) thickness gauging of rubber

easily devise arrays of capacitors to cover the width of the fabric. Thus, the sensor itself, aside from some simple mechanical issues, is rather trivial. The main difficulty with a sensor of this type is sensitivity. To estimate the sensitivity of a capacitive sensor, we can start with the mixing formulas in (5.1)–(5.4) and apply them to the space between two parallel plates of the same size as the resonant sensor, that is, each plate is 50 by 35 cm and are separated a distance 12 cm. From (5.3) and (5.4), the relative permittivity of a 1-mm thick fabric changes from 23.64 with the fabric containing 35% dip solution to 26.66 when the dip solution increases to 40% of the fabric volume, corresponding to 1% increase in solids retention. Using the mixture formula in (5.2), the effective relative permittivity of the space between the plates at these two levels of absorbed dip is

$$\varepsilon_{\mathrm{eff}} = \frac{\varepsilon_0 v_0 + \varepsilon_r v_f}{v_0 + v_f} \tag{5.10}$$

where v_0 is the volume of the space between the capacitor, v_f that of the fabric, ε_0 the permittivity of air, and ε_r the permittivity of the fabric. At a dip pickup of 35%, the effective permittivity is

$$\varepsilon_{\mathrm{eff}} = \frac{1 \times 0.119 + 23.64 \times 0.001}{0.12} = 1.1887 \tag{5.11}$$

At a dip pickup of 40%, the effective permittivity in the capacitor is

$$\varepsilon_{\mathrm{eff}} = \frac{1 \times 0.119 + 26.66 \times 0.001}{0.12} = 1.2138 \tag{5.12}$$

The capacitance at 35% dip pickup, assuming a parallel plate capacitor is

$$C(35\%) = \frac{1.1887 \times 8.853 \times 10^{-12} \times (0.5 \times 0.35)}{0.12} = 15.35 \quad (\mathrm{pF}) \tag{5.13}$$

At 40% dip pickup, the capacitance increases to

$$C(40\%) = \frac{1.2138 \times 8.853 \times 10^{-12} \times (0.5 \times 0.35)}{0.12} = 15.67 \quad (\mathrm{pF}) \tag{5.14}$$

This is a sensitivity of 0.32 pF/1% solids.

If the target is the ability to accurately detect a change of 0.02% in solids pickup, the sensor must be capable of accurately measuring capacitance increments of 0.0064 pF. This is not practical under the best of conditions and certainly not on the factory floor.

The calculations above are only rough estimates, and one can improve on this somewhat. The distance of 12 cm between the plates can be reduced, and the use of guard electrodes can improve things, but in the end, the capacitance is too low and effects of humidity and temperature are very likely to be much higher than the required sensitivity.

In the case of rubber thickness gauging, the capacitive sensor would take the form of a curved plate parallel to the calender as shown in Figure 5.46(b). In this

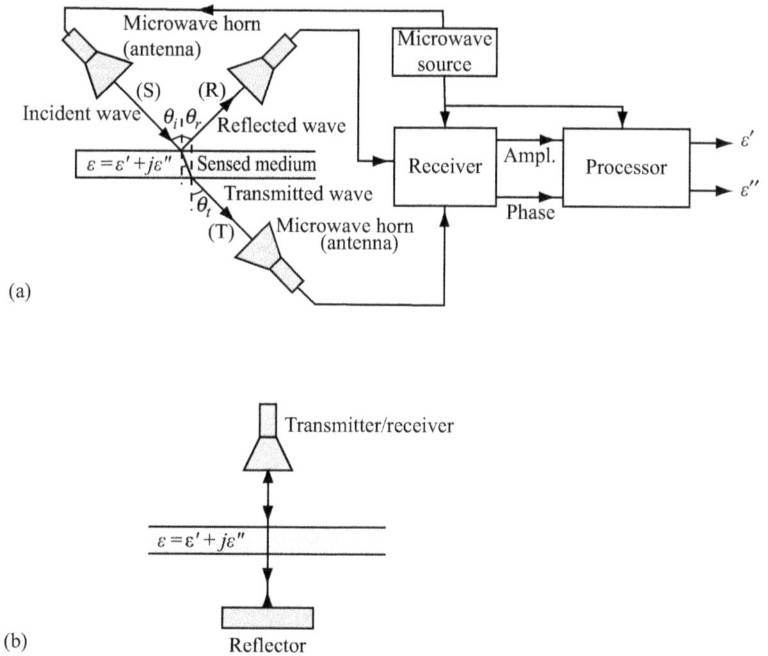

Figure 5.47 *(a) A general arrangement for reflection and transmission sensing and (b) simple arrangement for transmission and reflection sensing with a single antenna*

case, the plates are much closer together but the relative permittivity of rubber is much lower, leading to similar results. For these reasons, the capacitive sensing strategy was deemed insufficiently sensitive.

5.4.2 Reflection and transmission sensors

Reflection and transmission sensors at various frequencies are used in many applications including in what is sometimes called aquametry—the evaluation of moisture content in products such as wood, grain, foodstuff, foams, wool, etc. As such they could be adapted to the requirements of this work, at least in principle. The attraction of these methods is primarily cost effectiveness—they are likely to be less expensive but also the fact that the transmitters and receivers can be placed at considerable distances from the fabric or rubber sheet. The basic configuration is based on the reflection and transmission of electromagnetic waves from and through dielectric. The principle is shown in Figure 5.47(a) where an incident electromagnetic wave from source (S) propagates to a dielectric layer. Part of the wave is reflected (R) and part of it is transmitted (T). The reflected and transmitted components are dependent on the permittivity of the dielectric, and these in turn depend on the properties of the medium including density, moisture content,

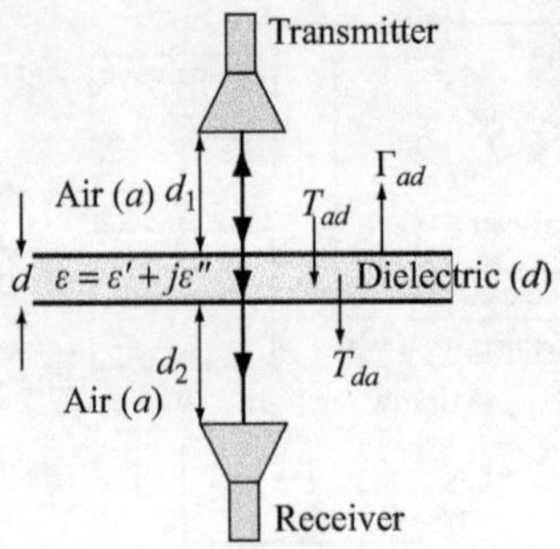

Figure 5.48 A simple transmission sensor

temperature, and sometimes even on strain or pressure. In a reflection sensor, the reflected wave is sensed, whereas in a transmission sensor, the transmitted wave is sensed. The waves can propagate at an angle as in Figure 5.47(a) or perpendicular to the sensed medium as in Figure 5.47(b). The methods in Figure 5.47 are often called bistatic methods because of the use of separate transmitters and receivers. Because of that, the transmission and detection can be continuous, so the methods can also be defined as bistatic continuous wave (CW) methods. It is also possible to use a single antenna in a pulsed mode as shown in Figure 5.47(b). This is an astatic (or monostatic) method that serves both as transmitter and receiver. A short pulse is transmitted and the return signal received by the same antenna. This is a transmission sensor except that the signal is transmitted through the dielectric twice.

The reflection from and transmission into a dielectric depends on the permittivity of the dielectric, angle of incidence, and the polarization of the wave. To get an idea on how these sensors operate and what is being measured, we will use the configuration in Figure 5.48 and assume plane wave propagation. Although the fields of the antenna only satisfy these conditions in the far field, and even there, only approximately, the propagation of plane wave affords a simple description of transmitted and reflected waves and hence a simple way of understanding the operation of the sensors. The reflection and transmission coefficients under these conditions are (see Section D.5 in Appendix D):

$$\Gamma = \frac{\eta_d - \eta_0}{\eta_d + \eta_0}, \quad T = \frac{2\eta_d}{\eta_d + \eta_0} \tag{5.15}$$

where η_0 is the wave impedance in air and η_d the wave impedance in the dielectric. The latter are given as

$$\eta_0 = \sqrt{\frac{j\omega\mu_0}{\sigma_0 + j\omega\varepsilon_0}}, \quad \eta_d = \sqrt{\frac{j\omega\mu_d}{\sigma_d + j\omega\varepsilon_d}} \tag{5.16}$$

where $(\varepsilon_0, \mu_0, \sigma_0)$ are the permittivity, permeability, and conductivity of air and $(\varepsilon_d, \mu_d, \sigma_d)$ the permittivity, permeability, and conductivity of the dielectric. The impedances can also be written as

$$\eta_0 = \sqrt{\frac{\mu_0}{\varepsilon_0(1 - j(\sigma_0/\omega\varepsilon_0))}}, \quad \eta_d = \sqrt{\frac{\mu_d}{\varepsilon_d(1 - j(\sigma_d/\omega\varepsilon_d))}} \tag{5.17}$$

where $\varepsilon_0(1 - j(\sigma_0/\omega\varepsilon_0))$ is the complex permittivity of air and $\varepsilon_d(1 - j(\sigma_d/\omega\varepsilon_d))$ is the complex permittivity of the dielectric. The first term is the dielectric constant of the medium and the second is loss associated with the medium. The term $\sigma/\omega\varepsilon$ is called the loss tangent of the medium. The complex permittivity is normally written as

$$\varepsilon = \varepsilon' + j\varepsilon'' \tag{5.18}$$

with ε' indicating the dielectric constant and $\varepsilon'' = \sigma/\omega\varepsilon$ the loss tangent.

Given an incident electric-field intensity of amplitude E_0, the reflected and transmitted electric- and magnetic-field intensities are

$$E_r = \Gamma E_i, \quad E_t = TE_i, \quad H_r = \Gamma\frac{E_i}{\eta_0}, \quad H_t = T\frac{E_i}{\eta_d} \tag{5.19}$$

E_t and H_t are the amplitudes of the electric- and magnetic-field intensities transmitted into the dielectric. To calculate electric-field intensity onto the receiver, the waves transmitted into the dielectric must transmit again across the opposite surface. Using Figure 5.48, we write

$$E_t = T_{ad}T_{da}E_0, \quad H_t = T_{ad}T_{da}\frac{E_0}{\eta_0} \tag{5.20}$$

where

$$T_{ad} = \frac{2\eta_d}{\eta_d + \eta_0}, \quad T_{da} = \frac{2\eta_d}{\eta_d + \eta_0} \tag{5.21}$$

In addition, because both air and dielectrics are lossy to a certain degree, the wave is also attenuated, and its phase changes as it propagates through air as well as the dielectric. The attenuation and phase constants in air and in the dielectric are

$$\alpha_a + j\beta_a = j\omega\sqrt{\mu_0\varepsilon_0}\sqrt{1 - j\frac{\sigma_0}{\omega\varepsilon_0}} \tag{5.22}$$

$$\alpha_d + j\beta_d = j\omega\sqrt{\mu_d\varepsilon_d}\sqrt{1 - j\frac{\sigma_d}{\omega\varepsilon_d}} \tag{5.23}$$

With these, and the notation in Figure 5.48, the field that reaches the receiver is

$$E_{\text{rec}} = T_{ad}\,T_{da}\,E_0\,e^{-\alpha_a(d_1+d_2)}\,e^{-\alpha_d d}\,e^{-j\beta_a(d_1+d_2)}\,e^{-j\beta_d d} \tag{5.24}$$

Or, in the time domain, assuming a sinusoidal form of the incident field ($E_i = E_0 \cos(\omega t)$):

$$E_{\text{rec}} = T_{ad}\, T_{da}\, E_0\, e^{-\alpha_a(d_1+d_2)-\alpha_d d} \cos(\omega t - \beta_a(d_1 + d_2) - \beta_d d) \tag{5.25}$$

The transmission coefficients, attenuation constants, and phase constants depend on the complex permittivity of the materials (air and dielectric). The properties of air are known a priori, and, if they are not, they can be evaluated directly through a calibration process without the dielectric. Hence, the two unknown values are the dielectric constant ε_d and the loss tangent $\sigma_d/\omega\varepsilon_d$ of the dielectric. These are evaluated by measuring the attenuation of the wave and the change in its phase, measured at the receiver relative to the transmitted wave.

Of course, in practice, the issues are more complex than this. The transmission coefficients T_{ad} and T_{da} are themselves complex values, and their product must be smaller than 1. This means that they contribute both an effective attenuation and a change in phase. For this reason, it is not sufficient to measure amplitude and phase but rather, one must also establish a calibration curve for the medium; otherwise, these effects contribute to errors in the measurement. Errors in measurements can also be introduced by reflections from structures in the sensing environment and by fields from external sources.

The method described here is akin to the free-space permittivity measurement described in section 4.11. In free space, permittivity measurement use is made of a carefully designed and calibrated sample holder, and the sample itself is carefully made to ensure accurate measurements using a network analyzer.

In some cases, especially if a reflection sensor is used, the configuration in Figure 5.47(a) is more convenient. Although the transmission and reflection coefficients are more complex and depend on the angle of incidence of the wave upon the dielectric as well as the polarization of the wave, the principle is still the same—one measures the phase and amplitude of the transmitted and reflected wave and correlates these with the dielectric constant and loss tangent of the dielectric.

In the context of the requirements of gauging of the fabric, neither the reflection nor the transmission sensor is appropriate. The fabric is very thin and hence both the attenuation and phase change within the fabric are low. On the other hand, the reflection coefficient is fairly large because of the high permittivity of the dip

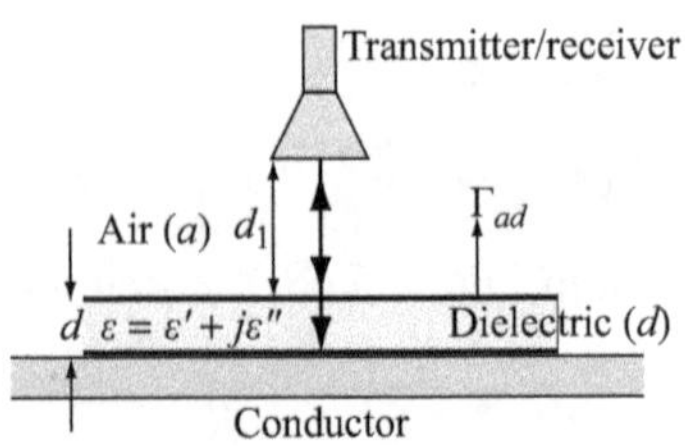

Figure 5.49 A dielectric sheet against a conductor backing

pickup in the fabric. This leads to very low sensitivity to moisture content although the method is very useful in thick dielectrics.

The method is much more appropriate as a rubber thickness sensor although it is nowhere near the sensitivity of the resonant method. In the case of the rubber thickness sensor, the most appropriate method is a pulsed reflection method as shown in Figure 5.49. Because the rubber sheet is located against the conducting calender, the reflection coefficient is an effective coefficient that depends on the dielectric. The wave received back at the antenna (assuming now a CW) can be written as follows:

$$E_{\text{rec}} = \left(\frac{\Gamma_{ad} - e^{-2\alpha_d d} e^{-2\beta_d d}}{1 - \Gamma_{ad} e^{-2\alpha_d d} e^{-2\beta_d d}} \right) E_0 \, e^{-2\alpha_a d_1} \cos(\omega t - 2\beta_a d_1) \tag{5.26}$$

where Γ_{ad} is the reflection coefficient at the air–rubber interface. The term in brackets is the effective reflection coefficient of the rubber–conductor combination, and the rest is the effect of propagation in air. This approximate formula takes into account the fact that the wave propagates twice through air and through the rubber and is perfectly reflected at the rubber–calender interface.

In practical measurements, the phase and amplitude of the reflected wave can be measured and correlated with the dielectric constant and loss tangent of the dielectric (rubber). In the pulsed approach, one can proceed in the same fashion, but the field is quite different including the fact that multiple harmonics will be present.

In reflection and transmission measurements, one can measure the power rather than the electric field, but since these are related quantities, the conclusions drawn here remain the same.

In addition to the methods described above, there are others as well as variations on these methods. One approach often used in measuring is an interferometric method. The method is particularly useful for thick dielectrics and requires the accurate evaluation of phases and amplitude. In that sense, it has the same properties of the transmission and reflections methods. The main difference between it and the reflection/transmission methods is that the total field rather than its reflected/transmitted components are measured.

Further reading

Some additional details of the initial prototype and on line measurements can be found in the following:

[1] S. DuFore, "Stripline Resonant Sensor Development for the Measurement and Control of Moisture in a Moving Web," M.Sc. Thesis, The University of Akron, December 2000.

[2] J. M. Madaras, K. M. Kot and P. M. Bujak, US Patent 6,565,914 Method for Controlling Deposited Polymer on a Substrate, May 20, 2003.

The design and optimization of the rubber thickness resonant sensor are detailed in the following:

[3] O. Bhuiya, "Design and Optimization of a Stripline Resonator Sensor for Measurement of Rubber Thickness in a Moving Web," M.Sc. Thesis, The University of Akron, December 2006.

Additional aspects of the simulation of the stripline resonators described in this work are described in the following:

[4] N. Farahat and N. Ida, "Open Stripline Resonator for Gauging in Industrial Applications," Review of Progress in Applied Computational Electromagnetics, ACES Conference, Verona, Italy, April 20–24, 2007, pp. 1846–1851.

[5] N. Ida and O. Bhuyia, "Design and Optimization of an Open Stripline Resonator for Rubber Thickness Gauging," Proceedings of the 11th International Conference on Optimization of Electrical and Electronic Equipment, Brasov, Romania, May 22–24, 2008, Vol. 4, pp. 97–100.

[6] N. Ida, "Open Stripline Resonator Sensors for Rubber Properties Gauging," Proceedings of MOMAG-08, Florianopolis, Brazil, September 7–10, 2008, pp. 1077–1083.

Some alternative sensors for moisture content and the more general issue of measuring complex permittivity may be found in the following:

[7] J. Musil and F. Zacek, "Microwave Measurements of Complex Permittivity by Free Space Methods and Their Applications," Elsevier, Amsterdam, 1986.

[8] J. R. King, "Microwave Sensors for Process Control Part I: Transmission Sensors," Sensors, Vol. 9, No. 9, 1992, pp. 68–74.

[9] A. Kraszewski, "Microwave aquametry—a review," Journal of Microwave Power, Vol. 15, No. 4, 1980, pp. 209–220.

[10] J. Thuery, "Microwaves: Industrial, Scientific and Medical Applications," Artech House, Norwood, MA, 1992.

[11] G. F. Engen, "Microwave Circuit Theory and Foundations of Microwave Metrology," IET, London, 1992.

[12] G. Roussy and J. A. Pearce, "Foundations and Industrial Applications of Microwaves," Wiley, New York, 1995.

Chapter 6
Evaluation of the sensors

6.1 Introduction

It is one thing to come up with a sensor but a whole different matter to come up with a sensor that satisfies the strict criteria required in a gauging application. The present chapter discusses the performance of the fabric coating sensor, its sensitivity, accuracy, calibration, and other parameters such as drift, sensitivity to environmental changes, and long-time stability. We start with initial performance tests on a prototype sensor whose purpose was to establish the viability of the sensor and to obtain data for further development that might be needed. Some of the tests were done in a laboratory environment but most were performed on an existing production line. The purpose of these tests, in addition to establishing the functional parameters of the sensor such as sensitivity and stability, was to also establish its viability in the industrial environment.

The first tests evaluate the empty sensor. The purpose is to verify the stability of the measurement and, as well, establish the effects of ambient temperature and humidity on the resonant frequency. These tests also have the advantage of comparison with accurate simulations since the effect is over the whole space occupied by the open cavity and the immediate environment. Following are laboratory tests that establish guidelines of what one should expect in actual measurements. As part of these, we define limits on shift in resonant frequency due to moisture in the fabric as well as verify the role of the even mode as the "sensing" mode, while the odd mode is seen to depend only on whole-cavity parameters such as environmental humidity and temperature.

Of course, what matters in the end are the actual measurements in the industrial environment. The prototype built for this purpose was tested over an extended period on a production line. The results given here show that all aspects of online testing can be identified from the signal obtained from the sensor. Even though the prototype sensor is very simple, stability is excellent and sensitivity is in line with the expected values. These measurements also pointed out to some possible modifications of the sensor, chief among them being the need for shielding it from nearby objects and personnel but also the need to stiffen the sensor against vibrations.

Any sensor must be calibrated before it can be used for any accurate measurements. The calibration method and some results used for the calibration process are described in some detail although the prototype sensor was only calibrated before installation. The production sensor is calibrated as part of the operation of the sensor at regular time intervals.

The sources of possible errors in measurement are explored as well by physically creating these errors and analyzing them from the signals obtained. These include the effects of fabric flutter, misalignment of the fabric in the sensor, sensor misalignment (offset), and others.

Although the present chapter deals exclusively with the fabric-coating sensor, the rubber thickness sensor has essentially the same properties. In both cases, only one mode is used for sensing, and since they are identical in dimensions and method of operation, including the method of motion, there is little difference worth reporting. In fact, many of the performance issues encountered in the fabric sensor, including the calibration process, are trivial in the rubber thickness sensor.

6.2 Empty sensor tests

After construction of the sensor's two halves, they were assembled on a temporary frame so that the distance between the center plates is 120 mm. This was the distance deemed sufficient for measurements on the production line, arrived at as a compromise between sensitivity (minimum distance) and requirements for clearance of splices in the fabric that, ideally, requires larger distances. The performance was first simulated using the finite-difference time-domain method and is shown in Figure 6.1.

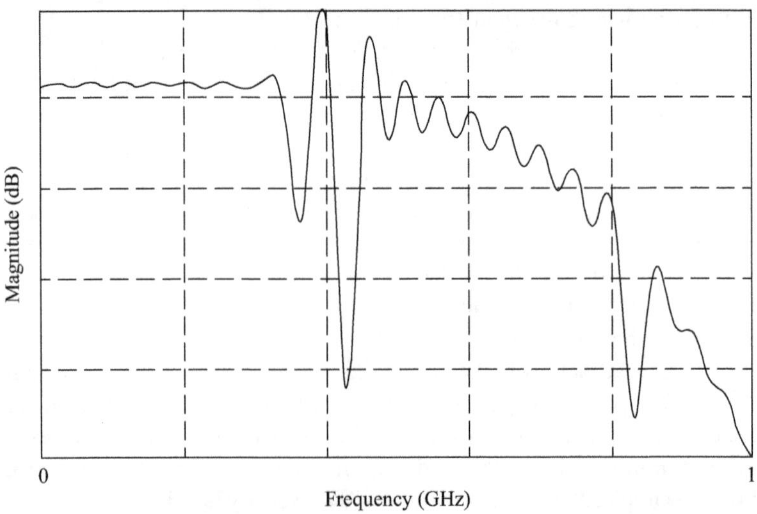

Figure 6.1 Finite-difference time-domain simulation showing empty resonator characteristics. The two resonant modes, even and odd, of the resonator can be clearly seen. The even-mode resonant frequency is at approximately 380 MHz, whereas the odd-mode resonant frequency is at approximately 420 MHz. The frequency span of the simulation is 1 GHz

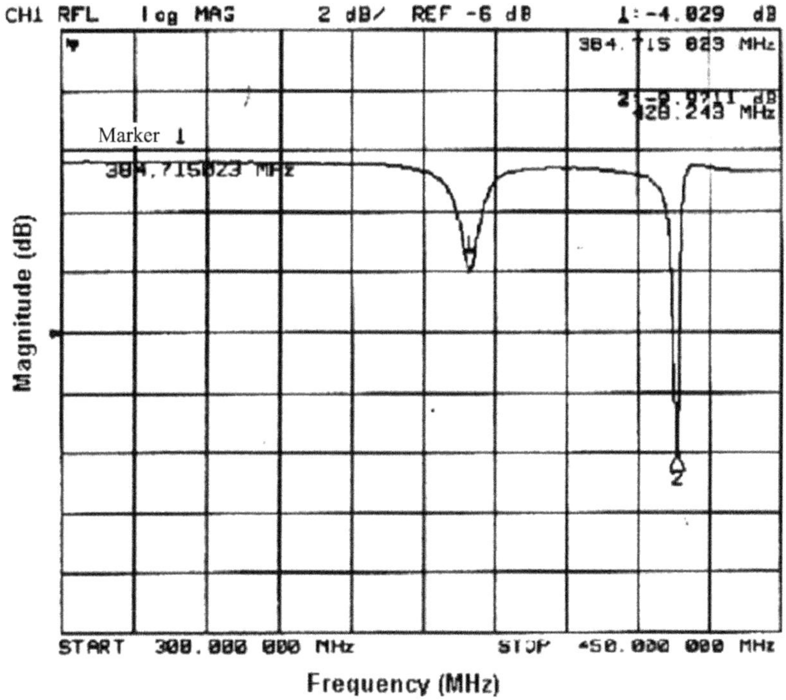

Figure 6.2 Network analyzer printout showing empty resonator characteristics. This graph also shows the two resonant modes, even and odd, of the resonator. The even-mode resonant frequency is at 384.71 MHz, while the odd-mode resonant frequency is at 428.24 MHz. The frequency span of the graph is between 300 and 450 MHz

The empty sensor resonates at 380 (even mode) and 420 MHz (odd mode). The network analyzer measurement in Figure 6.2 shows resonant frequencies of 384.7 and 428.2 MHz. The differences are small and are primarily due to small dimensional differences between the simulated and actual sensor. Figure 6.2 also indicates a smooth response and sharp resonances. These are indications that the signal-to-noise ratio and the quality factor of the cavity are high. This will be seen again in the online results to be discussed later in this chapter. Figures 6.1 and 6.2 are the same as Figures 5.4 and 5.5, respectively, but are reproduced here for convenience.

6.3 Laboratory tests

The primary function of the resonant sensor is to measure moisture content and, indirectly, the amount of solids left on the fabric after drying. Thus, the first tests were laboratory tests to verify the performance of the sensor with respect to moisture content in the fabric. Following calibration of the sensor (calibration is discussed in Section 6.6), and tests with the empty sensor, a number of tests were performed by wetting the fabric, weighing the fabric to obtain the amount of water, and then measure the resonant frequency with the wet fabric in the sensor. Multiple tests were performed using fabric samples similar to those that are to be monitored. Sample fabrics included fabrics of different densities, as well as bare and dipped fabrics. The testing procedure consisted of the following steps:

1. Weigh the dry-fabric sample.
2. Place the dry fabric in the sensor so that the plane of the fabric is lined with the plane of symmetry of the sensor.
3. Record even- and odd-mode resonant frequencies and weight.
4. Spray fabric with water using spray bottle.
5. Place moistened fabric in sensor as in step 2.
6. Repeat steps 3–5 until fabric is saturated (approximately 40% moisture content).

Table 6.1 shows sample data from a typical test. The fabric itself had a mass of 49 g. Water was added until the fabric was saturated. The final water weight was 62 g. This equates to a moisture content range from 0% to approximately 70%.

Table 6.1 Sample data from resonator test

Even mode (Hz)	Odd mode (Hz)	Odd–even (Hz)	Weight (g)
383,196,096	427,721,728	44,525,632	76.20
383,166,272	427,726,208	44,559,936	77.10
383,133,760	427,744,736	44,610,976	78.30
382,898,944	427,549,312	44,650,368	79.50
382,968,864	427,717,632	44,748,768	81.20
382,720,640	427,587,520	44,866,880	82.60
382,667,616	427,643,008	44,975,392	84.10
382,532,096	427,522,528	44,990,432	85.30
382,361,696	427,378,016	45,016,320	86.70
382,330,848	427,490,432	45,159,584	87.90
382,434,560	427,498,592	45,064,032	89.50
382,258,400	427,429,408	45,171,008	90.90
382,136,320	427,411,584	45,275,264	92.50
381,886,624	427,292,064	45,405,440	93.70
381,953,984	427,425,728	45,471,744	95.70
381,935,296	427,273,504	45,338,208	97.20
381,676,832	427,204,960	45,528,128	98.60
381,615,072	427,232,896	45,617,824	100.20
381,449,376	427,192,384	45,743,008	101.90
381,329,472	427,065,152	45,735,680	103.40
380,957,440	426,907,648	45,950,208	104.90
381,252,416	427,238,432	45,986,016	106.70
380,661,536	426,865,856	46,204,320	108.30
380,945,856	427,134,080	46,188,224	109.60
380,810,048	426,914,784	46,104,736	111.00

The typical moisture content on the web in the manufacturing process is approximately 35% and falls well within these limits. Table 6.1 shows, as predicted, the even-mode resonant frequency decreased from about 383.2 to 380.8 MHz as moisture content increased, while the odd-mode frequency remained relatively constant at 427 MHz.

Figure 6.3 shows four separate tests using the same fabric for moisture content. Both even- and odd-mode resonant frequencies were obtained and tabulated using the network analyzer. The difference between even- and odd-mode resonant frequencies was calculated and plotted versus the mass of the water on the fabric. It can be seen from the figures that the measurements are repeatable, and that, the change in frequency versus moisture content is linear. The reason for taking the difference between the even and odd resonant frequencies is that both are roughly equally sensitive to air humidity and to temperature, and therefore, this is a simple way of compensating for these effects. Using this information, the sensor can be modeled simply as a gain of approximately 13.3 kHz/g of water in the fabric.

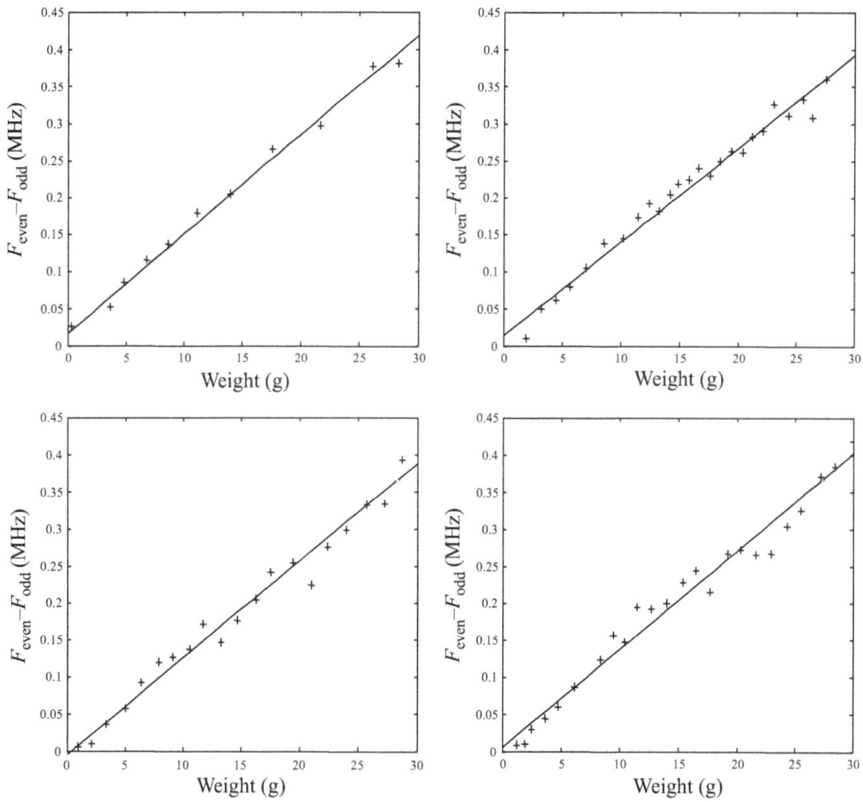

Figure 6.3 Four separate tests for moisture content measurement on the same fabric

The required accuracy of the sensor is specified by a maximum absolute error of 2% moisture content. Most of the observed data fall within this tolerance. The maximum error in moisture content encountered in this data is approximately 7%. However, the deviations from the linear characteristic appear to be randomly distributed, which implies that time-averaging techniques can eliminate the problem. The on-line testing results given later in this chapter confirm this idea. In addition, some of the error is due to alignment of the fabric and due to errors in establishing the water content, including drying during the preparation and measurements.

6.4 Online testing results

The preliminary tests were performed to insure that the sensor was a viable means for measuring moisture content and that the measurements are accurate, stable, and repeatable. For this purpose, the sensor was installed on an open frame made of aluminum so that the two halves of the sensor can be slipped into place without interfering with the production line. Figure 6.4 shows schematically how the senor is mounted on an open frame, allowing it to be moved in place without interfering with the moving fabric web. Figure 6.5 shows the sensor placed in position to measure the moisture content of a moving fabric web. The fabric is moving upwards. The two coaxial cables, seen in the forefront, connect the sensor probes to the network analyzer. Figure 6.6 shows the sensor before it is placed over the fabric to give a better view of the construction of the prototype sensor. The prototype was a fairly simple implementation of the sensor using 1.6-mm-thick aluminum plates for the ground planes and for the center plates. The center plates were held in place with plastic bolts, and the probes penetrated through the ground plane. The whole structure was attached to an aluminum frame that afforded some rigidity and

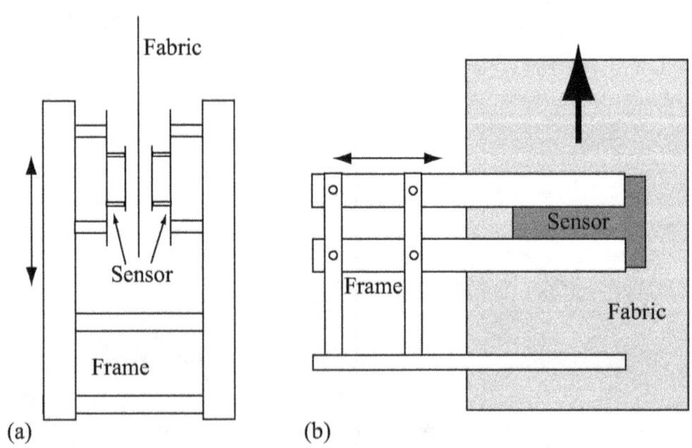

Figure 6.4 The two halves of the sensor attached to an open frame. The sensor is moved into position so that the web is centered. The web moves out of the page: (a) top and (b) side views

allowed the sensor to be placed over the moving fabric without interfering with production or stopping the motion of the fabric. In the prototype, the frame simply sat on the floor with the fabric moving upwards in the center of the frame (see Figure 5.3). Although a simple and relatively crude prototype, the results were extremely promising and predictable. These results were more than convincing as to the ability of a sensor of this type to measure and quantify all aspects of the fabric coating including amount of solids picked up and damage to the fabric.

Figure 6.7 shows schematically the relationship of the sensor to the other components of the dipping and curing process. The fabric is brought in on a roll (fabric letoff roll) and is then fed into the system by sewing the beginning of the letoff roll to the end of the last roll. The fabric then travels through a series of rolls called the festoon or accumulator. Then, the fabric is dipped in the dip tank. The fabric then travels through the dewebber system. The sensor was placed right after the dewebbers and just before the fabric travels into the drying oven. This way the final dip coating is sensed and kept at the required level.

Figure 6.5 Sensor measuring moisture content online (side view). The fabric can be seen on the left, moving up

Figure 6.6 The sensor before sliding around the fabric (seen on the right of the figure) (end view)

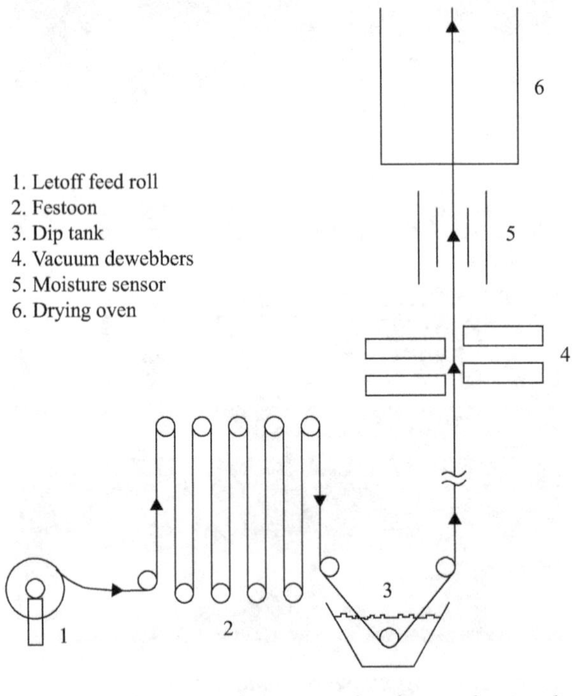

Figure 6.7 Conceptual diagram for the production line

The data was collected using a computer connected to the network analyzer. The computer communicates with the network analyzer via a National Instruments General Purpose Interface Bus interface card. A human–machine interface was created to enable an easy way to collect the required data. The interface allows the user to set the start and stop frequency scan. This is typically set to a narrow range that includes the expected resonant frequency but not too narrow to avoid missing the resonant frequency altogether. The defaults are set to 330 and 410 MHz but can be changed at will. The interface also sets the type of measurements to take including quality factor for the even and odd modes in both reflection and transmission configuration. Data can be collected at specific spots or in a continuous fashion. Additional functions can be added including buffering, storage and archiving, and transfer over the internet for remote monitoring.

Prior to the start of data collection, the gap between the center conductors was changed to 10 cm to get the best possible measurements. Data was collected by averaging five measurements together to get rid of noise effects and fluttering of the fabric. The even-mode resonant frequency of the reflected signal was the only measurement taken in these initial measurements. The odd-mode resonant frequency was considered to be constant after viewing some of the measurements taken, so it was dropped from the data acquisition. Figures 6.8–6.16 show the graphs of the even-mode resonant frequency over time, for several of the most interesting cases. Each graph shows some phenomenon happening in the system that could easily be used to better control the system. From these graphs, it can be seen that the sensor can detect minute changes in the moisture content. Also, the measurement appears relatively noise-free. No fabric weight data was taken on the fabric during the online testing trials, so comparative results could not be made.

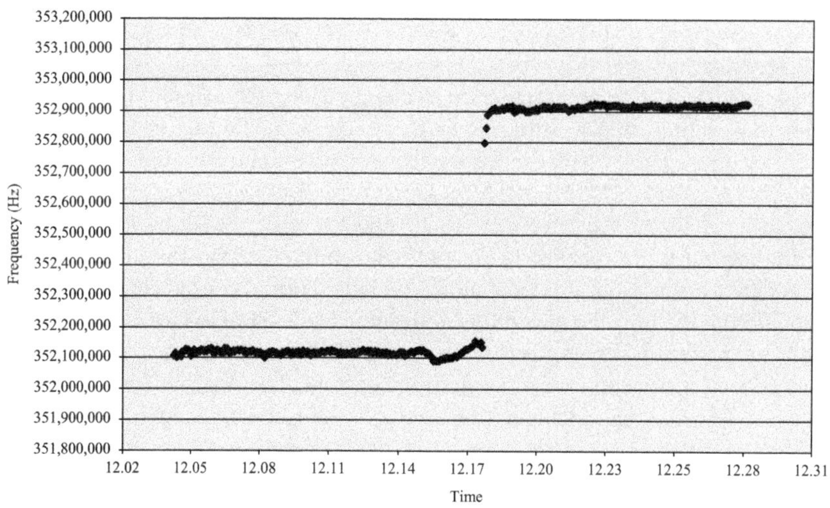

Figure 6.8 Resonant frequency versus time for fabric code (type) and speed change

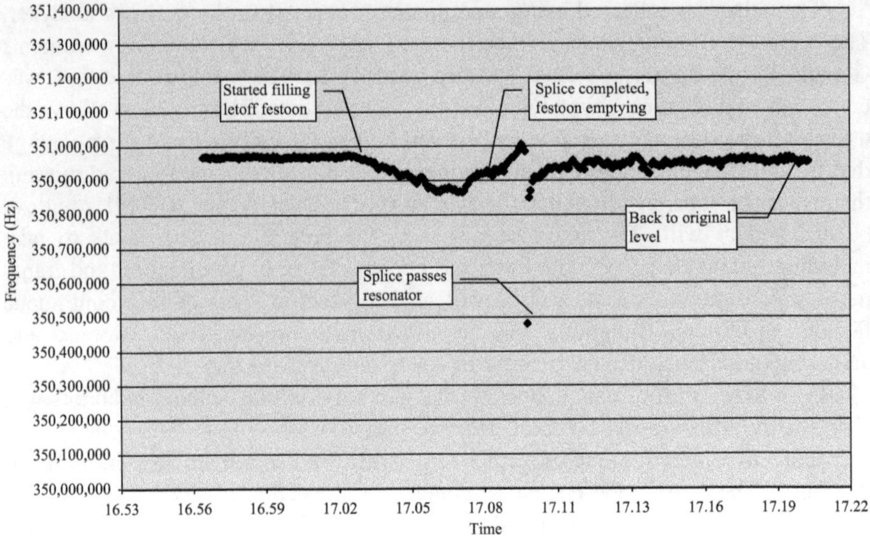

Figure 6.9 Resonant frequency versus time during web splicing. The splice is seen as the single dot at center–bottom

Figure 6.8 shows the response of the resonant sensor for change of the fabric materials as well as speed. Different fabrics (i.e., nylon or polyester) as well as fabrics of the same material but different density or even different batches have different properties; hence, they will pick up more or less dip. A decrease in speed allows the dewebbers to remove more dip hence the higher resonant frequency on the right hand part of the figure. In this case, the frequency increases by about 800 kHz. Note however that when the speed is constant, the resonant frequency is constant as well. The small variation in frequency along the constant speed sections are due to variations in dip distribution either because the fabric has picked more or less dip or, more likely, because the dewebbers have removed more or less of the material.

Figure 6.9 shows the effects of raising and lowering the festoon. When a new roll of fabric has to be attached to the end of a roll that is almost used up, a festoon or accumulator moves up to accumulate enough fabric to ensure continuous production during the time the new roll is being attached (sewn) to the roll that is being coated. As the fabric is being sewn, the new roll fabric is stationary, so the festoon lowers to feed the accumulated fabric to the system. The explanation to the effect seen in the figure is that raising and lowering of the festoon changes the tension on the fabric and that changes the amount of dip being picked up or that dip being removed by the dewebber changes. As can be seen in the figure, the resonant frequency decreases indicating more dip and then increases again indicating reduced dip pickup. After the festoon returns to normal, the resonant frequency returns to its previous value. Note also the splice in the fabric seen here as the point

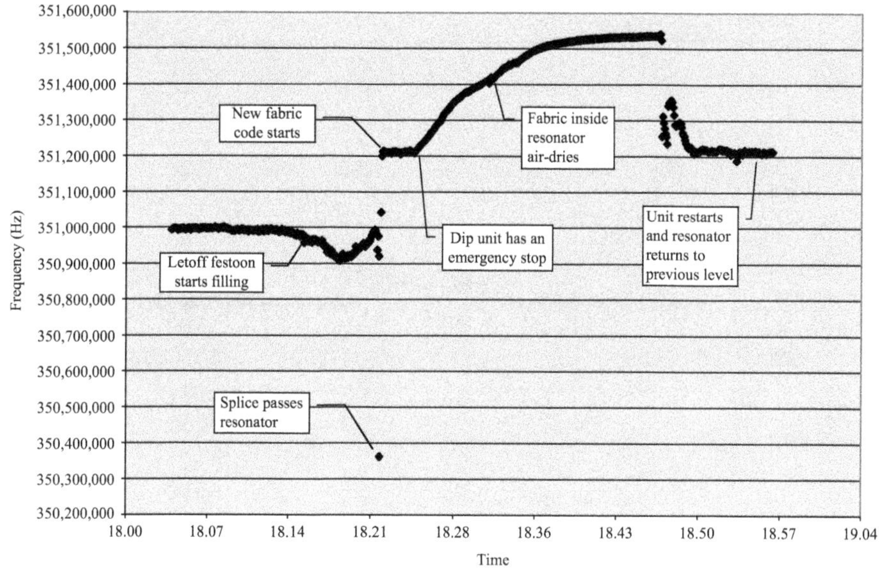

Figure 6.10 Resonant frequency versus time during a fabric change and line stoppage

in the lower middle part of the figure. As the splice passes through the sensor, the resonant frequency goes down considerably because of the larger amount of dip picked up by the splice. The added dip during the sewing operation represents a problem in quality assurance, but it can be eliminated by increasing the speed of the dewebbers to reduce the dip to the required value. This type of closed-loop control has not been implemented in the prototype system used for the measurements given here.

Figure 6.10 is somewhat similar to Figure 6.9. It shows the effect of the festoon movement up (left side of the figure), movement down (right side of the figure), and the splice passing through (lower left). However, the figure shows the response when a new fabric, different than the one in production, is attached. Since this fabric has lower permittivity (either because the fabric itself has lower permittivity or because it picks up less dip), the resonant frequency increases by about 200 kHz, from 351 to 351.2 MHz. A few minutes after the fabric change, the line stopped moving. The fabric dries off gradually as can be seen by the rising frequency during the time interval from 18.25 to 18.45. Once the line starts moving again, the measurement returns to the same value as before the line stop. The stability of the measurement guarantees that as long as the dip pickup and the dewebbing do not change, the resonant frequency remains the same. This results in particular but, to some extent the previous results as well, emphasizes the production-line-monitoring capabilities of the sensor. It can clearly detect and report such conditions as line stoppage, dewebber malfunction, lack of or reduction in dip

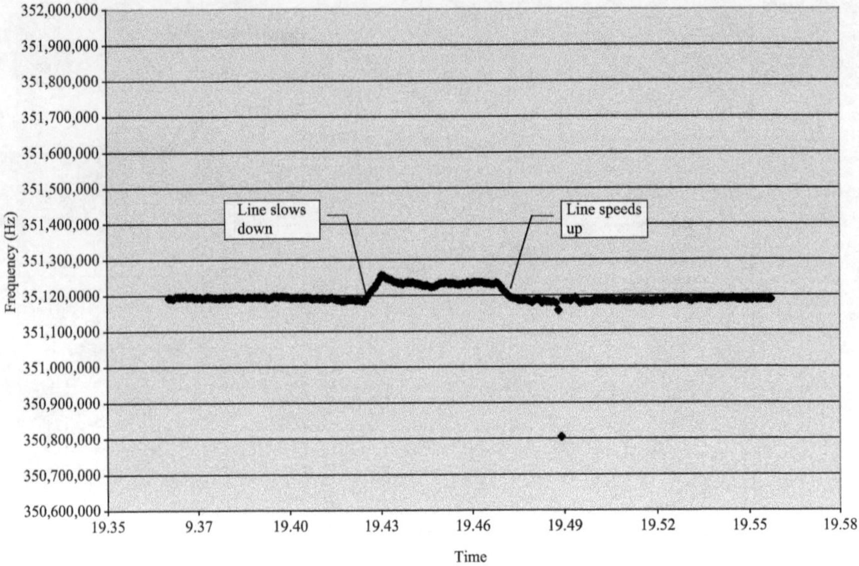

Figure 6.11 Resonant frequency versus time during line speed change

material, and many other conditions, some minor, some major. As such, it can help run the system unattended or with minimum operator supervision.

Figure 6.11 shows the line changing to a lower speed at 19.43 and then returning to normal speed at 19.47. The change in frequency is clearly noticeable during these times. Note as well that the change is gradual. As the line slows down, the dewebbers remove more dip off of the fabric; hence, the fabric is drier and the resonator frequency increases. It should be noted that the same effect would be obtained if the fabric were to pickup less dip or if the dewebber's speed were to increase.

One of the most important aspects of the resonator is the center plates. Since these were subject to optimization in Chapter 5, their effect was also tested in situ. Figure 6.12 shows the results of these tests. The first set of data points on the left part of the figure is for the thick rectangular plates, placed so that moisture was being measured and shows a resonant frequency of 351 MHz. Next, the sensor was pulled back from the fabric, and the center conductors were replaced with thin rectangular plates of the same dimensions. There was no appreciable change in resonant frequency. The small change in resonant frequency shown on the graph is due to the fact that the thin plates allowed a slightly larger distance between them and, since the effective permittivity is reduced slightly, there is a slight increase in the resonant frequency. These results are shown in the center of the figure. In the right section of the figure, we show the results after replacing the thin rectangular plate with butterfly-shaped plates with the same length and maximum width but with the middle of the conductor made narrow (see Section 5.3.1). This changed the

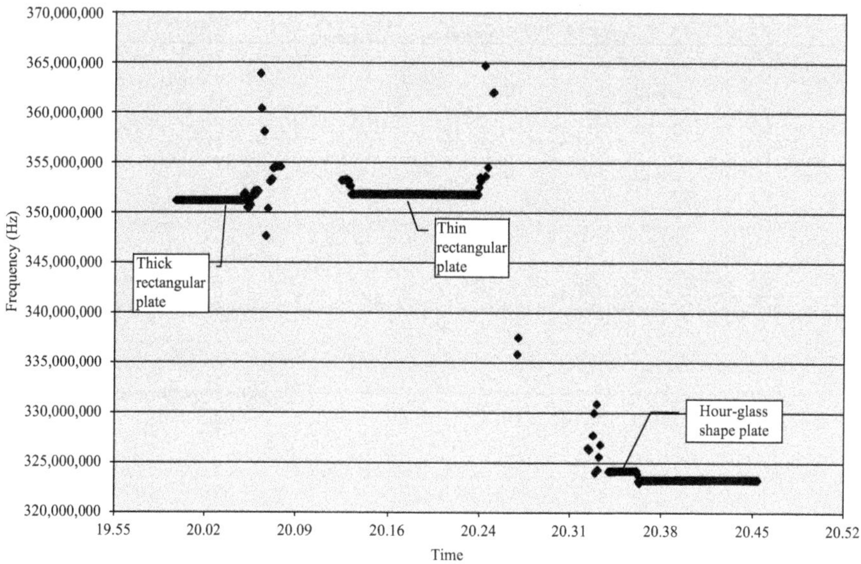

Figure 6.12 Resonant frequency versus time for three different resonator plates

resonant frequency by about 26 MHz. The step change in frequency after placing the butterfly plates on the sensor is due to moving the sensor from air back onto the fabric to be measured. Although each of the type of plate produces a stable resonant frequency, in the end, the thick plate was selected because it is sturdier than the thin plate, and the butterfly or hourglass-shaped plate has sharp corners that can snag the fabric. Since each produces a stable baseline frequency, it does not really matter which one is used other than the slight advantage of the thick plate. We note however that the butterfly plates produce a higher Q-factor in the cavity meaning that they are likely to produce higher sensitivity, which was not considered an important factor here since the sensitivity was more than sufficient.

Another functional parameter of the resonant sensor is the separation between the two center plates. As a rule, the wider the separation, the lower the danger that the sensor will interfere with the passage of splices and the lower the sensitivity of the sensor to misalignment of the fabric at the center between the plates. On the other hand, sensitivity suffers because of the reduction in the Q-factor of the cavity and reduction in the effective permittivity due to the mixture of air and the fabric/dip volume in the cavity. Another issue is the influence of external objects and personnel on the resonant frequency, which increases with the separation. To see the response of the sensor to separation between the plates, these were set to 100 and 200 mm to correspond to the minimum and maximum separation envisaged. Figure 6.13 shows the change in frequency due to changing the gap between the center plates. The resonant frequency changes approximately 4.5 MHz as the gap changes from 100 to 200 mm. The results were obtained with the butterfly-shaped

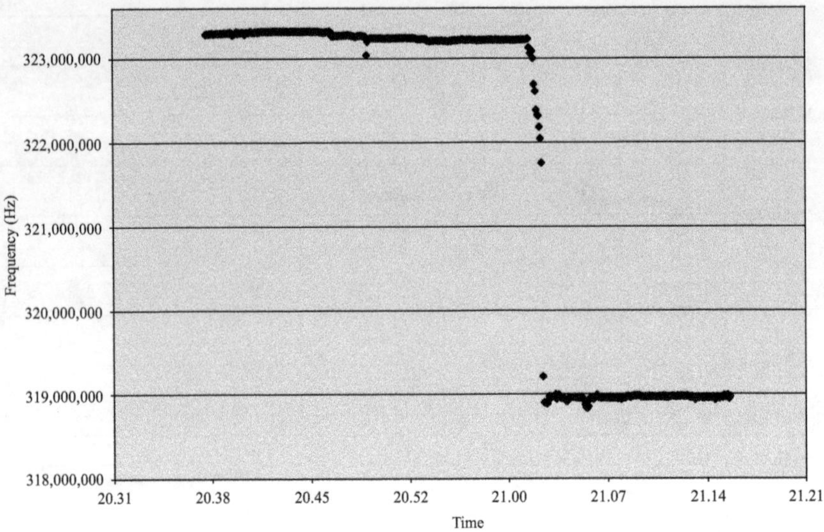

Figure 6.13 Resonant frequency versus time with a change in the distance between the center plates

center conductors. The lower separation shows a higher frequency because of the larger effective permittivity in the cavity (the fabric represents a higher fraction of the volume of the cavity).

The sensitivity of the sensor allows it to clearly distinguish the smallest variations in the test conditions. Figure 6.14 shows a slight shift in the resonant frequency caused by a change from a fabric to another of the same type but from a different supplier. The first fabric, whose response is on the left part of the figure, was slightly more absorbent than the second. The permittivity of the fabric is the same so the change must be due to a change in pickup of dip. The separation between the center plates was 200 mm.

A source of nonuniform dip coverage is either rubbing off of dip due to friction or slippage on the pull roll or due to streaking cause by contamination of the pull roll. Figure 6.15 shows a downward trend in the resonant frequency over time caused by accumulation of dip due to contamination of the pull roll. In this case, the dip kept building up and hence the reduction in the resonant frequency over time. Since streaking can occur anywhere on the fabric, it is important that the whole width of the fabric be monitored. The response due to streaking can be very gradual and hence it is important that the response be compared to the baseline resonant frequency continuously. Other "defects" in the coating or, indeed, in the fabric (such as tears) can be similarly detected although small area changes in properties may only show as a point or a very short line.

One purpose of the developed sensor is to serve as a feedback mechanism to control the dip and hence control the final dry coating of the fabric. This is done by using the resonant frequency to control the fans on the dewebbers and by so doing

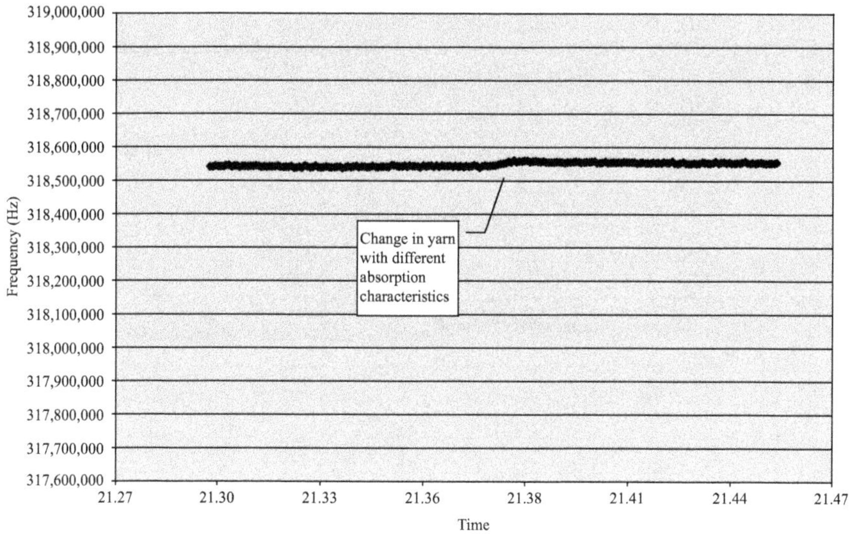

Figure 6.14 Resonant frequency versus time with the same fabric but a change in supplier

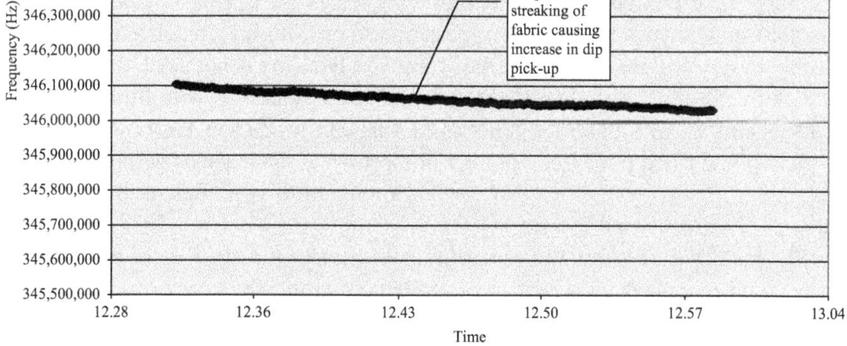

Figure 6.15 Resonant frequency versus time for streaking in the fabric due to pull roll stand contamination

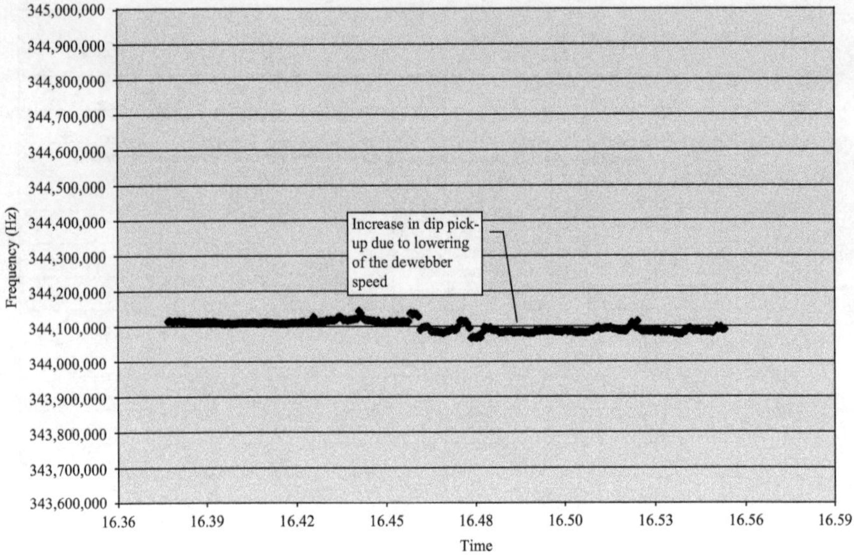

Figure 6.16 Resonant frequency versus time for fan speed changes in the dewebbers

keep the coating constant. This can be accomplished by increasing the speed of the fans if the resonant frequency decreases below the baseline resonant frequency and decreasing the speed if the resonant frequency increases. Figure 6.16 shows initial tests in establishing this control mechanism. Here, the speed of the fans was decreased from 3,200 to 2,800 rpm. The frequency went down by 30 kHz indicating an increase in dip remaining on the fabric. The control can be based on the feedback alone or it can be based on baseline values as well. That is, one can fix the resonant frequency for a particular type of fabric or even for a batch and change the dewebbers speed to maintain that frequency. If a baseline is not used, the frequency dip pickup can be changed so that the frequency is constant with time. However, the danger is that if a baseline is not used, one can produce a fabric with uniform dip pickup, but the dip may be too high or too low. A database of baseline resonant frequencies for various fabrics and various applications is sufficient to resolve this issue.

Figures 6.8–6.16 show the versatility, sensitivity, and stability of the sensor. It can be used to collect data and to control the coating thickness on the fabric to within very narrow margins. It is very effective in detecting variations in the amount of dip being picked up on the fabric but can also detect the festoon raising and lowering, splices, line stops, line speed changes, streaking, fabric code changes, fabric material changes, and fan speed changes. In that sense, it is more than a moisture sensor and can be used to monitor the production line functional parameters or, with proper additions, to also control the line. Certainly, simple alarms can be added or the line may be stopped if certain conditions (say, streaking) occur and an operator alerted to rectify the problem. The sensor's ability to operate in the industrial environment was clearly established from these simple measurements. There are also some performance characteristics that are not seen in these figures. For example, fluttering of the fabric, which in principle should be seen as changes in frequency, did not cause such changes primarily because the average position of the fabric remained centered but also because each resonant frequency shown is an average of multiple readings. Another concern was the possible contamination of the sensor with splattered dip, which would necessitate cleaning. This turned out to be an unfounded concern. The vertical motion of the fabric and the position of the sensor just before the drying oven prevented this from happening. As can be seen from the figures, the resonant frequencies are steady without observed noise. However, the sensor design, including coatings, used on the center plate took the possibility of contamination and the possible need for cleaning into account. Should that be necessary, cleaning can be undertaken by washing the sensor with water or with an appropriate solvent.

There were however some issues that needed addressing. One was the use of thin ground plates which can vibrate causing changes in the resonant frequency. In the industrial environment, there is also the danger of physical damage to the plates. Being made of aluminum, the thin ground plates can easily bend. In addition, the large separation between the ground plates meant that anything in the vicinity, and in particular, movement of personnel influenced the resonant frequency. As a result of these observations, a number of changes in the construction of the sensor were called for. These included use of much thicker plates and the bending of the ground plates into a partial shield. This was intended primarily to reduce external influences. A second recommendation was to coat the interior of the sensor with an easily cleanable coating such as Teflon (PTFE) similar to that used in nonstick utensils.

The second generation sensor has been described in Chapter 5 (see Section 5.2.5), and some of its characteristics have been optimized, including distance between the center plates, position of the probes, and many others. The main difference between this sensor and the prototype above is in the partial shielding of the sensor which makes it into a more closed sensor. This has far reaching consequences on the performance. By partially enclosing the sensor, the Q-factor of the cavity rises considerably improving sensitivity and signal-to-noise ratio. In addition, it reduces sensitivity to outside influences. The sensor was made of much thicker material to increase rigidity, and, in addition, the method of installation was modified to allow the sensor to move back and forth across the fabric to allow full coverage of the fabric. The implementation of the final sensor is described in the following chapter.

Following verification of the early prototype in the previous sections and production of the new sensor, a more detailed study of the performance of the sensor and the quantities that affect this performance was undertaken. Since the basic functioning of the sensor was established, the following sections discuss the more subtle performance characteristics of the sensor. These include variations in the gap between the center plates, off-centered fabric, position of the feed and load probes, and many others. The purpose of these tests was to establish limits of performance and to better understand the sources of errors that one can expect under normal operating conditions. Another important consideration is the calibration of the sensor and will be discussed here at some length.

6.5 Performance evaluation

6.5.1 Effect of distance from antenna tips to center plate

The purpose of the antenna probes is 2-fold. The feed probe, which is referred to here as the "reflection" probe, excites the cavity and sets it into resonance. It is connected to the source port of the network analyzer. We call it reflection antenna or probe simply because it is here that the reflection coefficient is measured through the S_{11}-parameter. The second probe is the load probe or "transmission" probe and is used to sense the transmitted power through the S_{21}-parameter. Since the coupling from the transmission probe to the center plate and from the center plate to the reflection probe is through the gaps between the probes and the plates one can expect that these gaps will influence the performance of the sensors. In fact, both the resonant frequency and the quality factor depend on these gaps as can be seen in Figures 6.17 and 6.18. Figure 6.17 shows that the resonant frequency

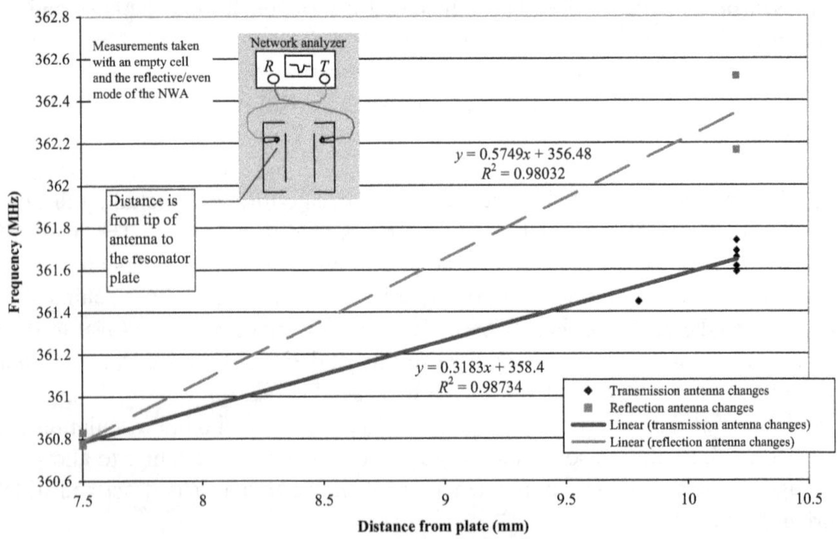

Figure 6.17 Effect of the distance between the tips of the antenna probes and the center plate on the resonant frequency of the cavity. The probes are on opposite sides of the resonator as shown in the inset. The cavity is empty. The straight lines show a linear least square plot

increases as either of the gaps increases. This may seem at first odd since the frequency should not be related to the coupling (i.e., to the power supplied to the cavity). However, the closer the probes are to the plate, the larger the influence of the conducting probe volume has on the effective volume of the cavity. In other words, as the conducting probes approach the plate, the perturbation due to their volume is larger (higher fields), effectively decreasing the volume of the cavity and hence the increase in resonant frequency. These results were also observed in simulations—a 1-mm change in the gap changes the resonant frequency by about 700 kHz. The surprisingly large change is due to the fact that this change occurs in a region of intense fields and hence has more influence on the resonant frequency that if it occurred in a region of low fields.

Figure 6.18 shows the quality factor of the cavity with respect to the position of the probes. The quality factor decreases as the source antenna moves further from the plate, because the energy coupled into the cavity decreases (while the losses are more or less constant). On the other hand, as the load probe moves away from the plate, the Q-factor increases since now there are fewer losses in the load probe as it encounters lower field values.

These results also show that the coupling is not critical and the cavity will operate over a range of positions of the probes. Of course, the resonant frequency changes are not important in the sense that once the distances are fixed, the resonator will operate around a fixed frequency. Any material in the cavity will then

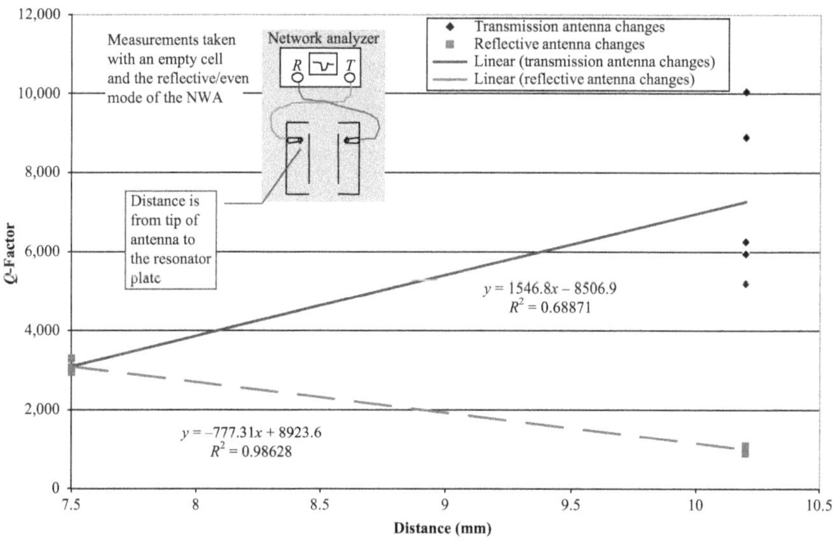

Figure 6.18 Effect of the distance between the tips of the antenna probes and the center plate on the Q-factor of the cavity. The probes are on opposite sides of the resonator as shown in the inset. The cavity is empty. The straight lines show a linear least square plot

lower the frequency to a value corresponding to the effective permittivity of the cavity. However, the *Q*-factor affects sensitivity and signal-to-noise ratio and, as a rule, the target is to increase the *Q*-factor as much as possible. From Figure 6.17, it is clear that the source probe should be closer to the center plate, whereas the load probe should be further away. Moving the tip from 7.5 to 10 mm increases the *Q*-factor by a factor of 2.

6.5.2 *Effect of flutter*

A wide, taught fabric, moving at a considerable velocity, is bound to flutter—that is, it will vibrate up and down around its axis of motion with larger vibration toward the edges. Since this flutter moves the fabric off the plane of symmetry of the sensor, there was a concern that errors may result from measurement of the fabric in the wrong position. To see how the sensor behaves under these conditions, the fabric (nylon) was tested at various position below and above the plane of symmetry of the sensor. With a separation between the center plates of 200 mm, this allowed positioning of the fabric at distances of up to ± 100 mm from the center plane. Figure 6.19 shows two curves. For the top curve, the probes were placed on opposite sides of the sensor (one above the upper center plate, the other below the lower center plate). The curve shows that as the fabric moves away from the center plane, the resonant frequency decreases. This is as expected since for the even

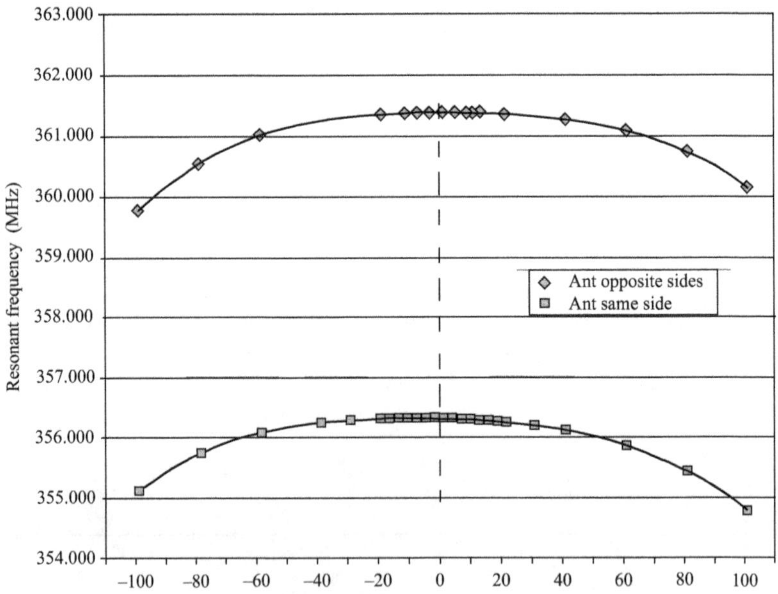

Figure 6.19 Effect of flutter of the fabric around the center plane of the sensor. The further away the fabric is from the center plane, the lower the resonant frequency and the lower the sensitivity of the sensor

mode, the electric field on the center plane is parallel to the fabric, whereas further away from the center plane, the parallel field decreases and the perpendicular component of the field increases. The center plane is indicated by the vertical dashed line. Because the fabric is thin, the main reaction is with the tangential component of the field which is highest at the center plane between the center plates. Hence, one can expect the sensitivity to go down as well. It is this loss of sensitivity that is the main concern, and it is the reason why the fabric must be kept as close as possible on the center plane. Nevertheless, for small deviations from the center plane, the changes are minor as can be seen from the flat section of the curve around the center. The lower curve in Figure 6.19 shows very similar results for the two probes installed on the same side of the sensor.

A flutter of a few millimeter is of no concern, but larger variations do occur. The simplest solution to this problem is to measure the resonant frequency a number of times (we chose five measurements) at the same position and average the values. Since the flutter occurs relatively fast, averaging produces a result very close to that without the flutter. Figure 6.20 shows the same two curves but on a narrower interval of ±40 mm. The variations are much smaller (about 150 kHz) justifying the averaging concept. It should be noted that the response is not perfectly symmetric especially with the two probes on the same side. This is simply

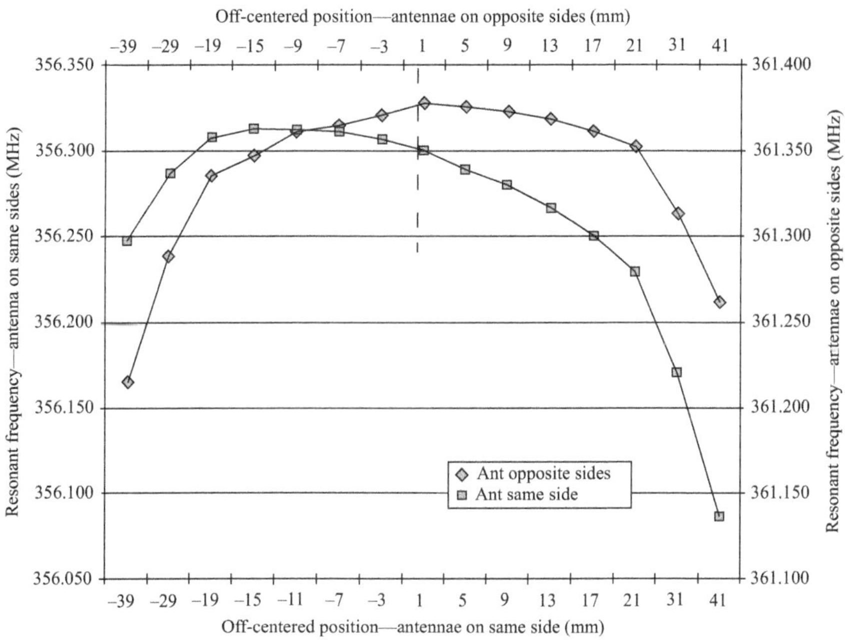

Figure 6.20 Effect of flutter of the fabric around the center plane of the sensor. This is the same as Figure 6.19 but with a flutter of maximum ±40 mm

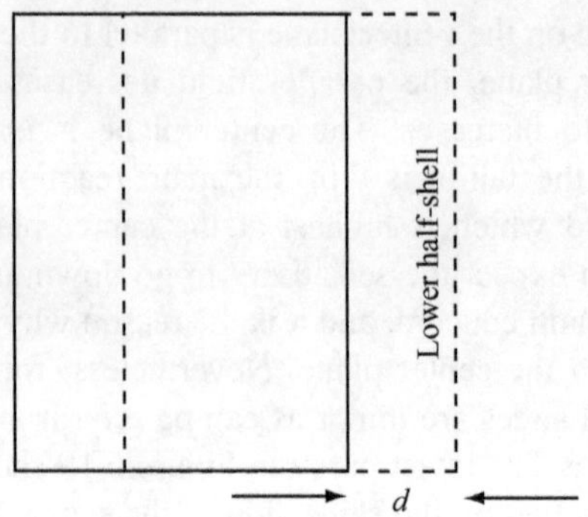

Figure 6.21 Definition of offset. The distance is measured on the outside of the shield between the upper and lower half-shells. The separation between the plates remains constant at 112 mm

from the fact that the feed is in one corner of one plate, and the fields around that corner are higher than at the other corners of the center plates. This asymmetry is not a concern especially if the fabric is maintained near the center plane.

6.5.3 Effect of cell offset

Because the sensor is made of two half-cells, connected to a frame, a relevant concern is the effect of any possible offset between the two half-cells. In particular, in the final implementation (to be discussed in the following chapter), the two half-cells move independently driven by screw drivers or by belts and there is a distinct possibility that cells may become either permanently misaligned or their relative position to each other may change during motion of the cells across the fabric. To evaluate the possible effects of offsets between the cells, the two half-cells were installed with deliberate offsets and the resonant frequency measured with the empty cavity (air) and with the copolymer calibration sheet. The offset is measured as the distance of one half-shell relative to the other as shown in Figure 6.21. Two types of center plates were used: one is coated with a smooth, black plasma coating, the other is coated with a rough plasma coating. The results for offsets of up to 35 mm are shown in Figure 6.22. In all four cases shown (air or copolymer in the cavity, each tested with either the smooth or the rough coating), the results show similar tendency: as the offset increases, the resonant frequency increases. This, of course, is due to the fact that the effective volume of the cavity decreases. However, for very small offsets, the changes are minimal indicating that small variations in position due to, say, vibrations should not pose a problem, and these small variations may not be detectable. The offset should also have an effect on the Q-factor. To test that the "depth" of the dip in the response of the sensor has been measured as an indication of the Q-factor—the deeper the dip, the higher the Q-factor. This is shown in Figure 6.23. As a rule, the Q-factor increases with the offset when the sheet is not present since the amount of lossy material in the cavity decreases. However, there is also an effect due to the coating on the center plates

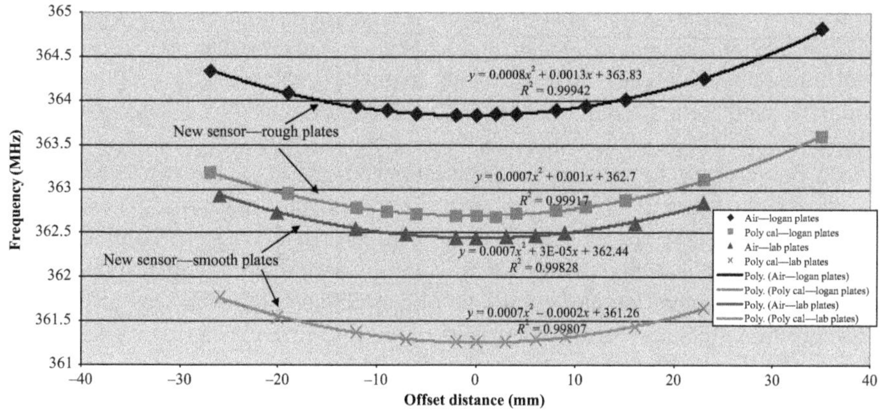

Figure 6.22 Resonant frequency as a function of offset between the two half-cells. Note the relatively flat section around the zero offset point. Four combinations of center plates and conditions in the sensors are shown

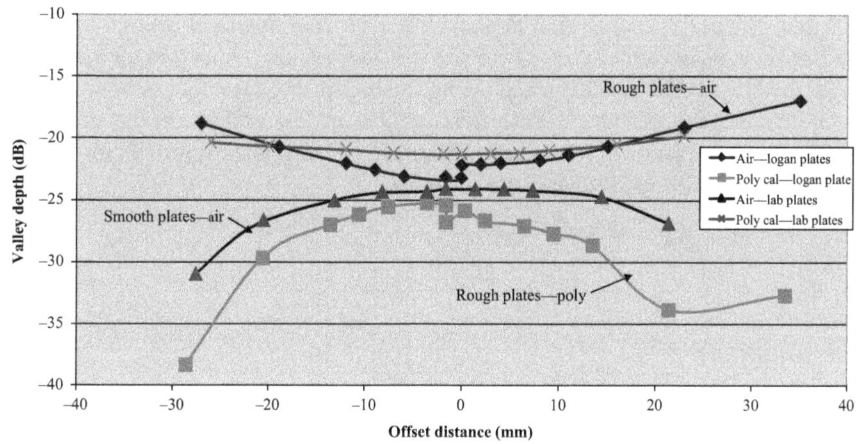

Figure 6.23 Estimate of the Q-factor from the depth of the dip in the response for different center plates and conditions in the sensor

since there are losses associated with these coatings. The dip level is not an accurate method of gaging the *Q*-factor since the width of the curve at the half-power points must be taken into account, but it gives some idea of the behavior of the *Q*-factor.

6.6 Calibration of the sensor

Both sensors described in this work rely on the change in resonant frequency due to changes in effective permittivity of the open cavity. As such, one can properly describe these as relative sensors, that is, one can easily say that a thicker material or one that contains more water will cause a larger shift in resonant frequency but one cannot, in general, associate a thickness or type of material or a certain moisture content directly to a resonant frequency unless the sensor has been properly calibrated. Calibration then is the establishment of the behavior of the sensor for the range of variations in the test material that one can expect in the production process. In the case of coated fabrics, the common target for production is the weight per unit area of the fabric. For example, the moisture content is measured in weight per unit area rather than, say, percentage of volume of the fabric. To be able to use the sensor to control the amount of dip pickup on the fabric, a clear calibration curve must first be established for each type of material (composition, thickness, etc.). In the case of the rubber thickness sensor, assuming the composition of the rubber is constant, the calibration curve relates the resonant frequency with the thickness of the material directly. One can establish theoretical curves from simulations or more practical curves based on measurements.

Figure 6.24 shows a calibration curve for thickness, correlated to the dip pickup measurements. The curve was obtained by using white Delrin sheets in lieu of the fabric. The white Delrin was used because its permittivity is roughly the same as that of the nylon material used in the nylon fabric. The sheets were 610 by 610 mm, one of thickness 1.588 mm and the other 2.38 mm used in combination—the first point is in air, the second for a single 1.588 mm sheet, the third for a 2.38 mm sheet, then two 1.588 sheets stacked up, and so on. As expected, the calibration curve is linear indicating that the electric field is more or less uniform in the small area around the center plane of the sensor. The small spread in the measurements is taken care of by a linear least square curve given as

$$f = -0.4262t + 359.74 \quad \text{(MHz)} \tag{6.1}$$

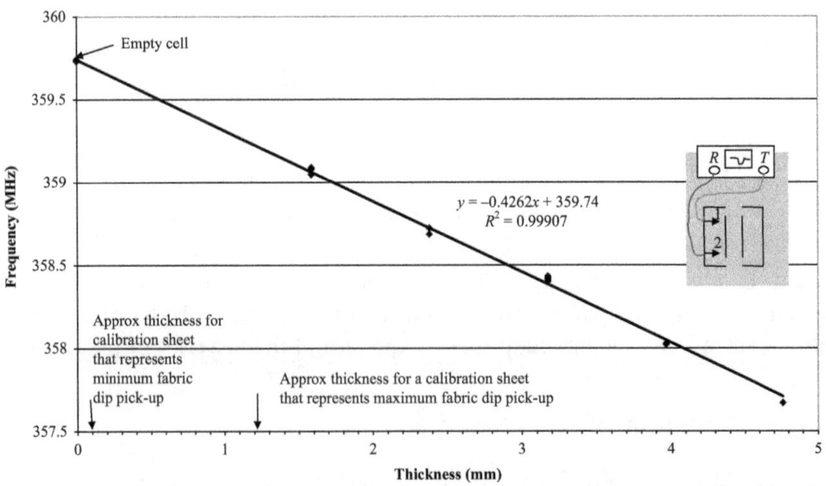

Figure 6.24 Material thickness calibration curve. Resonant frequency versus thickness for white Delrin

where f is the resonant frequency (shown as y in the figure) and t the thickness in mm (shown as x on the figure). R^2 is the correlation coefficient and indicates that in this case the linear least square representation is very close to the actual points ($R^2 = 0.99907$). The range of use for dip pickup is between 0.1 and 1.2 mm (indicated in the figure). These correspond to the minimum and maximum dip pickup. Note also the location of the probes connected to the source and load ports of the network analyzer. In all measurements, the sheets are centered between the center plates and the measurement is reflective (i.e., based on the S_{11}-parameter), measuring the even-mode resonant frequency.

Although the material used for the calibration is different than the fabric, because the curve is linear and because the limits expected for the dip pickup are known, the curve between the limits can be used for the dip pickup directly. The reason to use solid sheets for the calibration rather than, say, fabric with the correct dip is that these sheets are stable, do not stretch, and are unlikely to pick up moisture or other foreign materials and hence skew the measurements. We shall see in Chapter 6 that the use of solid sheets of material was needed so that the calibration can be done online before measurements begin and, periodically, during production runs.

Related to these measurements is Figure 6.25 which shows the Q-factor for the same material and thicknesses as in Figure 6.24. Although there is considerable spread for multiple measurements, the Q-factor follows an exponential curve with the Q-factor decreasing as the thickness increases:

$$Q = 6{,}731.3e^{-0.5818t} \tag{6.2}$$

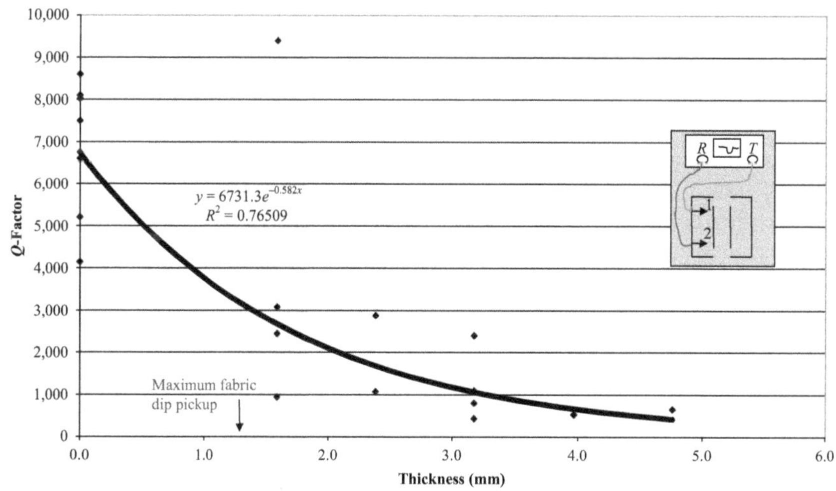

Figure 6.25 Q-factor as a function of material thickness corresponding to the measurements in Figure 6.24

This is exactly as expected—the thicker material introduces more losses and hence the decrease in the *Q*-factor. It should be noted however that at the maximum dip pickup, the expected *Q*-factor is above 3,000. Also, the relatively large spread of the points means that the correlation coefficient is lower ($R^2 = 0.76509$). Unlike the resonant frequency, the *Q*-factor as shown here applies to the dip pickup measurements only approximately since the loss tangent of the dip pickup is unlikely to be the same as that of Delrin. For this reason, the results in Figure 6.25 should be taken only as an indication of the behavior of the *Q*-factor.

In Figures 6.24 and 6.25, care was taken to ensure the Delrin sheets were centered in the cavity. As with the fabric flutter discussed above, any deviation from the center plane of the cavity will reduce the resonant frequency and introduce an error. The response of the sensor to off-centered calibration sheets is shown in Figure 6.26 for a 3.175 mm sheet of black acetal copolymer. The black acetal copolymer is used in lieu of the polyester fabric, again because of the similar permittivities of the two materials. The even-mode resonant frequency is shown as a function of position of the sheet between the center plates. The center plates are separated 112 mm apart. Zero distance means the sheet is at the surface of upper center plate (the plate next to the probes), and 112 mm means the sheet is at the lower plate.

The relation between plate thickness and resonant frequency shown by the solid curve was obtained from the measured points using a second-order least square fit, given by

$$f = -0.0008x^2 + 0.107x + 354.99 \quad \text{(MHz)} \tag{6.3}$$

The correlation factor of 0.9663 indicates a close fit.

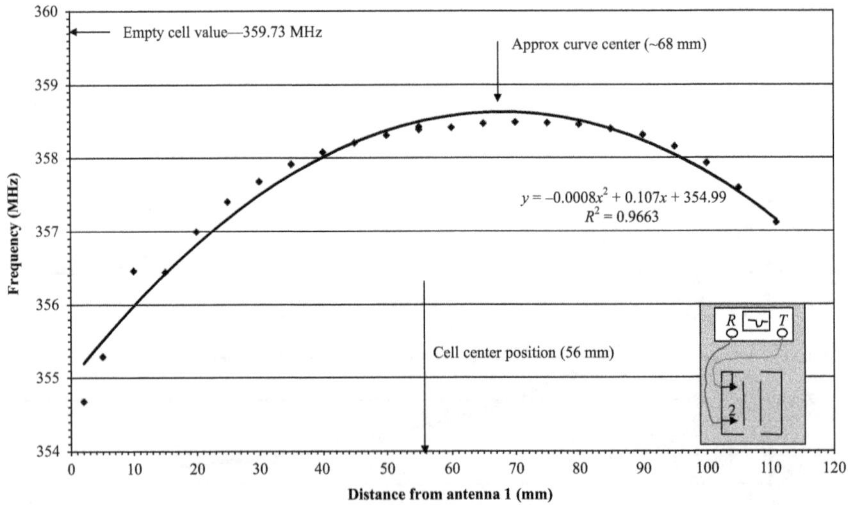

Figure 6.26 Resonant frequency due to a 2.175-mm-thick black acetal copolymer sheet versus distance from the left center plate in the inset. The plates are separated 112 mm apart

Although the information here is similar to that in Figures 6.19 and 6.20, it is interesting to see the asymmetry in the curve, that is, the change in frequency is larger as the calibration sheet approaches the upper plate because the fields are larger in that region. This also means that the physical center of the cavity and the electrical center are not the same. In fact, for normal operation, the fabric should move in a plane below the center of the cavity. In the case shown here, the geometric center is at 56 mm, whereas the electrical center is at 68 mm from the upper (driven) plate. This point is shown as the "curve center" in the figure. Around that point, the curve is flatter, and small deviations of the fabric from that point will cause minimal errors in the resonant frequency.

Figures 6.19, 6.20, and 6.26 all show an asymmetry in the response of the resonator, attributed to the location of the probes. To explore this issue further, the senor was equipped with four probes, two next to the upper center plate, two next to the lower center plate. This allowed for the source and load ports to be connected to two probes on the same plate or opposite plate in different combinations. This is shown in Figure 6.27. The inset numbers the probes for identification. The figure shows that for the reflection and transmission probes (source and load) on

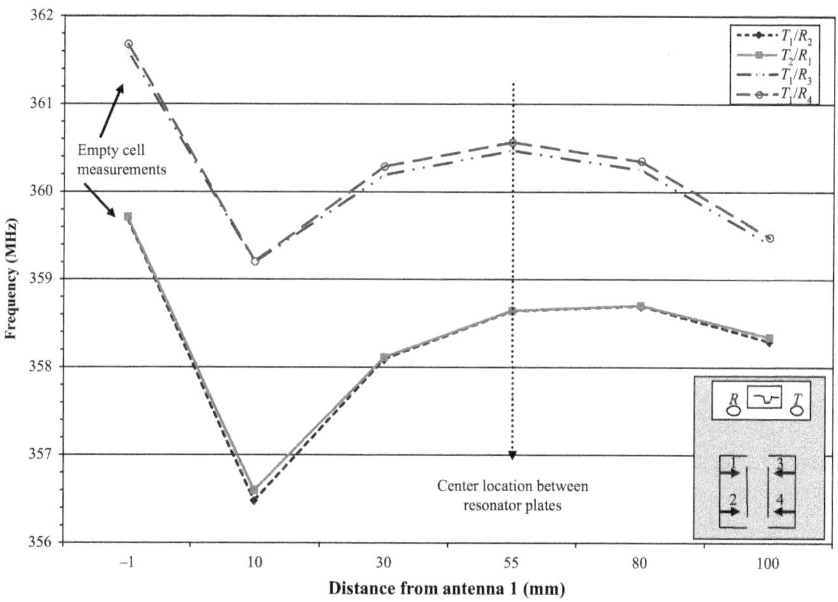

Figure 6.27 Effect of calibration sheet position on even-mode resonant frequency for different antenna locations configurations. T stands for transmission (load) port and R for reflection (source) port. The four configurations are shown in the upper right inset, whereas the ports and probes are shown in the lower right inset. The results were obtained for a 2.38-mm-thick white Delrin sheet

the same side of the sensor (T_1/R_2 or T_2/R_1), the response shows the asymmetry seen above in Figures 6.19, 6.20, and 6.26. These are the two curves in the lower part of the figure. Oppositely placed probes (T_1/R_3 or T_1/R_4) show a symmetric curve with the geometric and electric center coinciding.

In addition to improving the symmetry of the curve, the opposing connections improve the Q-factor as well as can be seen in Figure 6.28. The configuration T_1/R_3 has particularly large Q-factors, followed by the T_1/R_4 configuration.

Although one can argue that this asymmetry may not be important, since, once the fabric is located in the center of the cavity, it stays there, in fact symmetry in response is rather important. In particular, the method we used to eliminate the effect of fluttering assumes a symmetric response to deviations from the center plane. For this reason, the ideal configuration is to connect one probe as the source (say probe 1 in the inset in Figure 6.27) and connect the opposite probe on the same side, that is, probe 3 as the load.

All calibration measurements shown above were static, that is, the various sheets were placed in the sensor and the resonant frequency measured. This established the necessary calibration data and curves. In the following chapter, we

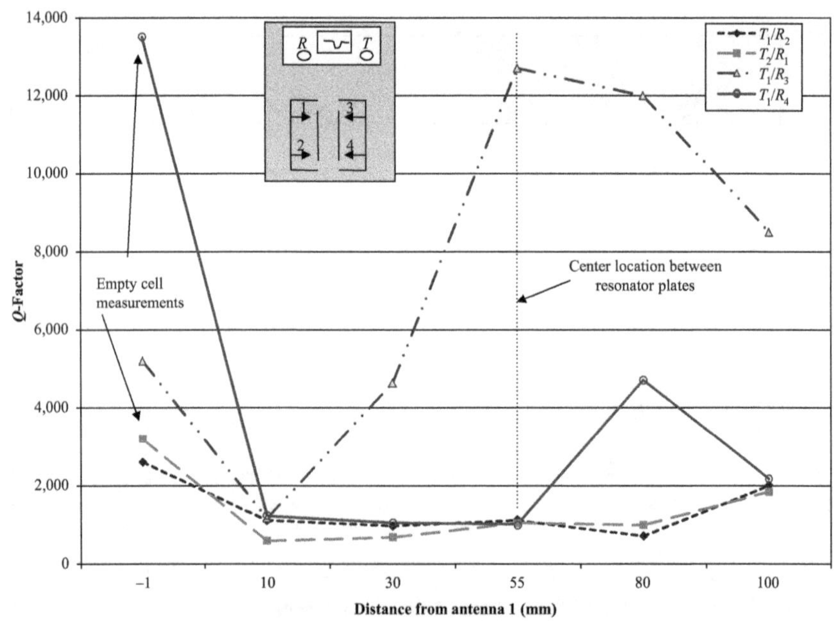

Figure 6.28 Effect of calibration sheet position on Q-factor for different antenna locations configurations. T stands for transmission (load) port and R for reflection (source) port. The four configurations are shown in the upper right inset, whereas the ports and probes are shown in the upper left inset. The results were obtained for a 2.38-mm-thick white Delrin sheet

revisit some of the calibration issues discussed here in the context of a production sensor. That will become necessary since the calibration sheets (which we will call "calibration standards") must be mounted in a frame to be held in position, and the frames mounted to the sensor structure. Influences of the frame and the motion of the cavity resonator must be taken into account and errors due to all of these factors established.

Chapter 7

Implementation and testing

7.1 Introduction

Following the design and simulation of the sensors, the implementation poses its own challenges, both mechanical and electrical. These have to be resolved, and their effects on performance must be evaluated. This chapter discusses the implementation of the sensors followed by testing. Much of the functional testing was reported in Chapter 6 and will not be addressed here. However, microwave systems are very sensitive to mechanical issues, and these will be addressed here. In particular, resonant systems, because of their high sensitivity to variations in volume of material, position of the tested material within the cavity, and, of course, variations in permittivity, require special attention to mechanical structures that support and facilitate the measurement.

As was discussed in the previous chapter, the prototype sensor, while certainly performing very well, was a fairly flimsy affair. The ground plates as well as the center plates were made of 1.6-mm aluminum sheet. The center plates were held in place with nylon bolts and the antenna probes held with rubber grommets through the ground planes. A much sturdier sensor is necessary for the industrial environment. In addition, the prototype sensor was stationary—it only evaluated a narrow strip of the fabric. The production sensor needs to cover the whole surface of the fabric.

The construction of the sensor must be sturdy while being nonintrusive, and the motion must be accurate and smooth. Some key details of the structure and its operation are described next for the fabric sensor. The rubber thickness sensor presents the same issues but since it is made of one half-cell, its implementation is essentially the same as that of the fabric sensor and there is nothing particularly different that requires a separate description.

7.2 The mechanical system

Following initial experiments with multiple stationary sensors and switches (see Section 5.2.5), it was realized that a single sensor moving back and forth across the fabric is a more practical solution even though, at first glance, it seems much more complex. There are a number of reasons that the single moving sensor was adopted. Multiple sensors are not likely to be identical leading to variations in

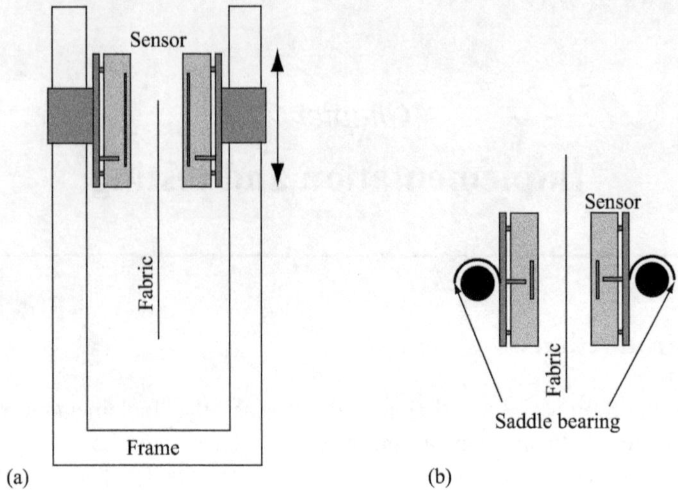

Figure 7.1 Schematic of the production sensing system: (a) top view with the sensor overlapping part of the fabric width and (b) front view showing the bearings

signals and switches are not only expensive but proved to be unreliable and short-lived. A moving sensor has the advantage of simplicity in sensing but with the addition of the complexity of the motion mechanism. This was deemed an acceptable disadvantage since it is an issue of design rather than maintenance, that is, once the mechanical design is complete and satisfactory, it should function properly with very little maintenance. On the other hand, the issues of multiple identical sensors (physical dimensions, separation of the cells, and variations in alignment) and in particular the cost and downtime related to replacement of failing switches are major maintenance problems.

The structure starts with a steel U-frame that allows the sensor to be placed around the fabric without the need to modify the production line [Figure 7.1(a)]. That is, the whole sensing mechanism can be built and then moved into position. It can also be removed as necessary without affecting the production line. Each half-cell is physically connected to a sliding bearing as shown in Figure 7.1(b). For stability and rigidity, the half-shells are connected to a thick 19-mm aluminum plate. The plates are welded to the sliding bearings, and these ride on the U-frame [Figure 7.1(b)]. A guiding structure keeps these plates parallel so that the distance between the center plates is constant throughout the travel. Each of the plates is driven by a screw mechanism rotated by stepping motors. The two half-shells are kept opposite to each other with zero offset by properly driving the two motors. The need for two separate drives is dictated by the need to keep the frame open so that it can be slid into place around the fabric.

In addition, two calibration sheets are included, one at each side of the fabric as shown in Figure 7.2. One is white Delrin and serves for calibration for the nylon

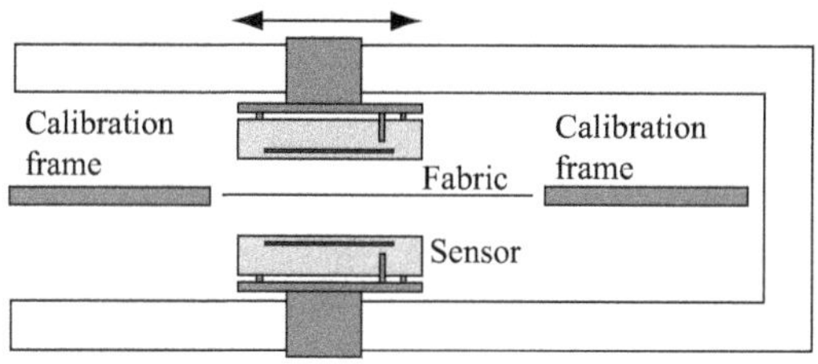

Figure 7.2 Top view of the sensing system with the two calibration frames included. The motion of the sensor allows calibration on either side depending on the fabric material

fabrics, whereas the second is black copolymer and serves as calibrator for the polyester fabrics (see Section 6.6). The travel of the sensor extends beyond the fabric on each side so that calibration can be performed on either calibration sheet. The sheets are thin (2.38 mm) and are 600 by 600 mm held in a nonconducting frame to ensure they are centered with the sensor. Since they are placed vertically, there is no concern with sag. Vibrations are well under control since the frames are attached to the same structure as the sensor mechanism, and calibration occurs after the sensors have stopped in the calibration position (centered with the calibration plates).

The sensor itself is a modification of the prototype sensor described in Chapter 6. The overall dimensions are kept the same, but the ground planes have been bent to form an open box (see also discussion in Section 5.2.5). In actual construction, the box was made of 6-mm-thick aluminum with the sides welded as can be seen in Figure 7.3 or Figure 7.4. Each half-shell was attached to a 19-mm plate (see Figures 7.1, 7.2, and 7.5) with four bolts. The purpose of the plate was to provide a convenient way to attach the sensor to the bearings [see Figure 7.1(b)] and guarantee the two half-shells are parallel to each other. It also added stiffness to the sensor as it moves and serves as a guide to ensure the half-shells stay parallel during the motion. In case service is needed, the sensors can be detached and reattached without interfering with the probes or the center plates by simply detaching the coaxial connectors and the four bolts holding the plate. The half-shells are removed together with their plates.

The center plates were of particular concern in the design as well as in the construction. Keeping the dimensions of the prototype, they were made much thicker (6 mm) and were plasma coated with either a smooth or a rough coating as mentioned in Section 6.5.3. The coating was added to facilitate cleaning of possible splatters of dip material. In tests, that was never an issue, primarily because the plates are placed in a vertical position parallel to the fabric. The plates are attached to a solid Delrin block (see Figures 7.3 and 7.6) while still allowing space for the antenna probes. The plates were attached with four recessed screws to the Delrin block and the block attached with two bolts to the ground plane (see Figures 7.6 and 7.7). Note also that the block sits in a channel (Figure 7.3) to ensure it does not shift. One of the concerns with the center plates was the possibility of the fabric

Figure 7.3 *Detail of the probe. Note the insulation at the aluminum shell. The probe is adjustable through a screw mechanism on the outside of the shell to allow for optimal coupling with the plate. Note also the rounding-off of the center plate designed to minimize the change of the fabric snagging on the plate. The channel is designed to keep the Delrin block in place with minimum chance of displacement should the center plate snag the fabric*

Figure 7.4 *The upper half-sensor and its attachment to the motion mechanism. Note the construction of the ground plane and partial shield afforded by the bent plates. The second half-shell has been moved away for a better view. The Delrin calibration frame can be seen in the lower part of the picture away from the half-shell*

Figure 7.5 *The two halves of the sensor placed over one of the calibration frames. The two calibration frames are located on the two extremes of the sensor's travel (see Section 7.4). Note how the two halves are attached to the motion mechanism at the top and bottom. The driving mechanism cannot be seen—it is hidden behind the guiding strips. One of the probes and the upper center plate with its Delrin support can also be seen above the calibration frame*

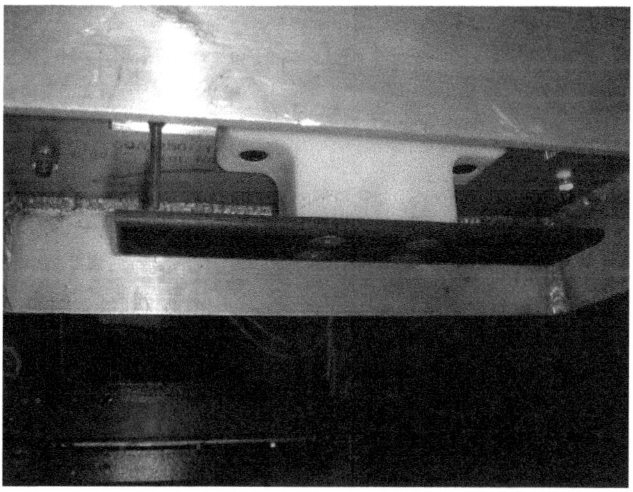

Figure 7.6 *Upper half of the sensor. The central plate is held in place on a Delrin block with four recessed screws to allow for a smooth surface. The Delrin block is solid and held in place in its channel with two bolts*

Figure 7.7 View of the upper half-shell from below showing the center plate attachment as well as two of the four bolts that hold the ground planes to the movement mechanism. Note the black plasma coating on the center plate. The center plate is held onto the Delrin block with four long bolts

snagging on its corners. Therefore, the edges were rounded (see Figure 7.3). The Delrin block seen in Figure 7.6 is fairly massive and the plate thick (6 mm) to help eliminate any possible vibrations of the central plate. Above the left edge of the plate is one of the probes (the other is on the opposite half-sensor). Note the location of the probes—it corresponds to the optimized location as discussed in Chapter 5. On the right side of the plate, there is a hole for another probe if the two probes are placed on the same half-sensor. Note also the gap between the end of the probe and the plate. This gap is adjustable from the top of the plate for optimal coupling. The gap is about 15 mm.

The antenna probes were mounted on opposite half-shells as can be seen in Figures 7.3 and 7.8. This configuration was found to be most beneficial primarily due to symmetry in response to off-centered fabric, a property used to minimize the effects of fabric flutter (see Section 6.5.2).

Figure 7.5 shows the two half-shells over one of the calibration frames. It also shows the guiding structure and the solid plates to which the half-shells are attached although the motors and the drive screws are hidden. The motors are also shielded to avoid any influence on measurements. These are stepper motors of considerable power necessitating shielding. Figure 7.9 shows a head-on view of the sensor over the calibration frame. Although the frame is thick (about 80 mm), the calibration sheet is only 2.38 mm thick and centered in the frame. Figures 7.3 and 7.6 show details of one of the half-shells complete with the center plate and the antenna probe. The complete empty cavity is shown in Figure 7.8.

Figure 7.8 The lower half-shell showing the lower center plate and the load probe under the left corner of the center plate. The support and construction is identical to that of the upper half-shell

Figure 7.9 Head-on view of the sensor with the calibration frame between the center plates. Note that the calibration sheet is very thin. The dark band at the center is the frame holding the calibration sheet

The sensor is connected to the network analyzer with two long cables that ride onto a flexible track to control the bending radius and to ensure minimum strain on the cables.

Overall, the requirements from the mechanical system are as follows:

Rigidity. Because the sensor detects any variation in its volume, any deviation from the rigid sensor (vibrations, offsets of any kind, changes in distance between the two half-shells) will change the resonant frequency and should be viewed as errors.

Accuracy. The motion must be such that the two drives move the two half-shells so they remain parallel to each other and that there is no offset between the two halves. This is normally taken care by the step size of the stepper motor, but it also involves any backlash in the screw driving mechanism. Any offset can be corrected by the stepping mechanism and that same mechanism allows introduction of a fixed offset if that becomes necessary, all under software control. Because the two half-shells are driven independently, they can be moved one at a time for purposes such as maintenance and cleaning.

Repeatability. Even more than issues like zero offset, repeatability is a key to a successful sensing system.

7.3 Evaluation of the mechanical system

Much of the usefulness of a sensor of the type described here relies on the rigidity and accuracy of the mechanical system especially since the sensor itself is relatively simple. The sensitivity of the sensor (Q-factor, frequency resolution of the network analyzer, etc.) contributes to this dependency on the accurate motion of the senor. Fortunately, the effects of the mechanical system on the output can be evaluated with the sensor itself, that is, the resonant frequency of the sensor can be used to measure the performance of the mechanical system. It suffices to run the sensor and look at the changes in resonant frequency and associate them with specific aspects of the mechanical structure. A number of effects can be expected, and these are quantified next. First, because the frame itself is not perfectly aligned, variations in the resonant frequency due to motion at different location along the frame can be expected. Then, the effects of ancillary structures such as motor/driver covers, brackets, and bearing housings are also likely to change the resonant frequency if they are in sufficient proximity to the sensor. In addition, the measurement of the resonant frequency can be done in one of two modes: one can stop the motion, measure the resonant frequency at a location across the fabric, then move to a new location and repeat the process until the fabric width has been covered and then repeat indefinitely. Instead, one can let the sensor run continuously at a constant speed and measure the resonant frequency at fixed or variable time intervals. Each method has its advantages although in the final product, continuous motion is more practical.

Figure 7.10 shows the sensor response as it moves in the region between the two calibration frames (where the fabric would normally be). In this test, the bearing housing is not installed and LMR195 Ultraflex cables are used. Two tests are performed. In one, the sensor is stopped every 50 mm to take a measurement as

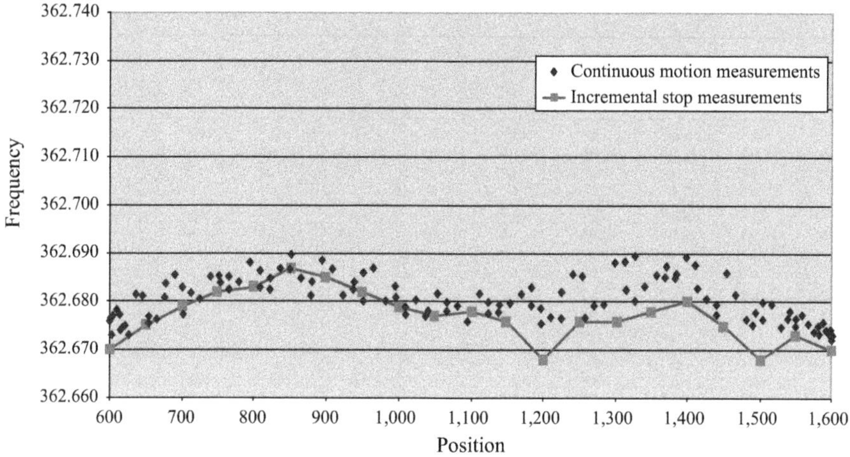

Figure 7.10 *Test showing the variation in resonant frequency due to imperfections in the mechanical system. The bearing housing cover is not installed. The cables to the network analyzer are LMR195 Ultraflex cables. Stop-and-go and continuous motion measurements*

shown by the continuous curve. The second test shows measurement taken at much shorter intervals while the sensor is in motion. In both cases, the maximum variation is 20 kHz, but the scatter in the continuous measurement is higher. This may actually represent the conditions that exist better than the stop and go measurement, that is, the variations along the test path, from all sources are more frequent than the continuous line shows.

Figure 7.11 shows the same test as in Figure 7.10 but with the lower bearing housing installed and with Thermax RG316 cables connecting to the network analyzer. The test shows three peaks with a maximum variation of over 50 kHz. Although the cables are different, the reason for the changes is not the cable since an identical test with the same cables as in Figure 7.10 (LMR195 Ultraflex) shows results similar to those in Figure 7.11. These are shown in Figure 7.12. It turns out that the changes are due to the metal lower bearing cover which influenced the resonant frequency. These variations can be seen in the stop and go measurements and in the continuous run measurements in both figures.

The variations seen in Figure 7.10 must come either from changes within the volume of the cavity, from outside influences, or from variations of the distance between the two half-shells of the sensor. Since the material in the cavity is air and after eliminating the possibility of outside influences, the distance between the half-shells was measured throughout the travel span of the sensor at intervals of 50-mm travel. The sensor was stopped and in each step the distance between the sensor's half-shells was measured at two locations (top and bottom) and at two different frequencies. The results are shown in Figure 7.13. The maximum variation in distance between the shells is 0.16 mm with a change in frequency of about 12 kHz.

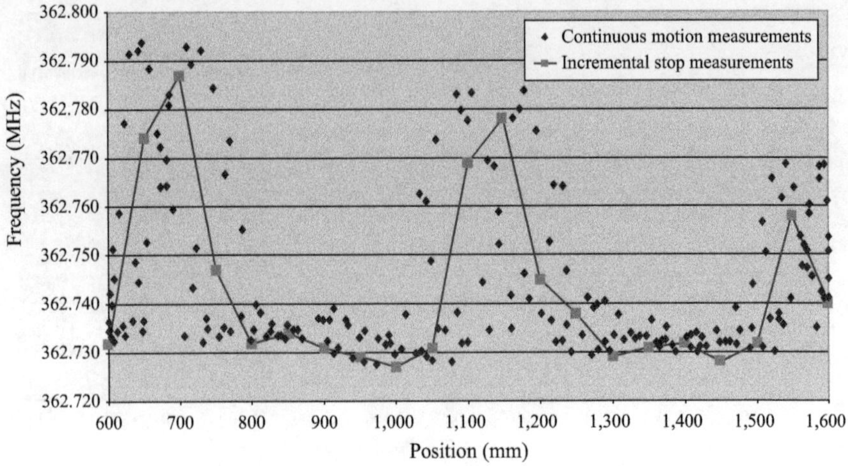

Figure 7.11 Test showing the variation in resonant frequency due to imperfections in the mechanical system. The bearing housing cover is not installed. The cables to the network analyzer are Thermax RG316 cables. Stop-and-go and continuous motion measurements

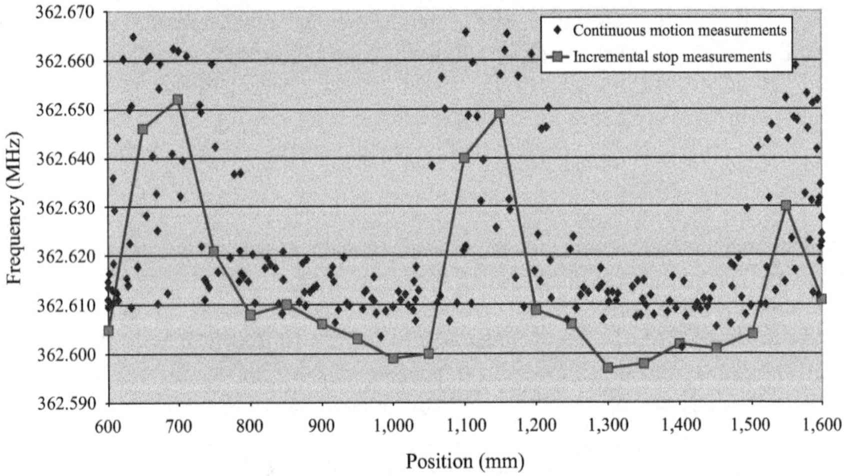

Figure 7.12 Repeat of the test in Figure 7.11 with the LMR195 Ultraflex cables to eliminate the cables as the source of the peaks. Stop-and-go and continuous motion measurements

This is lower than the result in Figure 7.10, which shows a change of about 15 kHz for the stop and go measurement. It is likely that the stop-and-go method introduces additional vibrations especially since the motors used to drive the system are high torque motors.

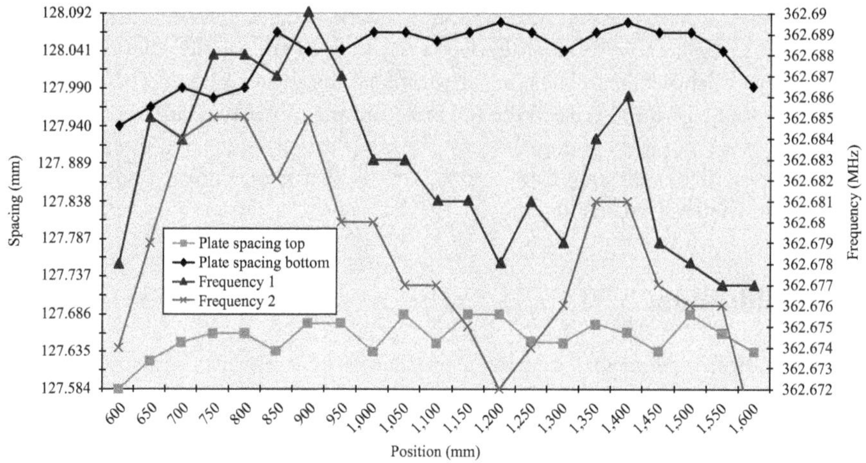

Figure 7.13 Variations in spacing between the top and bottom of the two half-shells and the resonant frequency variations along the path of the sensor's motion. Note the difference between the spacings but also that these variations remain more or less constant

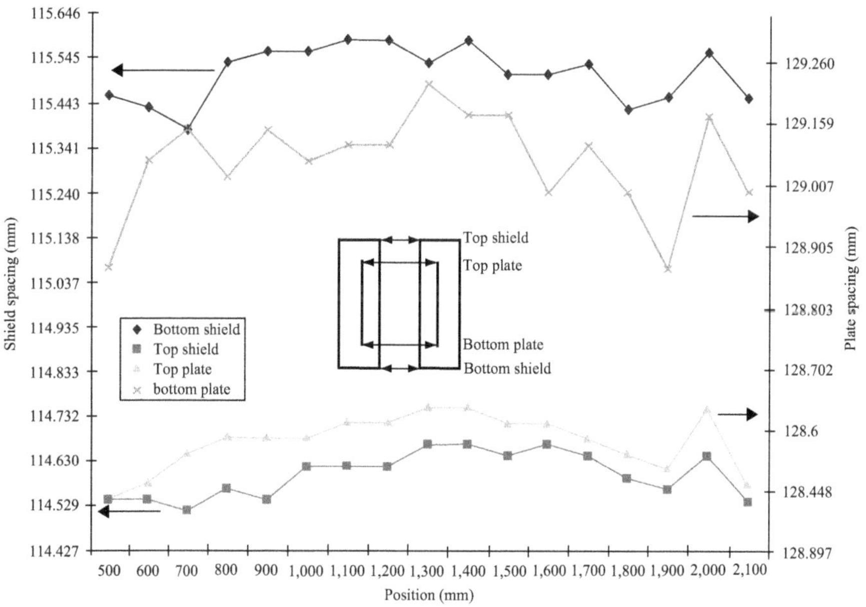

Figure 7.14 Measurement of the spacing between the two half-shells showing a difference of about 14 mm between top and bottom. The difference remains more or less constant along the path of motion

Figure 7.14 shows actual measurements of the spacing between the two half-shells. Measurements were taken between the tops of the shields, bottoms of the shields, tops of the center plates, and bottoms of the center plates. There is a difference of about 14 mm between the top and bottoms of the two half-shells, but the difference does not vary along the path. Since the variations are constant, they really do not matter although, of course, this is just a question of adjusting the positions of the two half-shells.

7.4 Calibration

The calibration method was discussed in Chapter 6. It consists of using two thin sheets of material to establish the calibration curves. One is a white Delrin sheet used for the calibration curve for nylon fabrics, the other a black copolymer for the calibration curve for polyester fabrics. In the production sensor, these "standards" were mounted in nonconducting frames centered with the sensor at each end of the sensor mechanism so that the sensor can be centered over one or the other standard (see Figure 7.2). However, since the frames are relatively small and hence in close proximity to the sensor, concerns of the influence of the frames on the resonant frequency had to be addressed. Furthermore, the frames had to be attached to the steel frame, and concerns of the influence of the steel frame at the end of travel of the open resonator sensor had to be addressed as well. The calibration standards were 610 by 610 mm so that with the length of the cavity equal to 500 mm, only 55 mm on each side of the cavity is available. The frame holding the standard took more than half of that leaving only 22 mm between the cavity shield and the standard's frame.

One of the effects of the calibration frame has on the performance of the system is shown in Figure 7.15. In this case, the far frame (the one closer to the metal structure in Figure 7.2) is placed with its near edge (the one closer to where the fabric would be) at 1,843 mm. The test, made in air, shows that as the sensor approaches the frame edge, its resonant frequency starts decreasing slightly about 50–75 mm before the edge of the sensor's shield reaches the edge of the frame. The reference scan, shown as the horizontal line with white diamond indication, is a scan in air without the calibration frame and its support. While the change is not large, indicating that the external influence on the resonant frequency is small, it does indicate that a minimum distance between the edge of the fabric and the frame of about 100 mm must be maintained. This also indicates that when scanning the width of the fabric, the sensor is only expected to produce accurate results for scans starting 100 mm from one edge and ending 100 mm from the other edge of the fabric.

The test in Figure 7.15 was done for a number of conditions as shown in the figure, including with a metal brace for the frame (used to ensure a minimum rigidity) as well as with antenna probes either on opposite sides of the sensor or on the same side.

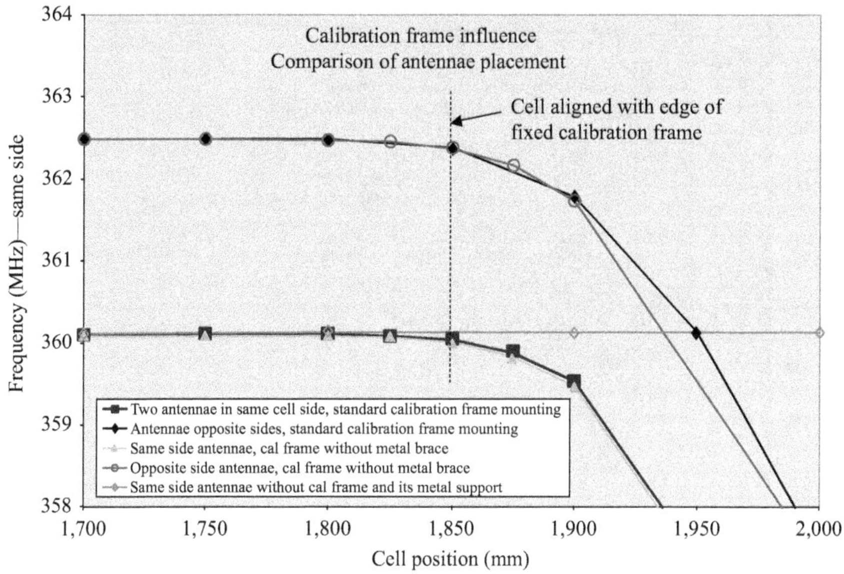

Figure 7.15 Effect of the calibration frame on the resonant frequency for various conditions. The reference frame edge is at 1,843 mm and starts affecting the resonant frequency within a range of about 50–75 mm

Whereas the metal brace used to hold the frame does not affect the resonant frequency as the sensor approaches the frame, it has a marked effect when the sensor moves over the frame as one would expect. Figure 7.16 shows the effect. With the metal brace, the frequency increases because of the reduction in the effective volume of the cavity. As the sensor advances, the effect changes because of the nonuniform distribution of the fields in the cavity with larger increases when the metal brace is under either edge of the center plate and lower increase when around the center of the center plates. Note that the figure only shows a small portion of the travel. The sensor is 350 mm wide and starts entering the frame at 1,843 mm. Its sensing position for calibration is in the center of the frame which corresponds to $1{,}843 + 350 + 50 = 2{,}243$ mm and the far edge is then at $2{,}243 + 350/2 = 2{,}418$ mm. The most important aspect of this test is that the frame must be either free standing or braced with nonconducting materials. The two arching curves show the effect of the frame itself (i.e., a freestanding frame) on the resonant frequency. This is the expected result indicating that the frame itself has a small effect once the edge of the frame has cleared the sensor. As with the results in Figure 7.15, the tests were carried out with both antenna configurations.

Figure 7.17 is an expanded view of the tests in Figures 7.15 and 7.16. The initial steep drop that starts at around 1,900 mm occurs as the sensor's edge passes over the frame. Once the trailing edge clears the frame, the test without the metal brace stabilizes at the actual resonant frequency of the standard in the frame

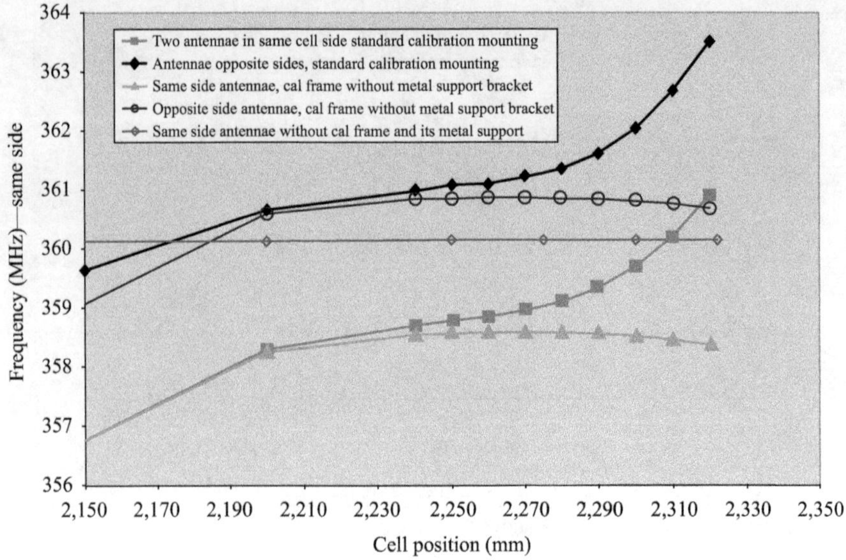

Figure 7.16 Effect of the calibration frame with and without the metal brace on the resonant frequency

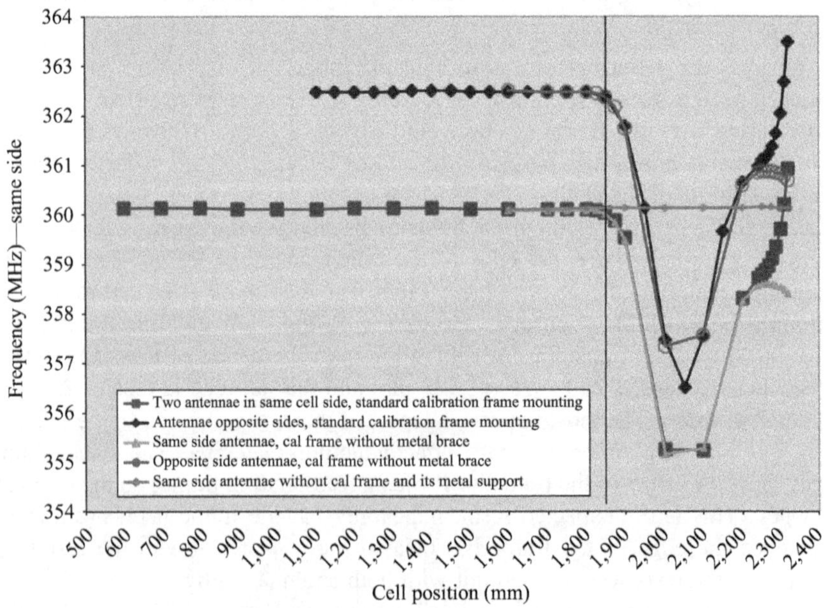

Figure 7.17 An expanded view of Figure 7.15 and 7.16. The steep drop in the resonant frequency is due to the frame encountering the leading edge of the resonator. The rise in resonant frequency occurs as the resonator clears the frame

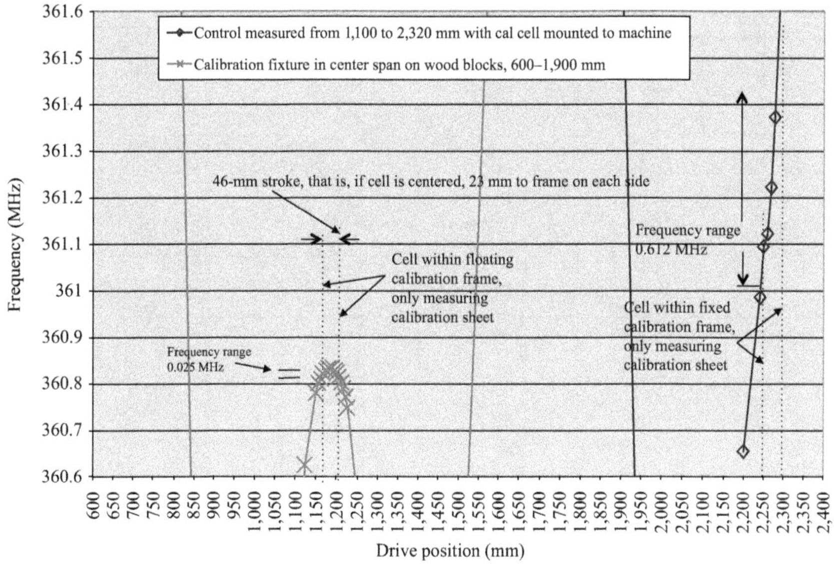

Figure 7.18 Errors in resonant frequency due to bracing of the calibration frame

(around 2,240 mm) for either the opposite antennae or the one-sided antennae. As the sensor continues past that point (the center of the calibration frame), the resonant frequency increases quickly if the metal brace is present or stays more or less constant, with a slight drop if the metal brace is absent (as can be seen more clearly in Figure 7.16).

To get a better feel of the effect of the metal brace vis-à-vis the unbraced frame, consider Figure 7.18. In this figure, the maximum change in resonant frequency as the sensor moves within the frame, that is, as the sensor moves from one edge of the frame to the other was measured. The gap between the sensor edge and the frame is only 22–23 mm on each side when centered. The figure shows the change in frequency within that span. For the unbraced frame (it simply lies on wood blocks or attached with nonconducting members), the maximum variation in resonant frequency is 25 kHz. This variation is due to the proximity of the frame to the sensor. With the metal brace, the maximum variation in the resonant frequency is 612 kHz. While it is clear from the results in Figures 7.15–7.18 that metal bracing or, for that matter, any metal in close proximity to the sensor produces unacceptable errors, nonmetal bracing also produces errors, albeit acceptable in magnitude and of the same order of magnitude as the errors produced by the motion over the length of the structure.

Another issue is associated with calibration repeatability. Because the sensor must be properly centered within the frame (so that it is at equal distances from the frame on all four sides, or at least on the long sides) and because there is at most 22–23 mm distance to the frame when properly centered, any deviation from these

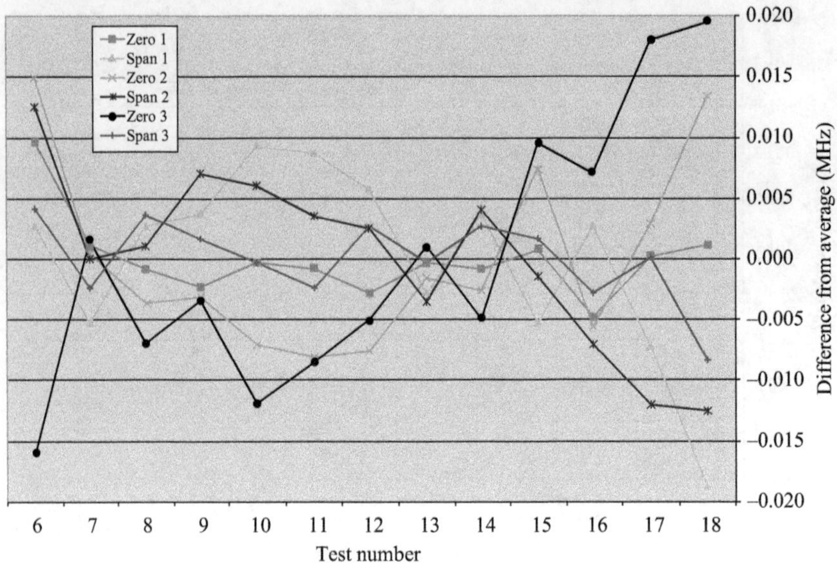

Figure 7.19 Repeatability test on the calibration standard

precise positioning requirements can change the resonant frequency. Since the sensor must be stopped for calibration and then started again, moving in the opposite direction, it is likely that some slack or mechanical hysteresis will be present introducing an error in the calibration. To evaluate any possible errors due to positioning within the calibration frames, three separate sensors were tested and each was tested 13 times to evaluate the repeatability of the mechanical system. The resonant frequency of the start and the stop of the calibration position were measured, and the average resonant frequency subtracted to obtain the change in resonant frequency due to the variations in the start and stop positions. These variations are attributed to minor slack (hysteresis) in the driving mechanism and, perhaps, in the position of the two half-shells. The results shown in Figure 7.19 indicate a maximum change in resonant frequency of less than 20 kHz in most cases. One test shows a maximum variation of 30 kHz. These changes are consistent with the changes seen for variations along the entire path of the sensor (see Figure 7.10).

In Chapter 6, we have seen that the resonant frequency is affected by the position of the antenna probes (see, e.g., the discussion on the effect of flutter). It is therefore necessary to also see if the position of the probes affects errors during the motion of the sensor. To do so the sensor was run with both probes on one half-shell and with opposing probes (one probe on the left half-shell, the second on the right half-shell). The tests were run in air and, separately, with the polyester calibration sheet. These are shown in Figures 7.20 and 7.21. Figure 7.20 shows the resonant frequency in air and, consistent with the results in Chapter 6, the changes

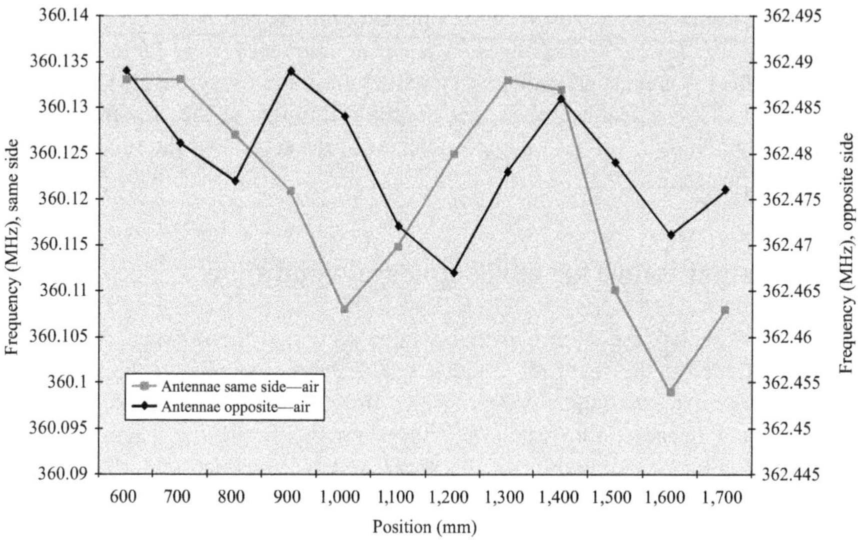

Figure 7.20 Sensitivity to position of the antenna probes, test in air. The variability is higher for the two probes on the same side of the sensor

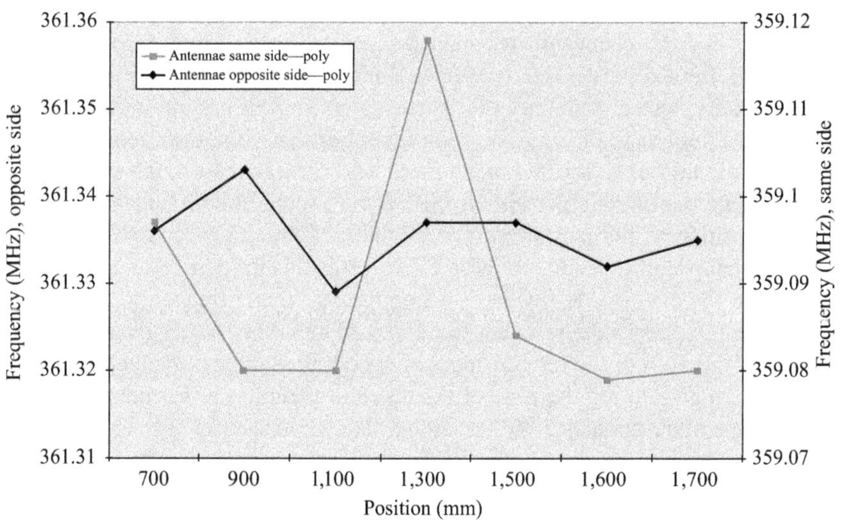

Figure 7.21 Sensitivity to position of the antenna probes, test with the polyester calibration standard. The variability is higher for the two probes on the same side of the sensor

in resonant frequency are lower for the opposing probes but, at the same time, the Q-factor is also lower. The changes in resonant frequency were larger when the polyester sheet was present, again with a reduction in the Q-factor. This sensitivity to location of the antenna probes was observed throughout the experiments and resulted in placing the probes on opposite sides of the sensor for the final design to reduce overall errors.

7.5 Compensation for environmental conditions

As can be expected, the environmental conditions within the resonant cavity must have an effect on the resonant frequency. Of these, the relative humidity and temperature are of particular concern since they are expected to fluctuate continuously and because both can affect the resonant frequency. There are other effects including contamination of the space and the conducting surfaces, all of which will affect the resonant frequency. The effect of relative humidity is fairly obvious—water and water vapor have a high relative permittivity and hence will change the resonant frequency. The effect of temperature is due to the change of the relative permittivity of air (see Appendix B) and of water vapor.

To see the effect and to quantify the errors, the resonant frequency of the cavity was measured over a period of time, while, at the same time, the temperature and humidity were recorded and correlated with the resonant frequency. Both the even and odd frequencies were measured since the odd-mode frequency can be used to compensate for the change in resonant frequency due to environmental effects. In Section 5.2, we discussed the fact the odd-mode electric field is perpendicular to the fabric being tested, and since the fabric is very thin, it has a minimal effect on the resonant frequency of the cavity. On the other hand, the even mode, which is parallel to the fabric, is affected much more, and for this reason, the even mode is used to gauge the fabric's permittivity. Both the even and odd modes are roughly equally sensitive to the permittivity of the bulk of the cavity and will change in roughly equal proportions due to these effects. It is this property that allows one to compensate the even-mode frequency using the odd-mode frequency.

Figure 7.22 shows the resonant frequency of an empty cavity over a period of 15 h out of the 16 h of the test, during which both the temperature and the relative humidity change. During the time of the test, the variation in humidity is 2% and that in temperature about 1.5 °C. Assuming that these are the only environmental changes during that period (no other change has been observed during testing), the change in resonant frequency can be attributed to these changes alone. The resonant frequency changes by a maximum of 186 kHz during the 16-h period. The measurement was taken at 5 s intervals and showed some spread although it is rather small—the maximum spread is about 20 kHz. One can safely assume that some of this spread is due to variations in conditions but some of it seems to be random.

To separate the effects of temperature and relative humidity, it is useful to inspect Figure 7.23. In this figure, the resonant frequency, relative humidity, and temperature were plotted, each normalized with respect to its maximum value in

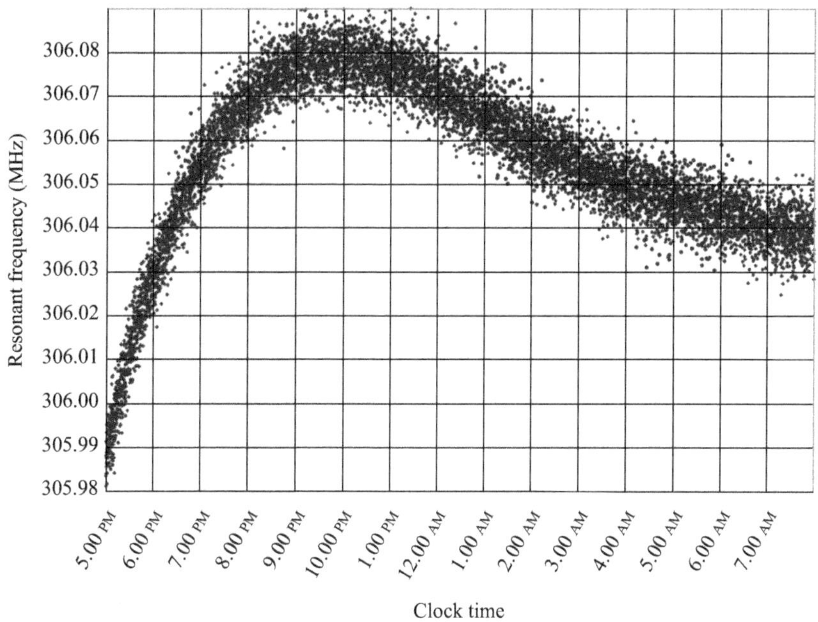

Figure 7.22 The even-mode resonant frequency due to changes in temperature and relative humidity measured over a period of 15 h

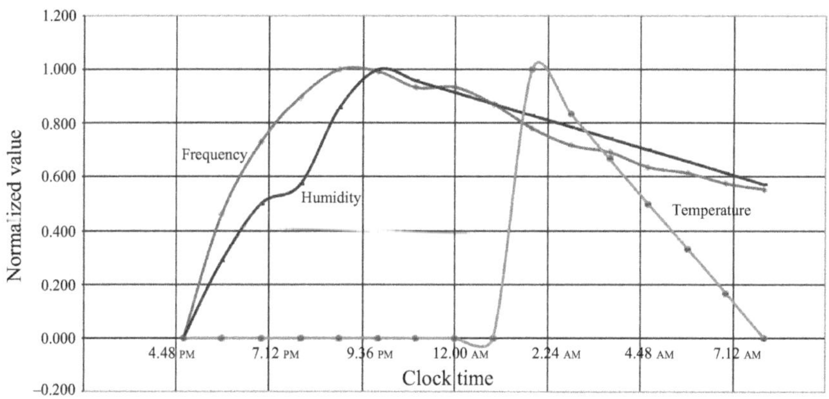

Figure 7.23 Resonant frequency, relative humidity, and temperature over the test period. The values are normalized with respect to maximum value recorded and a third-order polynomial fitted over the values to obtain a continuous curve

the test period. It is clear from the figure that the resonant frequency follows the change in relative humidity exactly, whereas the change in temperature does not seem to have any visible effect. This of course is expected based on the relatively high relative permittivity of water and the relatively weak dependency of the permittivity of air on temperature.

Note: in Figure 7.23, the change in resonant frequency was modified from negative to positive to show the trend, that is, as the relative humidity increases, the resonant frequency decreases, whereas in the plot, they trend the same way.

The error in the resonant frequency is approximately 186/2 = 93 kHz/%RH. This is a very large error and cannot be left alone. A change of 10% in relative humidity would change the resonant frequency by almost 1 MHz. This is particularly concerning because one can easily expect relative humidity to vary between, say, 30% and 80% at any time of the year and even more in extreme cases. Variations can occur in short periods of times or may be slow, changing over long periods of time. For this reason, it is critical that the odd-mode measurement be used to compensate for the effect of the relative humidity and any other bulk effect within the cavity.

7.5.1 Compensation method

There are two basic ways one can compensate for the bulk effects in the cavity. One is the use of the odd-mode resonance mentioned above. We will use this method here. It is however possible to generate a set of data similar to the data above which indicates the change in resonant frequency due to humidity and temperature and subtract that change from the measured resonant frequency. This is very simple and can be easily handled by the computer. It does, however, require the measurement of relative humidity, a measurement that can only be done outside the cavity. There are however many disadvantages of this method. First, it cannot take into account any other effects such as, say, contamination of the cavity or even as simple an effect as dew forming within the cavity or on its walls. Further, with time, the response of the cavity to environmental effects may change due to changes in alignment, etc., and this would require generation of a new curve for relative humidity. Finally, humidity measurements are slow and likely to lag behind the measurement of frequency.

On the other hand, the use of the odd-mode resonance for the purpose of compensation is almost ideal in the sense that it can compensate for any condition in the bulk of the cavity and there is no need for either external sensors or recalibration. Its only disadvantage is the need to measure the odd-mode resonance and that requires a wider scan of the frequency range since the odd-mode resonance is at a frequency higher than the even-mode (see Figure 6.2). This is a small price to pay, and the network analyzer is well equipped to handle this task.

Figure 7.24 shows plots of the even- and odd-mode resonant frequencies taken over the same period of time, and each sample was taken at identical times. Clearly, the two resonant frequencies behave almost identically although, because the odd-mode frequency is higher, the variation in the resonant frequency is also higher

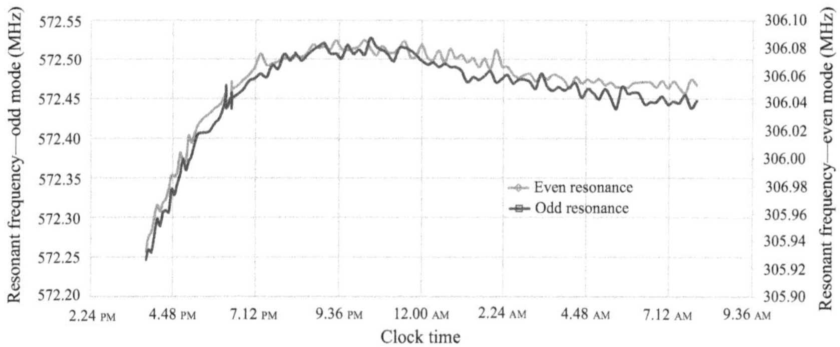

Figure 7.24 Odd- and even-modes resonant frequencies over time. The changes in the resonant frequencies are due to environmental conditions within the cavity

since, when the equivalent permittivity in the cavity changes, the resonant frequency changes by a factor proportional to the permittivity. In fact, the maximum change in frequency over the period of measurement for the even mode is 186 kHz, whereas for the odd-mode resonance, it is 322 kHz. However, if one scales this change by the ratio of frequencies, that is, by the ratio 305/572, one obtains 322 * 306/572 = 172 kHz. Although the two frequencies do not change by exactly the same ratio, this is close and hence this change can be used for compensation of environmental conditions in the cavity.

The compensation itself can take many forms. The simplest method is to take a fixed value as reference, say the first reading in the measurement sequence in Figure 7.22 or in Figure 7.24. Suppose that value is k. Now the compensated even-mode frequency is written as

$$f_{ec}^{n} = f_{e}^{n} - \left(f_{o}^{n} - k\right) * \frac{f_{e}^{n}}{f_{o}^{n}} \tag{7.1}$$

In this relation, n refers to the current measurement, c to the compensated value of the measurement, o to odd mode, and e to even mode. Applying this simple calculation to the data in Figure 7.22 and plotting the compensated values results in the plot in Figure 7.25. The maximum change in frequency over the test period is 48.8 kHz, resulting in an error of approximately 24.4 kHz/%RH. This is roughly 1/4 of the error in Figure 7.22. Looking at the averaged value along this plot, the maximum variation is about 20 kHz with an averaged error of about 10 kHz/%RH.

In practice, this error can be reduced further by updating k at regular intervals so that the compensation is done on a smaller range. This can be done in conjunction with the periodic calibration of the sensor (see Section 7.4). In this approach, k is updated immediately after the calibration is performed and the sensor starts measurements. For example, by updating k every 10 min, the maximum error is less than 4 kHz/%RH and is, in fact, smaller than the spread observed in

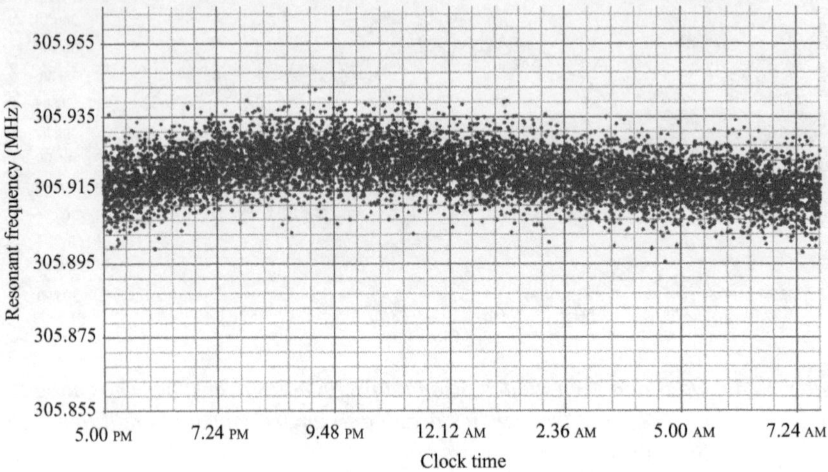

Figure 7.25 Compensated resonant frequency. The effects of relative humidity and temperature were reduced using the odd-mode resonant frequency

Figure 7.22. The method adopted in this work makes use of this compensation procedure to update the value of k periodically (approximately every half hour or so). Although it may seem that the choice of k is arbitrary, that is, that its value is affected by the relative humidity in the cavity, the use of the calibration frames removes this uncertainty since the relative humidity during calibration is taken into account at the calibration step. That is, after the calibration, the value of k is devoid of effects of relative humidity or any other bulk effect in the cavity since the permittivity of the calibration sheets is known exactly.

Although this method of compensations seem to be almost ideal, it is not. First, it assumes that the odd-mode resonant frequency is not influenced by the fabric but only by the bulk volume. That is only approximately true. Indeed, the odd-mode is largely insensitive to the fabric but not entirely so. This can be seen in Figure 7.25. We have also shown in Figure 5.6 that the odd-mode resonance changes with permittivity of the fabric, especially when the moisture content of the fabric is high (high permittivity). Therefore, the compensation method shown here can only be seen as a method of improving accuracy rather than eliminating the effects of environmental conditions. The second limitation of the method is that it cannot be applied to the rubber thickness measurement since there the odd-mode resonance has no meaning.

There are of course many more methods that can be used for this purpose, but the method described here is sufficiently accurate for the purpose.

Chapter 8
The network analyzer

8.1 Introduction

As was indicated in the introduction, one of the features of the current work is the incorporation of a network analyzer as part of the overall sensing system. Although a network analyzer is a very expensive piece of equipment often associated with laboratory work, its use in a system of this type is justified on a number of grounds, not the least being development time, accuracy, and stability. In the industrial environment, it also affords a single unit that can be replaced or repaired quickly without the need of specialized personnel, saving downtime, and, ultimately, costs. Of course, there are alternatives to this approach. These include separate oscillators and voltage-controlled oscillators (VCOs) that can then be related to the shift in resonant frequency. Although less expensive than the network analyzer, these alternative methods are also less accurate and in the context of this work were deemed insufficient. It should also be remembered that development of equipment is a lengthy process and even when individual building blocks exist, the integration and testing of the equipment is tedious. For these reasons and, of course, because of the accuracy afforded by the network analyzer, we decided to use a vector network analyzer (VNA) as a single integrated test piece in spite of the initial cost and the fact that only a small portion of the analyzer's capability is actually used.

Because of the central role the network analyzer holds in the sensors described in this work, this chapter describes briefly the structure of network analyzers and the measurements one can perform. The measurements described are primarily those needed for the present work and hence this description should not be viewed as a tutorial on the use of network analyzers.

8.2 What is a network analyzer?

The network analyzer is, perhaps, the most sophisticated and versatile (and expensive) measuring instrument for radio frequency (RF) analysis in existence and, as such, is not normally used or known by the average practitioner. However, at its very basis, it is a trivially simple instrument. To understand these statements, consider first the circuit (i.e., low frequency) approach to analyzing a network. Figure 8.1 shows a one-port network and the necessary instrumentation to analyze the network. Since all we have access to is the port, we can measure the current into

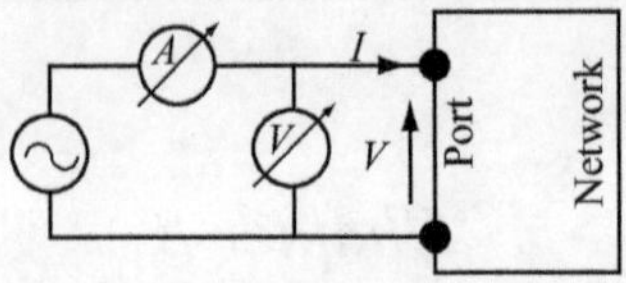

Figure 8.1 Measurement of input voltage and current on a one-port network. Impedance and power are calculated from the basic measurements

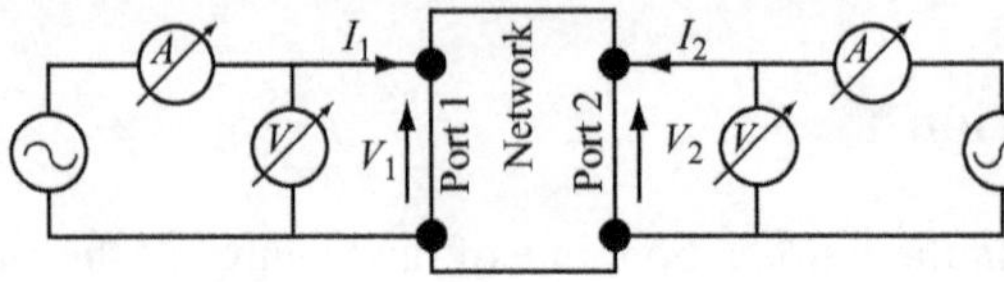

Figure 8.2 Measurement of input voltage and current on a two-port network

the port and voltage across the port. From these, we can calculate the impedance of the network and, with some additional effort, the phase between current and voltage, real power entering the port, reactive power, and so on.

If one considers a two-port network, then identical measurements can be performed on the output port (see Figure 8.2). Indeed, one can extend this to an *N*-port system—all that is necessary is to place appropriate measuring instruments on each port and, perhaps, have a facility to read them automatically and perform the analysis. Indeed, we can call the system in Figure 8.1 or Figure 8.2, a network analyzer provided that we add to it analysis tools such as a piece of software or, if one were to stick with the basics—a pencil and paper. It should be noted that the measurements in Figure 8.2 can be done with a single set of measuring instruments and a single source by switching them from ports 1 to 2. However, doing so, one cannot measure them at the same time and hence it is not possible to measure, for example, the voltage on port 2 when port 1 is driven. For this reason, network analyzers are specifically designed as two-port (or more) systems.

Of course, this is not what one would pay a prince's ransom for. A network analyzer is much more than this, but at the fundamental level, the picture drawn in the previous paragraph is useful to keep things simple and in context. At high frequencies, one cannot measure voltages and currents directly as we have discussed in Section 4.2. Rather, we can measure waves, or more specifically power, propagating, reflecting, and transmitting into and out of a network. Network analyzers, therefore, do not measure voltages and currents but rather power. That is, they make use of power detectors to measure incident, reflected, and transmitted power and from these evaluate the *S*-parameters as well as other quantities. Thus, as a second level of complexity, a network analyzer can be viewed as in Figure 8.3. The source is a variable frequency generator that can scan the frequencies of interest under controlled conditions of amplitude, phase, frequency, and spectral purity. The generator may be a true variable frequency source such as a VCO, but,

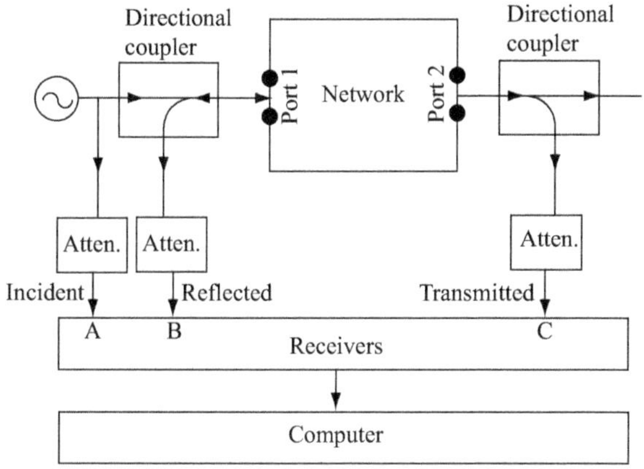

Figure 8.3 A simplistic block diagram of a network analyzer capable of detecting the incident, reflected, and transmitted signals due to a two-port network

more often, it is a synthesized source. The signal from the generator is split into a forward wave and an incident or reference wave. The forward wave is fed to the port through a directional coupler, whereas the reference wave is fed to the receiver (detector) directly to define the incident power. The directional coupler is used to allow the incident power to propagate to the input port but prevent the reflected power from the port to propagate back to the generator. Rather, the reflected power is coupled into the receiver through the directional coupler to detect the reflected power. On the transmission side, the transmitted power is fed into a second receiver (or a second receiving channel) to detect the transmitted power, again, through a directional coupler. The power detection is usually performed through a balanced bridge detector, a diode detector, or, in some cases, a small thermistor (see Section 4.5.2). The blocks indicated as attenuators allow for adjustment of signal levels as needed by the measurement. The detected signals are then analyzed in the onboard computer that can calculate the *S*-parameters and from them any other quantity calculable. The term "network" is a generic term, and it may represent any device or circuit analyzed. In one-port networks, only the incident and reflected signals are detected. Of course, things are much more complicated than this, and the designers of network analyzers had to worry about many issues including noise, losses, and accuracy, but the user does not need to worry about all of that. A two-port network analyzer is shown in Figure 8.3 and the extension to *N*-ports is shown in Figure 8.4. Note that the source is common to all ports and switched between them for measurements although, for two-port network analyzers, one can also use a split source rather than a switched source. Each port is served by two channels, a reference channel and a measurement channel. The ADC (analog-to-digital converter) and digital signal processing units process the data locally before transferring to the computer.

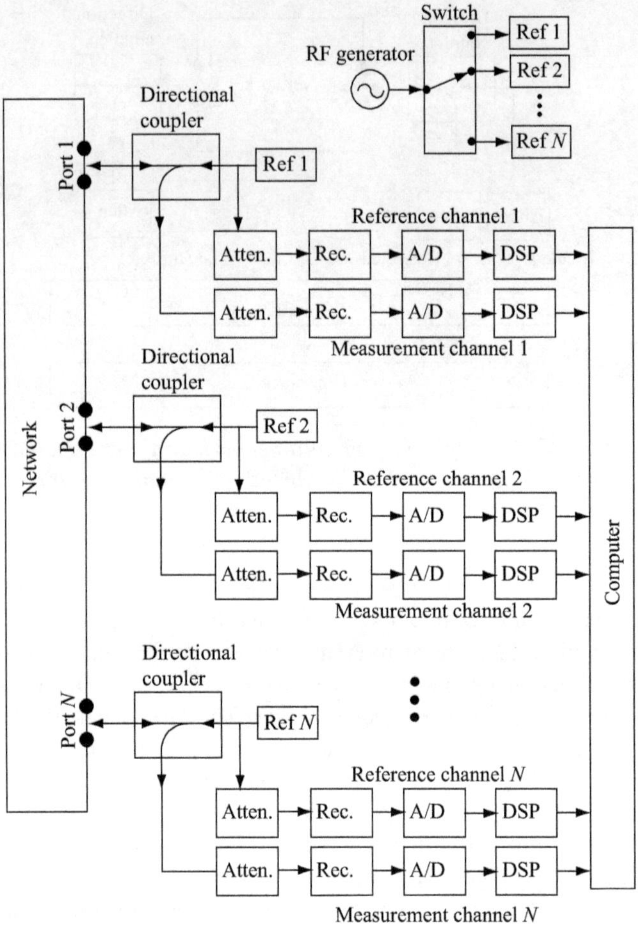

Figure 8.4 Block diagram of an N-port network analyzer

As was alluded to previously, the instrument is much more complex than the description above reveals. Issues of signal losses, mutual coupling between ports, directivity and isolation of signals, frequency range, speed of analysis, signal purity, attenuation, phase changes, and many others had to be resolved before a network analyzer could become the sophisticated instrument it is today. Also, all control, analysis, and computation of parameters is done with an integrated computer in the network analyzer so that one can compute, analyze, and display results in many forms as well as store and archive results and testing parameters as needed. However, these are issues that are specific to particular instruments, whereas here our intention is to discuss the use of the network analyzer rather than its construction and its many features.

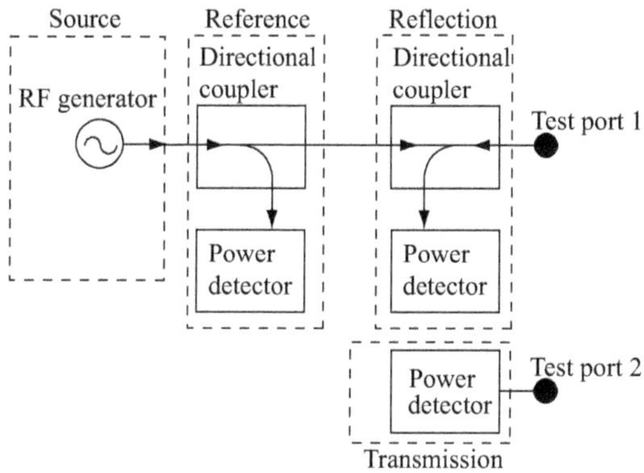

Figure 8.5 Simplified block diagram of a scalar network analyzer (SNA)

8.2.1 Scalar and vector network analyzers

There are two basic types of network analyzers—scalar network analyzers (SNAs) and vector network analyzers (VNAs). In very simple terms, an SNA only measures the amplitude of waves, whereas the vector instrument measures both amplitude and phase. Behind this broad distinction, there are many important differences that ultimately define the way these instruments are used. A simplified diagram of the scalar analyzer is shown in Figure 8.5. The reference detector measures incident power, whereas the reflected and transmitted powers are measured separately. The directional couplers are used to separate between incident and reflected powers. Obviously other functions are incorporated in the instrument including attenuators, compensators, mixers, ADCs and processors, a computer, and many more. This diagram shows what is the main attraction of the SNA—its simplicity, and therefore its relatively lower cost. In general, the analyzers use a broadband signal that is then downconverted to a low-frequency AC or even to DC to measure the power in the signal. The power detectors themselves are relatively simple and may include diodes or thermoelectric devices, which are themselves broadband devices. This means that the power receivers do not require retuning as the frequency changes, that is, a frequency sweep only requires tuning of the source and measuring the power at each frequency. This leads to fast frequency sweeps and quick measurements.

While simplicity, speed, and cost are important, the SNA is not always the best choice for measurements. It is limited by the broadband noise in the detectors as well as external noise, and its calibration is usually not as accurate as more advanced VNAs. Nevertheless, in many applications, especially if the frequency range is narrow, the SNA is sufficient for all but the most demanding measurements.

The VNA differs from the SNA not only in that it measures both amplitude and phase but also in the way it does so. Probably the most important general distinction is that the measurement is narrowband and hence the VNA does not suffer from broadband noise and has a much higher dynamic range than the SNA. It accomplishes this by using heterodyning—a downconversion technique that

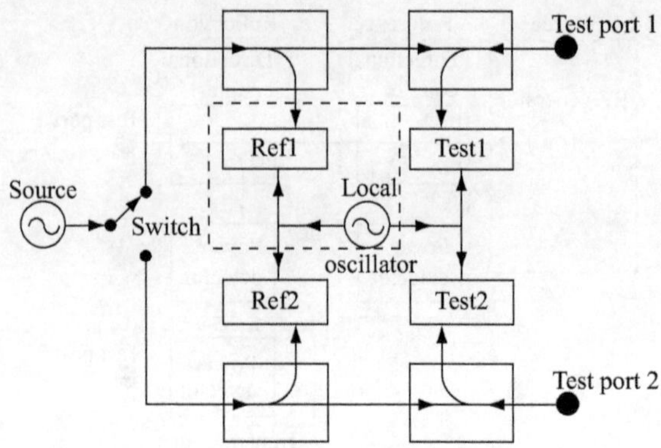

Figure 8.6 Simplified block diagram of a two-port vector network analyzer (VNA). The section within the dashed lines is expanded in Figure 8.7

employs a local oscillator in tune with the scanning source to produce a fixed intermediate, lower frequency for the power measurements. As a consequence, the power measurement itself is more accurate and, in general, the frequency steps the instrument is capable of are smaller. All this of course leads to a much more complex architecture and consequently to a more expensive instrument but with the added benefit of a more accurate measurement for amplitude and phase and of more flexibility in measurements it can handle. In operation, the VNA stimulates the network with a swept-frequency continuous wave (or, in certain applications with a pulse) and measures the phase and amplitude of the traveling waves at the stimulated port and at all other ports of the network, while these are terminated at their characteristic impedances (usually at 50 Ω). A simplified diagram of the VNA is shown in Figure 8.6. The source generator supplies the signal for measurements. In most analyzers, this is a synthesized source with tight control on frequency and power and can be swept over a wide frequency range as well as over a range of output power. In older VNAs, especially those with a narrow frequency range, the generator may be a VCO. The output is typically a pure sinewave although in some VNAs it is also possible to produce multifrequency signals as well as modulated sinewaves for more advanced measurements. Aside for accuracy of amplitude and frequency, the VNA source must be extremely agile, capable of sweeping rapidly over a wide range of frequencies or a range of power so that the response of the network can be obtained in a reasonable time. Figure 8.6 shows a two-port analyzer and hence the signal is switched (sometimes split) between the two so that each port can be analyzed. The source is followed by attenuators to control the amplitude fed onto the ports. Each port is fed through directional couplers so that the incident (reference) signal and the reflected signal can be measured separately. The incident and reflected signals are then mixed with the signal from a local oscillator to produce the IF signal that the receivers detect. Since each signal is detected by a separate receiver, both the amplitude and phase can be measured. The mixers, receivers, and all other functions needed are included in the blocks marked as *Ref*

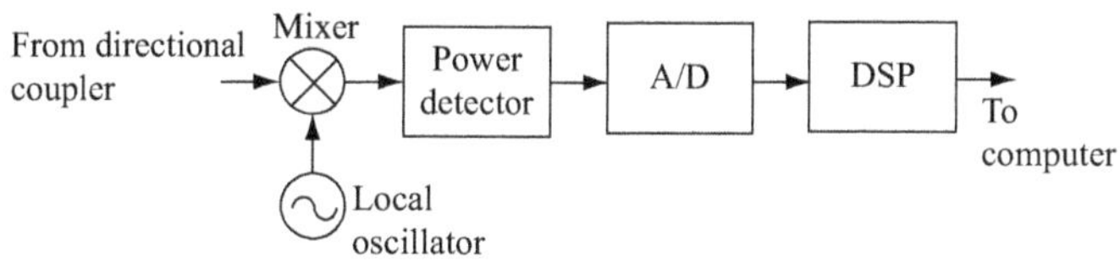

Figure 8.7 Block diagram of one of the receivers in Figure 8.6

and *Test* but are not shown explicitly (see Figure 8.7). As was discussed above, additional functions are needed to make this diagram practical, but these are not shown.

Figure 8.7 shows a simplified diagram of one of the receivers, say for the incident wave of port 1 (see dashed lines in Figure 8.6). After passing through the directional coupler, the signal is mixed with the local oscillator signal before entering the detector. Once the power of the signals is detected, it is typically converted to a digital signal using a high-speed ADC and fed to a digital signal processor for further analysis, display, and archiving.

At the end of the measurement process, there is a computer that performs all signal analysis, controls the measurement process, the display, and all other functions necessary for the measurement. A more general diagram of a multiport VNA is shown in Figure 8.8. This is essentially the diagram in Figure 8.4 with the addition of a local oscillator and a mixer to each channel. The N-port network is connected to the VNA directly. The source is switched between the ports, attenuated, split into a reference (incident) signal, and fed to the port through a directional coupler to measure the reflected signal, as described above. The local oscillator is common to all ports. The analysis is, of course, done in the computer following digitization and processing. It should be noted that although the N-port VNA has no theoretical limit on the number of ports, the complexity of the system, associated cost, and the speed at which the N-ports can be analyzed limit the number of ports to a relatively modest number. Four- and six-port analyzers are available but anything beyond that becomes very expensive and with limited utility.

The description above gives enough details to understand the principles and, perhaps, also hints to the technical difficulties faced by the designers of this marvelous instrument. It is not possible to discuss these in the context of this work, nor is there a particular need to do so. Nevertheless, it should be borne in mind that all components of the system introduce some errors that the instrument must compensate for and that much of its complexity is due to these circuits. Clearly, for example, a directional coupler presents an insertion loss, and the coupling is not perfect. The directivity of the coupler and its isolation from inputs that are not supposed to be coupled are not ideal and so on. All these and more must be taken into account and compensated to produce the accuracy needed. The generator and the local oscillators are again not ideal, their spectral purity is finite and stability of amplitude and frequency must also be taken into account. Indeed, there are many more issues that must be resolved including external noise, impedance of ports, even thermal issues (often, e.g., the network analyzer must be turned on for a

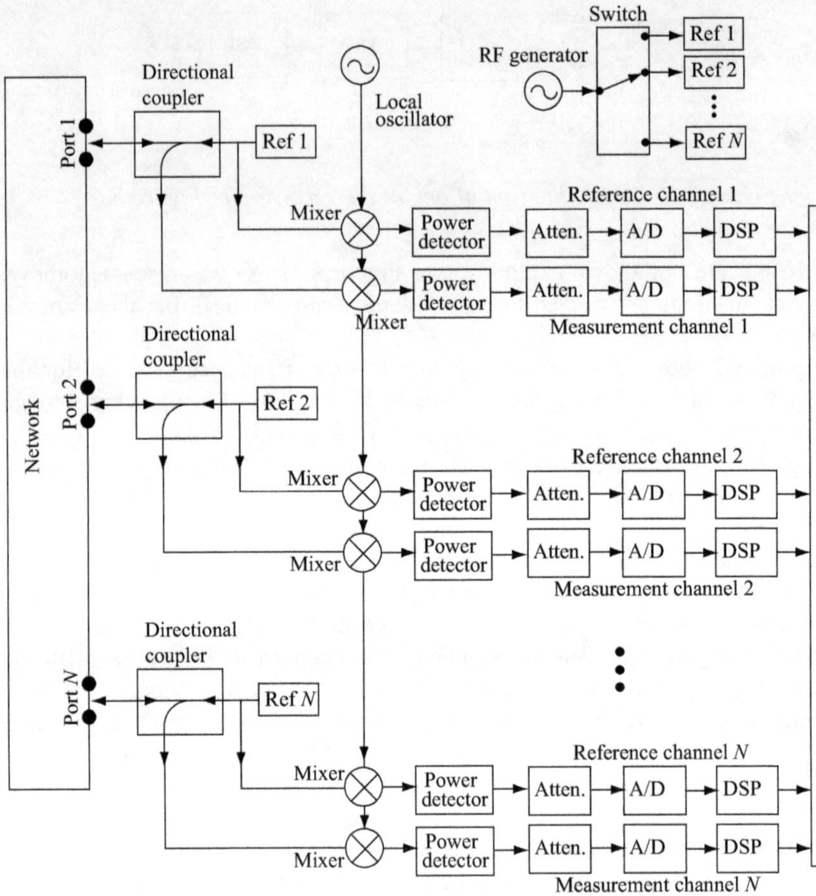

Figure 8.8 Structure of an N-port vector network analyzer (VNA)

considerable time before measurements commence to ensure it is in thermal equilibrium). Because of these issues, one of the most important aspects of measurement with network analyzers is the calibration process that must, necessarily, precede measurements.

It should also be noted here that, in general, VNAs are slower than SNAs primarily because of the need for frequency scanning, whereas in SNAs, the measurement is broadband. However, because of their other attributes, primarily their versatility and accuracy, VNAs are more often used in spite of their higher cost.

8.3 The measurement process

Most network analyzers measure the linear response of networks. However, there are also instruments that can extend some of the measurements to nonlinear

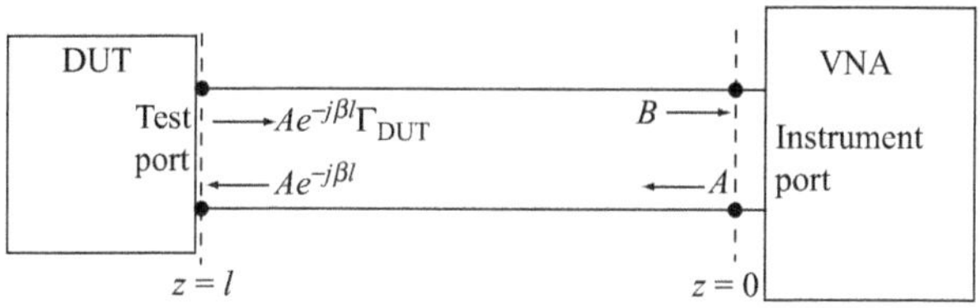

Figure 8.9 The change in phase between the measurement plane and the DUT plane

networks and devices, and, as mentioned in the introduction, they can also be used for measurements in space through the intermediary of antennas. In fact, the present work uses exactly this facility, that is, the network analyzer couples power to the stripline resonator through a probe (a short antenna) connected to one port of the analyzer and detects resonance through a second antenna connected to the second port of the analyzer.

There are a number of fundamental measurements that are often undertaken, but in almost all cases, these involve the determination of the S-parameters of the network (or device). The VNA does so by accurately measuring the incident and reflected amplitudes and phase on each of the ports of the network and doing so accurately. From these, one can then determine, through the use of the onboard computer, other quantities including losses, impedance, and even time-domain quantities such as delays.

8.3.1 Calibration

Much of the accuracy in measurements depends on the calibration of the instrument. In the case of the network analyzer, there are two types of calibrations that are necessary to achieve the high accuracy one demands of them. The first is what is often called "factory" calibration. This consists primarily of calibration of the source power, frequency, sweeping steps, attenuators, frequency response of the receivers, and the like. In essence, this calibration ensures a properly functioning instrument with known parameters so that the user can proceed with its use. Because of the sensitivity of measurements, it is not unusual for a network analyzer to need periodic recalibration. In addition, many measurement procedures specify calibration before each measurement.

More important from the user's point of view is the calibration of the instrument with the text fixture in place. This calibration is necessary to take into account the particulars of the measurement and ensure minimal errors. As a simple example, a device under test (DUT) must be connected to the network analyzer through a cable (transmission line) of some length. Even if we can safely assume the transmission line to be lossless, the length of the cable introduces a delay, which manifests itself through a change in phase. Using Figure 8.9, and given a forward propagating wave A, suppose we are trying to measure the reflection at the DUT (actually the S_{11}-parameter). The reflected wave B received back at the network analyzer is delayed because the waves must propagate to the DUT and back.

We referred to this effect in Chapter 4 (see Section 4.2.6) as the shift in the reference plane when we discussed the generalized *S*-parameters. The wave reflected back into the test port of the network analyzer is

$$B = Ae^{-j2\beta l}\Gamma_{\mathrm{DUT}} \tag{8.1}$$

where Γ_{DUT} is the reflection coefficient at the DUT. The reflection coefficient at the test port of the network analyzer is therefore

$$\Gamma_{\mathrm{TEST}} = \frac{B}{A} = \Gamma_{\mathrm{DUT}}\, e^{-j2\beta l} \tag{8.2}$$

Or, since $\beta = 2\pi/\lambda$ and $\lambda = v_p/f$ where v_p is the phase velocity along the line, we can write this result in terms of frequency f and the delay along the line, $\tau = l/v_p$ as

$$\Gamma_{\mathrm{TEST}} = \Gamma_{\mathrm{DUT}}\, e^{-j4\pi f\tau} \tag{8.3}$$

Clearly then, the phase of the reflection coefficient changes depending on the length of the transmission line, l, and the phase constant of the line, β, although the amplitude remains the same. Given that at high frequency even a very short delay (line length) can introduce significant errors in the phase, it is imperative that the phase difference be known and therefore removed from the measurement. This is done during the calibration process.

This particular problem was discussed in Chapter 4 as the reference plane in evaluation of *S*-parameters and is an obvious issue in the use of VNAs. But it is not the only one. Many of the errors that can be removed through calibration are systemic errors of the instrument and include test-port match errors, directivity errors in the directional couplers, frequency response, and isolation at the test port.

Port match errors on both the source and load ports are always present. Even though the ports' nominal impedance is 50 Ω (some network analyzers used for communication work may have a 75 Ω impedance), the actual impedance varies somewhat. In addition, every port has a return loss, and these combine to reduce the accuracy of the test. These must be compensated prior to actual measurement.

Directivity errors refer to the directional couplers. In an ideal directional coupler, the signal travels through the coupler from the input to the output and coupled to the coupled output as shown in Figure 8.10(a). In a real directional coupler, some signal may travel in the opposite direction and into the coupled arm of the coupler shown in Figure 8.10(b) by the dashed path. This error in directivity can also be compensated for in the design of the analyzer or by proper calibration.

Frequency-response errors refer to the fact that each receiver in the VNA has a slightly different frequency response because of slight variations in circuits and paths of signals. These are often referred to reflection and transmission tracking.

Finally, isolation refers to the isolation between ports. Ideally, the ports should be perfectly isolated, but in practice, some coupling between the receivers of the

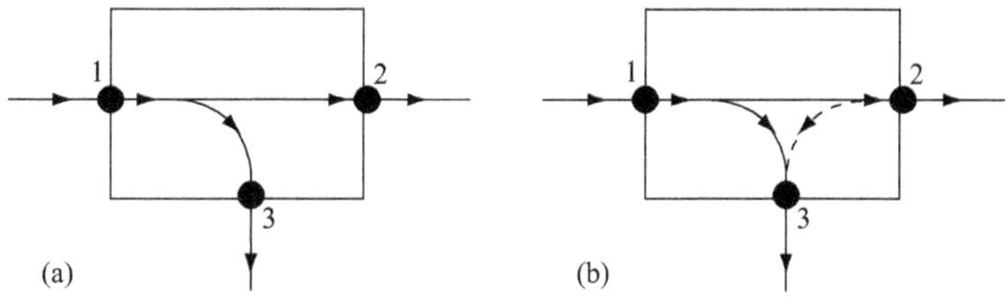

Figure 8.10 (a) An ideal and (b) real couplers showing back coupling through the dashed path

ports must exist. This crosstalk between different channels is also compensated for during calibration.

Actual calibration is done by connection of calibration standards to the reference planes (the location where the DUT would otherwise be connected) and following a procedure prescribed by the type of calibration required. Calibration standards are devices with exactly characterized properties including impedances. With these, the VNA measures the incident and reflected waves and uses these measurements to compensate for the mismatches and errors that exist. The calibration method depends on what is being measured. The most common methods of calibration are the short-open-load-through and involve first connecting the insertable standard between the reference planes and then connecting the two calibration ports (reference planes) together. This is particularly simple when the DUT is a coaxial device and hence can be inserted directly at the reference planes. Otherwise, special connecters and adaptors are needed to use with non-insertable devices such as in the case of characterization of electronic components. Calibration methods are based on error models, and, as one would expect, there are quite a few error models to choose from, each with a certain level of complexity. Error models refer to what errors are being compensated for. For example, if one compensates at one port for coupler directivity, reflection tracking, and source match, then this becomes a three-term model. If this is done on a two-port system, there are three additional errors due to load match, transmission tracking, and isolation. If these six errors are applied to each port (forward and reverse model), the result is a 12-term error model. Error models range from 3 to 16 (for two-port analyzers). Which model is to be used depends on the measurement and the accuracy needed.

There are of course other methods of calibration, some simpler, some more complex, and there are variations of almost all of them. Most methods of calibration are documented and their advantages and disadvantages given by the manufacturer. In all cases, standards and connectors are available from manufacturers, and basic calibration components are usually included with the instrument. Many more standards, connectors, adapters, and test fixtures are available to fit any calibration and testing need. The instrument manual provides data for basic use and calibration, while additional publications are available from the manufacturer for other applications. Manufacturers also offer automatic (sometimes called electronic) calibration units that simplify the calibration process to that of connecting the unit and running the calibration process. Two- and four-port automatic calibration units are available for VNAs, but these are specific to the manufacturer of the VNA.

In addition to systemic errors that careful calibration can minimize, there are other errors that can be caused by various components of the measurement system or by the network analyzer itself. One of the most obvious and common source of problems in measurements is the cables connecting to the DUT or the test fixture. Bending of cables, damaged or poor shielding, and poor connection can result in errors or even in false measurements due to reflections within the cables. These errors are particularly annoying because they often appear to be random. Similarly, poor-quality connectors or loose connections at the connectors or within test fixtures can similarly lead to poor measurements. Operation at the very limits of the analyzer such as at maximum power can also lead to nonlinear behavior, whereas at the limits of frequency response, there may be issues with tracking and with the local oscillator. In general, one should remember that it is unreasonable to expect good measurements with poor quality testing equipment. A quality instrument such as the VNA can easily be reduced to a useless instrument by use of poor cables, improper connections, or test fixtures or by improper measurement procedures.

Finally, one more point: the network analyzer is a complex instrument that requires time to master. A first encounter with it is daunting even for experienced practitioners. The instrument itself is expensive and everything associated with it, from connectors to test fixtures, is equally costly. It is therefore well advised to take the necessary time to learn the instrument and to practice its use before embarking on actual measurements to avoid unnecessary expenses and, even more importantly, misleading measurements. In the end, it is the practitioner who must judge the measured results in spite of the sophistication of the instrument.

8.3.2 Measurements

The primary measurements of network analyzers and in particular VNAs are the *S*-parameters of a network. A network for the purpose of this discussion is seen as anything that can be connected either as a one-, two-, or multiple-port quantity. It may be a simple passive or active component such as an inductor or a diode (one-port devices or networks), or it may be a transmission line section, an amplifier, or a cavity resonator (two-port devices or networks). In all cases, the *S*-parameters are evaluated and are independent of the properties of the network analyzers provided that proper calibration has been performed. *S*-parameters characterize the linear behavior of the network and, as we have seen in Chapter 4, almost all useful characteristics of the network can be obtained from the *S*-parameters. Through the use of cascaded measurements, one can characterize even more complex circuits.

VNAs can also be used to characterize materials in application as simple as measurement of dielectric constants and loss tangent or as complex as characterization of biologic properties and interactions or scans for tumors since, with proper antennae connected to its ports, the network analyzer can be transformed into a low power, short-range radar, or a microwave microscope. With its onboard computer and powerful signal-processing capabilities, it becomes a powerful imaging tool.

Since VNAs measure the reflected power as well as transmitted power as a function of frequency, these results can be easily transformed into the time domain. Hence, some of the important uses of VNAs are in time-domain reflectometry.

The primary purpose of the present work is detection and quantification of moisture content in open transmission line resonators; hence, we will limit the discussion of VNA measurements to those that are useful to that end. These include detection of resonance, measurement of quality factor, measurement of permittivity, and measurement of losses as well as some other quantities that are useful in determining the primary quantities.

The network analyzer measures S-parameters internally but, from the point of view of the user, the S-parameters are usually not "visible." For example, suppose one wished to measure the standing wave ratio (SWR). To do so, the network analyzer measures the S_{11}-parameter [see (4.68)]. The SWR is then

$$\mathrm{SWR} = \frac{1 + |S_{11}|}{1 - |S_{11}|} \tag{8.4}$$

The network analyzer will display the SWR, and the user never actually knows what S_{11} is unless, of course, he or she chooses to measure the reflection coefficient, which is S_{11}. All that means that the instrument is set to perform the necessary calculations to obtain the measured result the user asks for but internally it measures S-parameters.

In the following, we will describe the processes by which the analyzer obtains the results, but it should be remembered that much of this process is part of the algorithm for particular measurements.

8.3.2.1 Measurement of the *S*-parameters

Given a two-port network as in Figure 8.11, the S-parameters are defined from the forward waves a_1, a_2 and backward waves b_1, b_2. As was discussed in Chapter 4, the relations between the forward and backward waves are

$$b_1 = s_{11}a_1 + s_{12}a_2 \tag{8.5}$$

$$b_2 = s_{21}a_1 + s_{22}a_2 \tag{8.6}$$

The reference planes Z_0 are at the DUT, that is, it is assumed that the phase difference between the DUT ports and the instrument port have been compensated for.

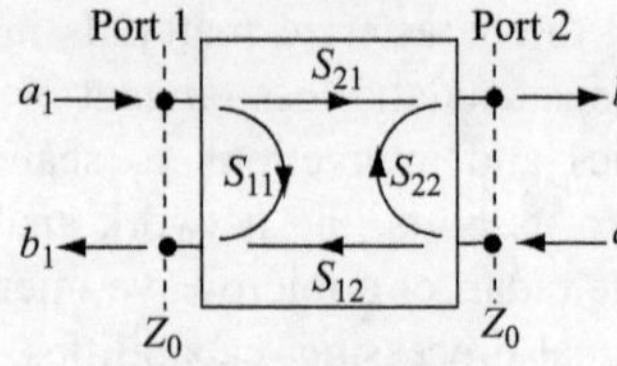

Figure 8.11 Relation between forward and backward waves at the ports of a network

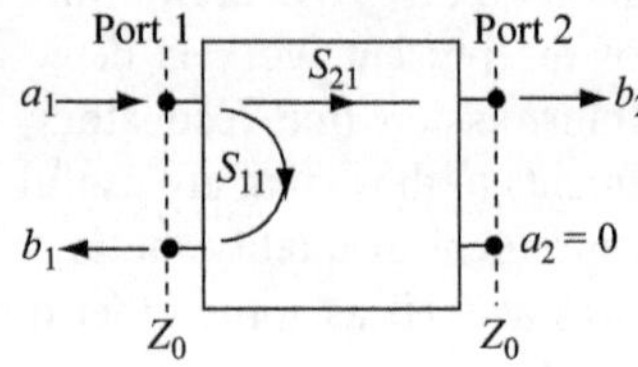

Figure 8.12 Forward measurement to obtain S_{11} and S_{21}

The *S*-parameters are then obtained by selectively setting one of the forward waves to zero. In the network analyzer, this is done by properly terminating one of the ports with the characteristic impedance as follows:

By terminating port 2 with its characteristic impedance, we force $a_2 = 0$ and obtain

$$S_{11} = \left.\frac{b_1}{a_1}\right|_{a_2=0} \quad \text{and} \quad S_{21} = \left.\frac{b_2}{a_1}\right|_{a_2=0} \tag{8.7}$$

Similarly, by now terminating port 1 with its characteristic impedance, $a_1 = 0$, and we obtain

$$S_{22} = \left.\frac{b_2}{a_2}\right|_{a_1=0} \quad \text{and} \quad S_{12} = \left.\frac{b_1}{a_2}\right|_{a_1=0} \tag{8.8}$$

That is, the sequence in the network analyzer is to terminate port 2 with its characteristic impedance (usually 50 Ω), excite port 1 with a wave, and measure the reflected and transmitted waves b_1 and b_2 (Figure 8.12). The S_{11}- and S_{21}-parameters are then calculated as the ratios between the backward waves b_1 and b_2 and the incident wave a_1 as in (8.7). This measurement is referred to as the forward measurement.

Now the process is repeated on port 2 by terminating port 1 and measuring the backward waves again to obtain S_{22} and S_{12} (Figure 8.13). This measurement is referred to as the reverse measurement.

It should be noted here again that these measurements are only as accurate as the termination is exact, and the systemic errors have been removed. Hence, a two-port calibration prior to evaluation of the *S*-parameters is absolutely essential,

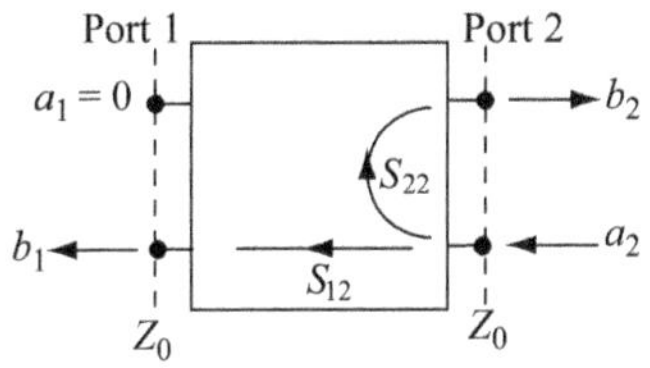

Figure 8.13 Reverse measurement to obtain S_{22} and S_{12}

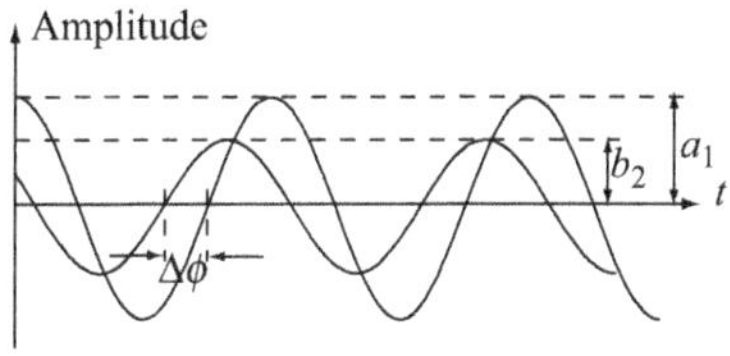

Figure 8.14 Measurement of magnitude and phase of the S_{21}-parameter

otherwise the errors in the S-parameters can be significant, and these will carry over to subsequent measurements based on the S-parameters.

The reduction of this process to a single port is obvious since in that case, $a_2 = 0$, and only the parameters S_{11} and S_{21} exist. The extension of an N-port network is also obvious (see Section 4.2.2): each port is analyzed as above, by setting all incidents except for that on the measured port to zero by terminating the ports with their (compensated) characteristic impedance.

The S-parameters are obtained over the range of desired frequencies, and hence they represent the frequency response of the network over that range.

The S-parameters are in general complex values. An SNA can only measure the differences in amplitude between the wave quantities and hence can only obtain the magnitude of the parameters. A VNA also measures the phase differences between the waves and hence measures the complex S-parameters. It should be noted that in most VNAs, the phase is referenced to the input wave on a port (the reference wave) although some VNAs can actually measure absolute phase. Figure 8.14 shows how the S_{21}-parameter and its phase with respect to the incident wave a_1 are obtained from measurement of the delay between the two waves. The phase difference is therefore the argument of S_{21} (in this measurement), whereas S_{21} is the ratio between the amplitudes ($|S_{21}| = b_2/a_1$).

Although in most cases the S-parameters are used for subsequent measurements, their magnitude and phase can be displayed either on a Smith chart screen display or as plots. The magnitude is usually displayed in dB versus frequency, whereas the phase is displayed in degrees versus frequency.

8.3.2.2 Measurement of the reflection coefficient

Perhaps the most fundamental measurement is that of the reflection coefficient. This was explained in the previous section and, indeed, the S_{11}-parameter is the overall reflection coefficient. That is, the reflection coefficient as defined in transmission lines or in space equals S_{11} if the reflecting section is infinite or if the reflection section is matched at the load port. If the reflecting section is not infinite or not matched, S_{11} is the sum of all reflections at the port including those due to internal reflections. However, its knowledge has far reaching implications for the network or device. Since the reflection coefficient is a measure of mismatch between the transmission line leading to the port and the port itself, the S_{11}-parameter can be used directly to match the network. It is often used in this fashion to match loads such as antennas or to match between components and transmission lines in circuits. In the network analyzer, impedances are normalized with respect to the test-port impedance; hence, the reflection coefficient can be written as

$$\Gamma = S_{11} = \frac{1 - Z/Z_0}{1 + Z/Z_0} \tag{8.9}$$

Since S_{11} is obtained over a frequency range, Z, the impedance of the port is also some function of frequency. By modifying Z of the port (e.g., by adding passive components to an antenna), matching is obtained when $Z = Z_0$ either in a range of frequencies or at a particular frequency, as is often the case with antennas that operate at a fixed frequency. If matching is undertaken, it is often done with the aid of the Smith chart (see Section 2.12). The ideal condition of course is for S_{11} to be zero. In practice, one tries to reduce the reflection coefficient as much as possible. If the display is in Smith chart mode, the goal is to bring the trace to converge to the center of the chart, a point that corresponds to zero-reflection coefficient. The measurement is usually used to evaluate the required impedance Z for matching. Then the impedance of the device is modified to equal the matched impedance and then the network analyzer is used again to verify that matching occurs.

We have mentioned the SWR in (8.4). The network analyzer can compute the SWR directly from (8.4) after the S_{11}-parameter has been found. The SWR can also be used for matching, but now the goal is to reduce the SWR to SWR $= 1$, a value corresponding to $S_{11} = 0$. As the mismatch increases, so does the SWR with SWR tending to infinity for shorted (zero impedance load) or open (infinite impedance load).

The S_{11}-parameter can also be used for detection of resonance in circuits. Since at resonance the impedance of a network is purely real, resonance also corresponds to minimum reflection at series resonance and maximum reflection at parallel resonance. Thus, the S_{11}-parameter can be used as a resonance indicator directly.

In this section, we discuss the measurement of the S_{11}-parameter using a VNA and its application to two important functions. One is the matching of a load to an antenna (or transmission line) (see Section 2.9.1 for a discussion on matching), and the second is the detection of resonant frequency in a number of configuration as they were used as part of the present work. Because these applications make use of the Smith chart and the Smith chart is a numerical tool, we use specific numerical examples. However, the methods are totally general.

Matching of a load to a 50-Ω antenna

Consider Figure 8.15(a). The load in this case is an energy harvesting board connected to a 50-Ω antenna. The board is designed to harvest power from a 915-MHz transmission. At 915 MHz, the impedance of the board is $12.5 + j115\ \Omega$, as measured with a network analyzer. Under these conditions, the reflection coefficient can be calculated directly:

$$\Gamma = \frac{12 + j115 - 50}{12 + j115 + 50} = 0.6368 + j0.6737 \qquad (8.10)$$

Clearly, this is the same as the S_{11}-parameter

$$S_{11} = 0.6368 + j0.6737 \quad \text{or} \quad S_{11} = 0.9270\angle 46.6^\circ \qquad (8.11)$$

The mismatch on the line produces an SWR [see (8.4)]

$$\text{SWR} = \frac{1 + |S_{11}|}{1 - |S_{11}|} = \frac{1 + 0.9270}{1 - 0.9270} = 26.4 \qquad (8.12)$$

Measurement of resonant frequency

This can be done with the network analyzer over a range of frequencies. Figure 8.16 shows the S_{11}-parameter over the frequency range 905–920 MHz.

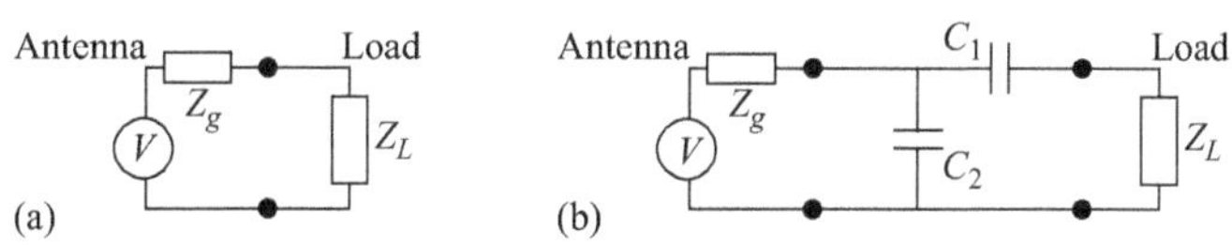

Figure 8.15 (a) A load connected to a 50-Ω source (antenna) and (b) the matching network needed to match the load

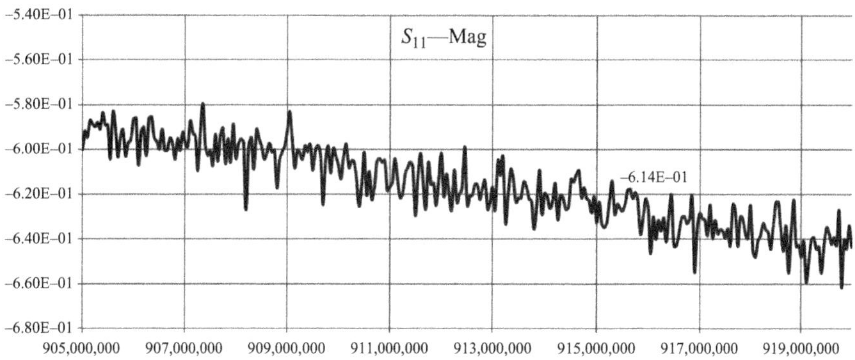

Figure 8.16 The S_{11} for the circuit in Figure 8.15(a). S_{11} is shown in dB

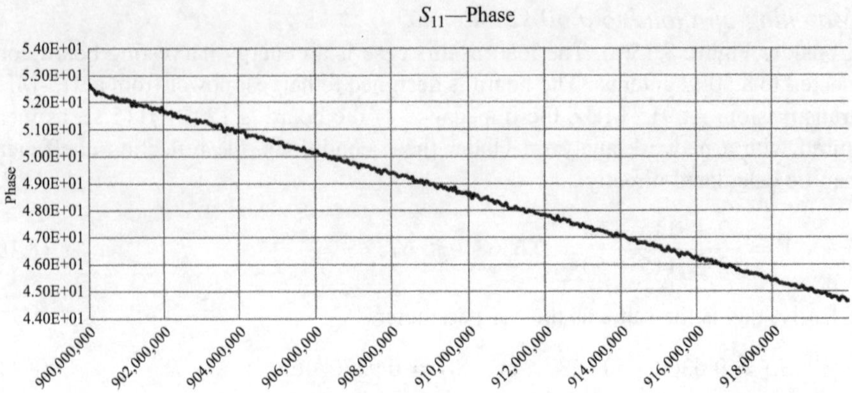

Figure 8.17 The phase angle of S_{11} for the circuit in Figure 8.15(a)

The S_{11}-parameter is in dB. The phase of S_{11} is shown in Figure 8.17 over the same range. At 915 MHz, the network analyzer provides the following:

$$S_{11} = -0.6275117\angle 46.3^\circ \tag{8.13}$$

If we convert this to magnitude using (4.77), we have

$$|S_{11}| = 10^{(-0.6275117/20)} = 0.93 \tag{8.14}$$

or:

$$S_{11} = 0.93\angle 46.3^\circ \quad \text{or} \quad S_{11} = 0.6425 + j0.6723 \tag{8.15}$$

Note that there is a slight difference between the calculated values and those obtained from the Smith chart. This is because of the ripple on the S_{11} curve in Figure 8.16.

To better match the board to the antenna, a network consisting of a series and a shunt capacitor is added as shown in Figure 8.15(b). The choice here is obvious—since the impedance of the board is inductive, the network must be capacitive. The process as performed on the network analyzer in matching mode is shown in Figure 8.18. A series capacitor equal to 1.84 pF moves the impedance from its original location at point P [where the normalized impedance is $(12 + j115)/50 = 0.24 + j2.3$] to point P_1 at which the normalized impedance is $0.24 + j0.43$. The shunt capacitor, equal to 6.14 pF, moves the impedance the center point at which the normalized impedance is $1 + j0$. As can be seen from the figure and from the inset, the S_{11}-parameter now is

$$S_{11} = 0.01\angle 180 \quad \text{or} \quad S_{11} = -0.01 \tag{8.16}$$

That is, the reflection coefficient has been reduced to -0.01 indicating an almost perfect result. The impedance seen by the antenna is virtually 50 Ω and SWR = 1.

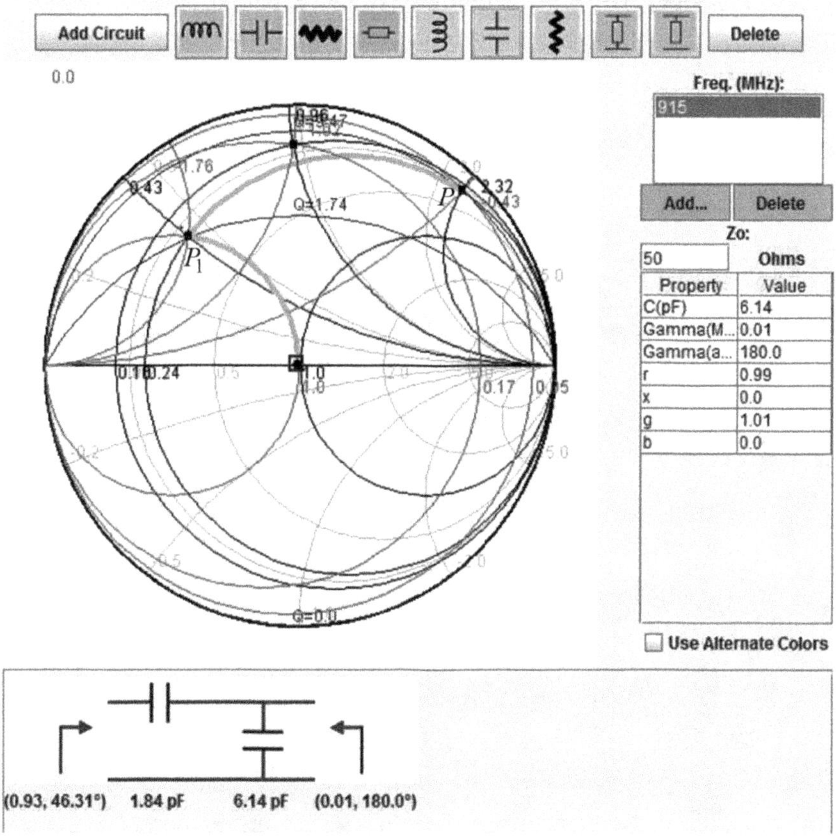

Figure 8.18 Screenshot for the design of the matching network in Figure 8.15(b)

In practice, it is not possible to use the capacitor values given. Using standard capacitors with values close to those indicated in the design will produce some mismatch. For example, a series capacitor of 1.5 pF and a shunt capacitor of 5 pF produces an impedance of $86 + j28\ \Omega$ instead of $50\ \Omega$, with $|S_{11}| = 0.3$ and SWR $= 2$. This is of course not "perfect," but it is much better than the starting point.

Figure 8.18 shows one of the many modes of a VNA—in this case a matching mode.

Since the present work deals primarily with resonance, an example is useful. A coaxial resonator of length 30 cm was designed as in Figure 8.19. The dimensions of the resonator were designed to produce a nominal characteristic impedance of $50\ \Omega$ when air-filled. The purpose of the device was to measure the permittivities of water. Polyester, nylon, and the solution used for coating of the nylon and polyester fabrics. The dialog allows the user to build the network, select components, and to compute and display the results. The image shown is the final display for a scan over a range of frequencies.

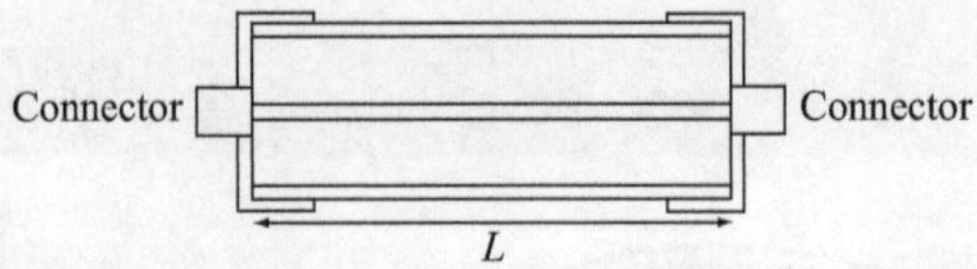

Figure 8.19 A coaxial section used as a resonator

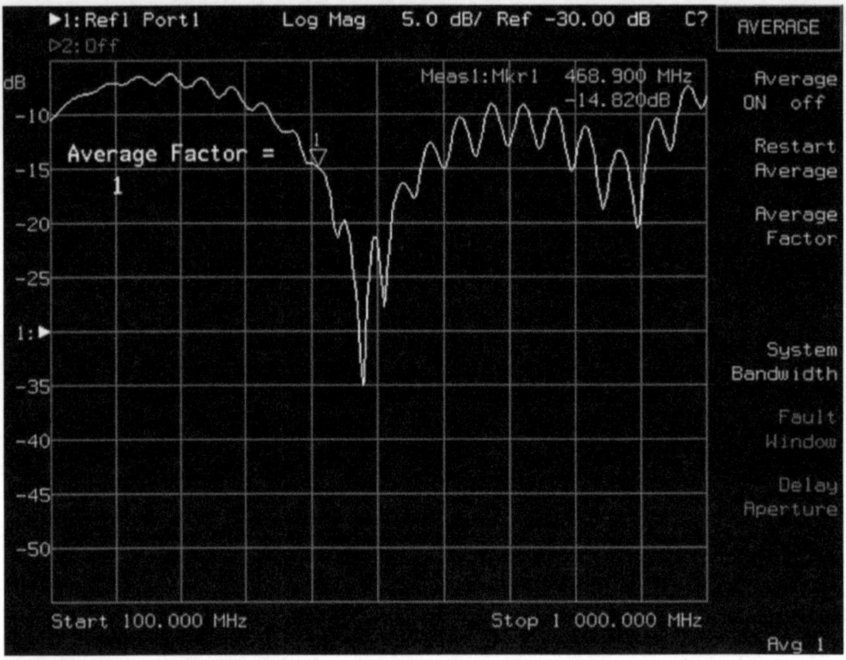

Figure 8.20 Reflection measurement on port 1 of the device in Figure 8.19 (air-filled)

Figure 8.20 shows the frequency response of port 1. A clear resonance can be seen at a frequency of 530.4 MHz with minor variations in the reflection coefficient at other frequencies, indicating variations in impedance. The frequency is scanned from 100 kHz to 1 GHz. The second port is not used (off), and the magnitude is given in dB with 5 dB/division. Note also the reference signal specified at −30 dB.

Figure 8.21 shows the same measurement after a polyvinyl chloride (PVC) pipe was inserted in the chamber. The PVC pipe fills the chamber lengthwise but only partially radially since its wall thickness is smaller than the air space available. The resonant frequency has now moved to a lower value of 468.9 MHz because of the higher permittivity of the PVC. Although the purpose of this example is simply

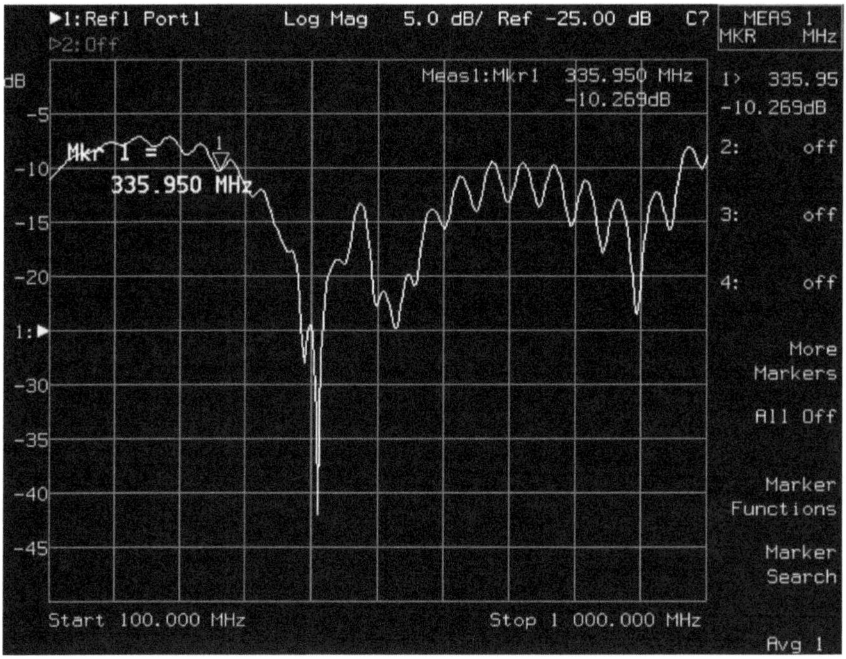

Figure 8.21 Reflection measurement on port 1 of the device in Figure 8.19 (PVC pipe in the chamber)

to show the utility of the S-parameters in detection of resonance, it should be noted that the effective permittivity of the chamber can be easily evaluated as

$$\varepsilon_{\text{eff}} = \left(\frac{530.4}{468.9}\right)^2 = 1.284 \tag{8.17}$$

This, of course, is not the permittivity of the PVC pipe but rather the effective permittivity duc to the partially filled cavity (a mixture of air and PVC).

Figure 8.22 shows the reflection measurement when the cavity is filled with water. Now the resonant frequencies moved to even lower frequencies as expected from the higher permittivity of water. Resonances can be seen at 144, 218, 290, 363, 435, 510 MHz, and so on. It should be noted that there are resonant frequencies lower than 100 MHz as can be seen from the curve that tends lower at 100 MHz, but its measurement would require a scan staring at a lower frequency. The many resonant frequencies are not surprising since the high permittivity of water increases the electric length of the cavity.

8.3.2.3 Measurement of the transmission coefficient

The measurement of the S_{21}-parameter, or the transmission coefficient, was described above. As with the reflection coefficient, it can be used for a number of applications. One of the simplest applications is its use in detection of resonance in

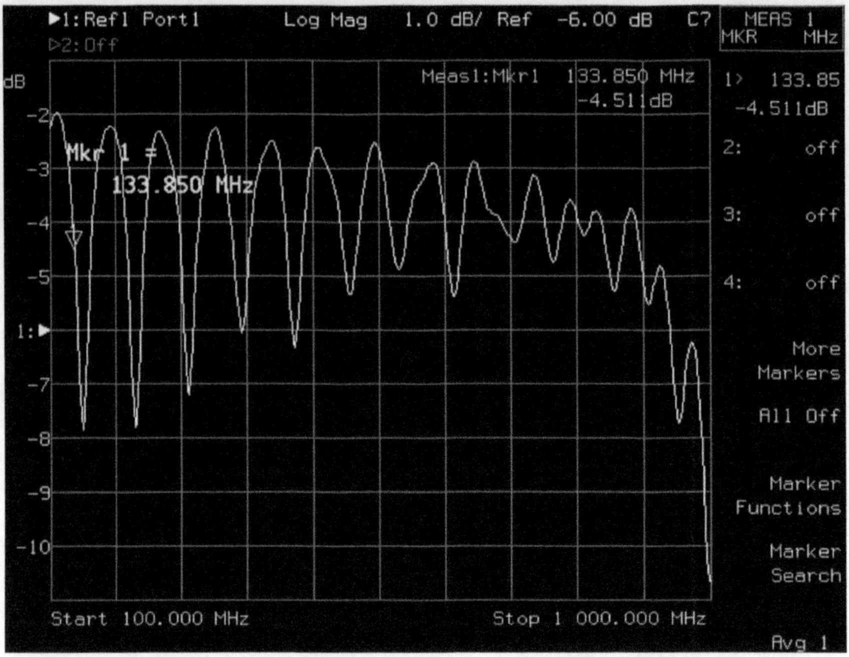

Figure 8.22 Reflection measurement on port 1 of the device in Figure 8.19 (chamber filled with water)

a manner similar to that of the reflection coefficient except, of course, that when the S_{11}-parameter is minimum, the S_{21}-parameter is maximum. Thus, in detection of resonance with the S_{21}-parameter, one looks at peaks rather than valleys in the response. This is not usually a recommended use because unlike valleys, which are very sharp, peaks tend to be flat, and hence the decision on what the resonant frequency is tends to be less accurate. To see this, consider Figures 8.23–8.26.

Figures 8.23 and 8.24 are both scans from the empty cavity of Figure 8.19. Figure 8.23 is a plot of the transmission coefficient S_{21} and Figure 8.20 of the reflection coefficient. The scans are between 300 kHz and 1.3 GHz. The first thing to note is that the minima in Figure 8.24 are at the exact location of the maxima in Figure 8.23 as can be seen by comparing the markers. Clearly also, the maxima in the transmission scan are broad and flat, whereas the minima in the reflection scan are narrow and well defined. It is also interesting to compare Figure 8.24 with Figure 8.20. Both of them are reflection scans with the cavity air-filled. In an ideal case, these should be identical, but in fact the first resonant frequency in Figure 8.20 is 530.4 MHz, whereas in Figure 8.24, it is 534.26—a difference of almost 4 MHz. The two scans were taken at different times, and the difference is likely due to changes in properties of air—temperature, humidity, and pressure which all change the permittivity and hence the difference. It is also possible that one or the other scan was done after a "better" calibration or that the instrument

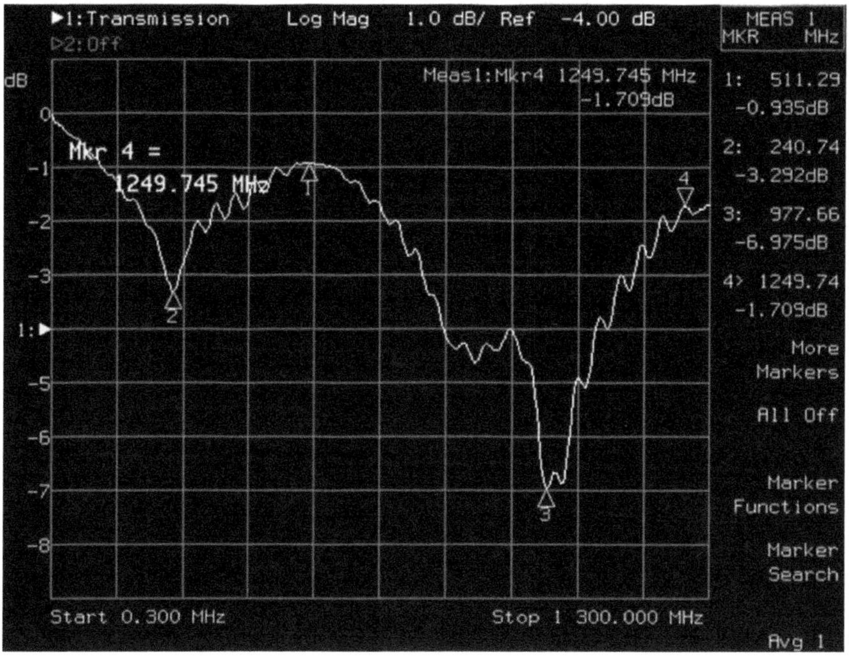

Figure 8.23 Transmission test on port 1. The marker at the peak between markers 2 and 3 shows the resonant frequency (maximum transmission, minimum reflection)

may not have been in thermal steady state. Although the difference is small (less than 1%), it is significant and it represents the testing conditions during the actual tests. Note also that the trace along the scan is different as well with different variations. These can usually be attributed to the test cables (i.e., reflections due to bending) and/or due to imperfections in the manufacturing of the resonator and are not significant in this test (i.e., we are only interested in the resonant frequency, not in the frequency response of the test cables).

The tests in Figures 8.23 and 8.24 were repeated with a polyester core, for the purpose of evaluating the permittivity of polyester. The results are shown in Figure 8.25 for the transmission scan and in Figure 8.26 for the reflection scan. The resonant frequency reduced from 534.26 for the air-filled cavity to 514.98 MHz with the polyester core. To be noted again is the fact that the minima in Figure 8.26 correspond to the maxima in the transmission test in Figure 8.25 although the peaks in the transmission test are quite difficult to identify.

8.3.2.4 Measurement of quality factor

Measurement of quality factor of a resonator is often needed in evaluation of circuits or, in the case described here, to evaluate the accuracy of measurement of resonant frequencies. Indeed, the higher the quality factor, the sharper the

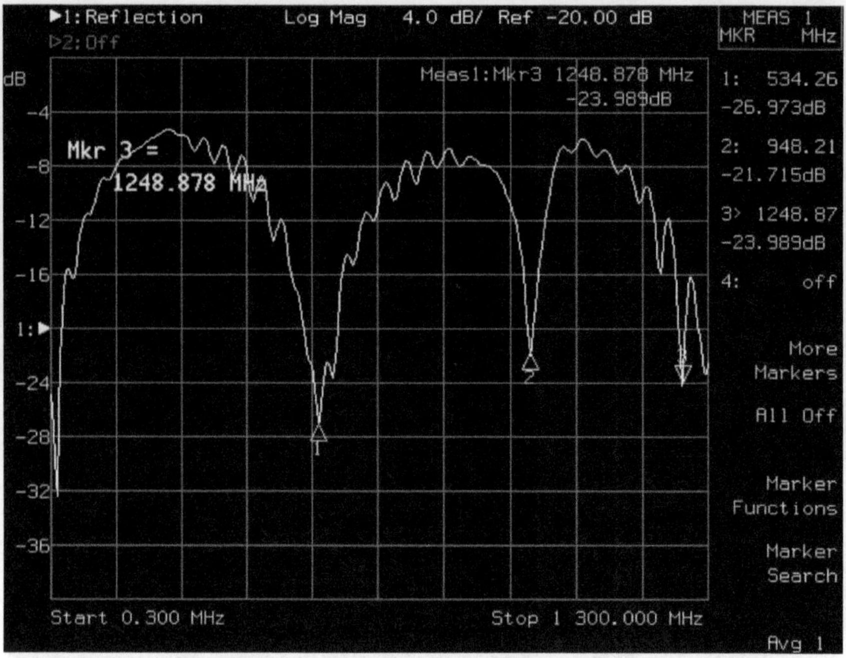

Figure 8.24 *Reflection scan for the same conditions as in Figure 8.23. The resonant frequencies at 534.26 and 848.21 MHz correspond to the peaks in Figure 8.23*

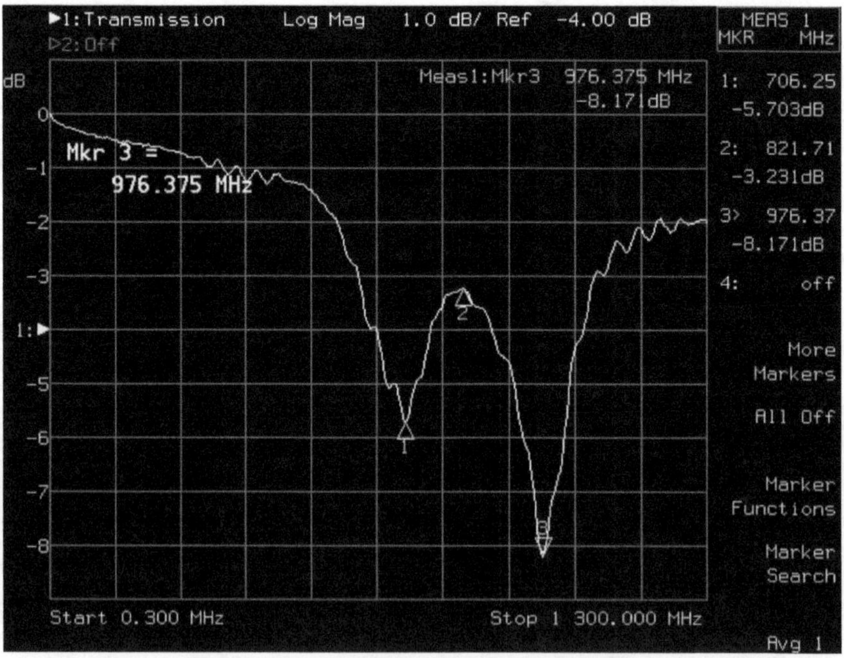

Figure 8.25 *Transmission test with a polyester core filling part of the space in the coaxial resonator*

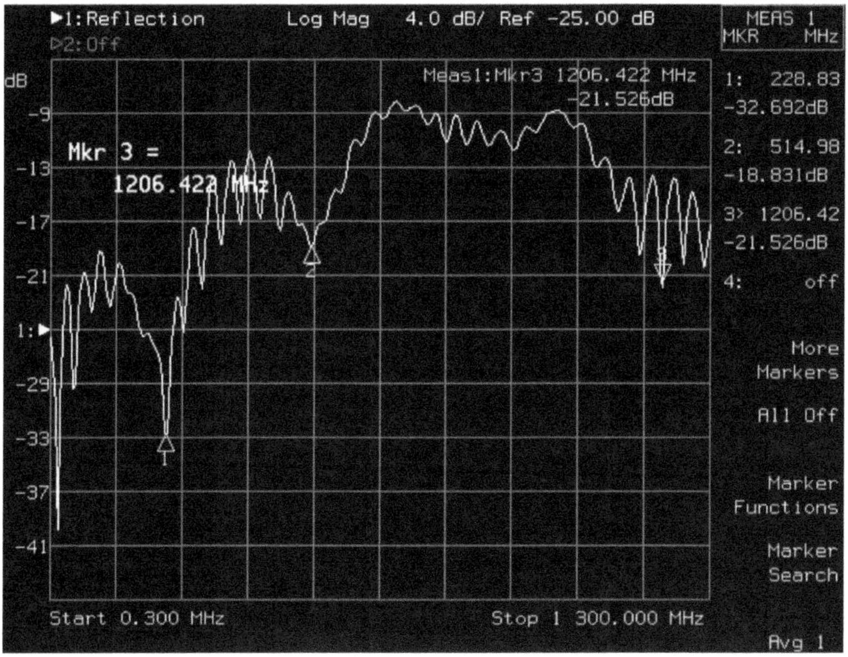

Figure 8.26 Transmission test with a polyester core filling part of the space in the coaxial resonator

resonance and the more accurately it can be measured. There are a number of methods suitable for this purpose. The easiest method to understand the measurement of the loaded quality factor is based on the fact that the quality factor is the ratio between the resonant frequency and the bandwidth between the two half-power points [see (2.208)]:

$$Q_L = \frac{\omega_0}{\mathrm{BW}} = \frac{f_0}{f_u - f_l} \tag{8.18}$$

Once the resonant frequency has been identified on the scan, the 3 dB points below (f_l) and above (f_u) the resonant frequency can be identified automatically and the quality factor displayed. Other algorithms that accomplish the same purpose exist and may be used, but the method above is simple and accurate. However, if the unloaded quality factor or the external quality factor is required, we must resort again to the *S*-parameters. Section 4.6 discussed the theoretical details of this measurement. At resonance, the unloaded quality factor is [see (4.101)]:

$$Q_0 = \left(\frac{1}{1 - S_{21}}\right) Q_L \tag{8.19}$$

The measurement of Q_0 requires a couple of steps. First, we must find the resonant frequency to evaluate the loaded quality factor using (8.18) and the process of evaluating the half-power frequencies. These are usually evaluated from the S_{11}-parameter in a reflection test. This is followed by a transmission test to find the S_{21}-parameter at the resonant frequency. The simple calculation in (8.19) is carried out by the network analyzer to obtain the unloaded quality factor. The external Q-factor is also available from the S-parameters [see (4.102)]:

$$Q_e = \frac{Q_L}{S_{21}} \tag{8.20}$$

Although the external quality factor is not often needed, (8.20) shows the ease with which additional, secondary values can be measured without additional measurements and the associated necessary calibrations.

Of course, there are other methods of measurement for the Q-factor since there are different methods of evaluating the losses in a cavity. For example, one can measure the loaded Q of a cavity from the reflection coefficient or, more conveniently, from the SWR since at the half-power points, the magnitude of the real part of the impedance equals the real part of the resonator impedance. Either the phase or the magnitude may be used for this purpose. Other methods use the transmitted power. However, in modern measurements, all these can be done using a network analyzer, and hence the method described above is sufficient and has the advantage of clarity and simplicity.

8.3.2.5 Measurement of impedance

Measurement of impedance is a fundamental process in a VNA and is related to the S_{11}-parameter as discussed in Chapter 4. Impedance of a device connected to a transmission line produces a reflection coefficient:

$$\Gamma_L = \frac{Z_L - Z_0}{Z_L + Z_0} = |\Gamma_L|e^{j\theta_\Gamma} \tag{8.21}$$

where Z_L is a load impedance to be measured and Z_0 either the characteristic impedance of a line or the impedance in space. One has to measure both the magnitude of the reflection coefficient and its phase θ_Γ, something that is inherent in VNA measurements. Once the S_{11} coefficient has been measured, one can evaluate the impedance of the device as

$$Z_L = Z_0 \frac{(1 + S_{11})}{(1 - S_{11})} \tag{8.22}$$

This of course is a complex value as one would expect.

The measurement of impedance is fairly simple except of course that one must take the normal precautions associated with VNA measurement including proper calibration.

8.3.2.6 Measurement of insertion loss, return loss, attenuation loss, and reflection loss

These losses were defined in Section 4.3.3.3 in terms of the S-parameters. In all cases, it is sufficient to measure the S_{11}- and S_{12}-parameters (or, in some cases, the S_{22}- and S_{21}-parameters). The various losses are calculated from the *S*-parameters as

Reflection loss

$$RL = 10\log\left(1 - |S_{11}|^2\right) \tag{8.23}$$

Attenuation loss (transmission loss):

$$AL = 10\log\frac{|S_{12}|^2}{1 - |S_{11}|^2} \tag{8.24}$$

Insertion loss:

$$IL = 20\ \log|S_{12}| \tag{8.25}$$

Return loss
that return loss is a nonnegative number for reflection from a passive network:

$$RTL = 20\ \log\frac{1}{|S_{11}|} \tag{8.26}$$

8.4 Measurement of complex permittivity and loss tangent

The measurement of complex permittivity was discussed extensively in Chapter 4. All methods discussed relied on the measurement of the *S*-parameters either in a waveguide, a transmission line, or in free space. In addition to the *S*-parameters, the methods required extensive computation and, in some cases, special calibrations and compensation techniques. All these of course can be done with a network analyzer and either a standard fixture or, in some cases, a specially made fixture. Sample preparation is necessary for most methods. Of particular interest are the transmission line methods and the cavity perturbation method. These methods were used in the present work to measure the resonant frequency of solid sheets of nylon and delrin for sensor calibration purposes and of the liquid solution to verify the numerical simulations. Although in general the dielectric constant is known, measurements often become necessary either because of variability between suppliers or even batches of the same material from the same supplier or because the permittivity is not available at the required frequency. Also, the loss tangent is often not known and requires measurements.

8.4.1 Resonant methods

The use of a coaxial resonator for measurement of permittivity was discussed in Sections 8.3.2.2 and 8.3.2.3, in which the effective permittivity was calculated from the resonant frequency of the cavity with and without the sample in the cavity [see (8.10)]. As was indicated in Section 8.3.2.2 and as discussed in Section 3.5, the use of resonant methods can only measure the permittivity of the cavity due to all effects that may occur in the cavity, hence the term effective permittivity. However, under one of two conditions, the measurement can provide the actual permittivity of the sample:

1. The material being tested fills the entire resonator.
2. The sample is small and placed at a location where the electric-field intensity is maximum and constant throughout the sample.

The condition in (1) can be satisfied if a sample can be machined to fill the cavity exactly or if the cavity is filled with a liquid permittivity of which is required. In some cases, such as measurement on gases or in relating permittivity due to humidity, the condition is satisfied exactly. It is not, however, possible to do so if the material comes in sheets or if it is bulk material such as, say, grains. The condition in (2) can be satisfied theoretically for almost any cavity and many solid materials that can be machined. It is not suitable for liquids, and the placement of the sample in the cavity can be problematic. In some cases, it requires some kind of support such as a suspending wire, and these issues must be taken into account.

Once the sample has been made and properly placed, the measurement is straightforward. The resonant frequency of the empty cavity is measured first. Then the measurement is repeated with the sample in the cavity.

8.4.1.1 Small sample in cavity—cavity perturbation

The cavity perturbation method was described in Section 3.5 and will not be repeated here. However, from a practical measurement point of view, there are a number of issues that have to be resolved for the measurement to be valid. The first of these is the type of cavity to be used. The most common are rectangular cavities and cylindrical cavities. The latter are particularly useful because the connections to coaxial cables are usually easier to handle and result in better matching than in rectangular cavities. Figure 8.27 shows the sample placement in a cylindrical cavity. If the cavity is driven in the TM_{010} mode, the sample is a thin rod of any length (up to the axial length of the cavity), ensuring that the field is constant in the

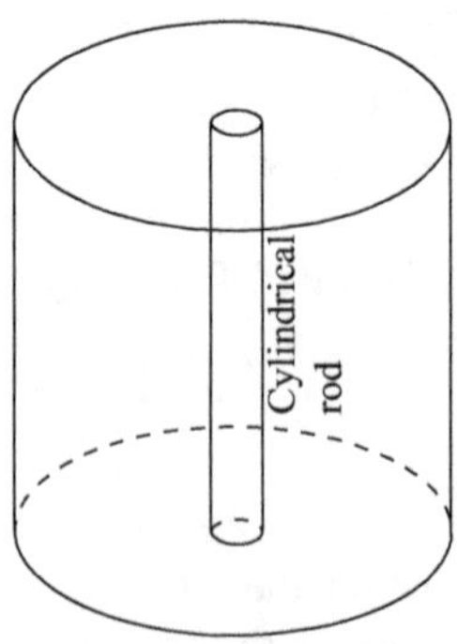

Figure 8.27 Cavity perturbation measurement in a circular cavity driven in a TM_{010} mode

sample. Under these conditions, the perturbation equation in (3.84) for the shift in resonant frequency and (3.108) for the Q-factor produce the following:

$$\varepsilon_r' = 0.539\left(\frac{V}{V_s}\right)\left(\frac{f_0 - f}{f_0}\right) \tag{8.27}$$

$$\varepsilon_r'' = 0.269\left(\frac{V}{V_s}\right)\left(\frac{1}{Q_s} - \frac{1}{Q_0}\right) \tag{8.28}$$

where V is the volume of the cavity, V_s the volume of the sample, f_0 the resonant frequency of the empty cavity, f the resonant frequency of the cavity with the sample, Q_0 the Q-factor of the empty cavity, and Q_s the Q-factor of the cavity with the sample.

In a rectangular cavity driven in the TE_{103} mode, the sample is also a rod and is placed at the center of the cavity as shown in Figure 4.30. The real and imaginary parts of the permittivity were given in (4.148) and (4.149) and repeated here for convenience:

$$\varepsilon_r' = 1 + 0.5\left(\frac{V}{V_s}\right)\frac{f_0 - f}{f} \tag{8.29}$$

$$\varepsilon_r'' = \left(\frac{V}{4V_s}\right)\left(\frac{1}{Q_s} - \frac{1}{Q_0}\right) \tag{8.30}$$

Clearly then, the process is identical and the results are very similar.

One first measures the resonant frequency f_0 of the empty cavity followed by the measurement of Q_0, the loaded Q-factor of the empty cavity as described in Section 8.3.2.4. Then the process is repeated for f and Q_s, and the permittivity calculated from the equations for the specific cavity used.

The measurement of the resonant frequency is typically done through the S_{11}-parameter (for series resonance) or from the S_{21}-parameter (for parallel resonance). The loaded Q-factor is measured from the half-power points in the frequency response of the cavity.

There are a few precautions that should be borne in mind to ensure accurate results.

1. The cavity must be a low-loss cavity to ensure that losses in the cavity walls do not affect the loss tangent. If the sample is very low loss, the losses in the cavity walls may introduce a large error in the calculation of ε''. In very low-loss materials, the sample fills a larger section of the cavity or even the whole cavity. In such cases, (8.27)–(8.30) cannot be used since these were specifically derived from the condition that the sample is thin. In such cases, one has to go back to the perturbation equations and recalculate ε' and ε''.
2. Matching to the cavity is very important, otherwise the S-parameters may be in errors affecting the results.

3. The sample must be machined accurately and must be as small (thin) as possible since, strictly speaking, the field is only constant on the axis of the circular cavity or the center of the rectangular cavity. The thicker the rods, the less accurate the measurements. Nevertheless, it is possible to come up with calculations for samples that are large—for example, disks the radius of the circular cavities if this is necessary.

8.4.1.2 Large sample in the cavity

If the dielectric properties of low loss, or low-density materials (such as foams or gases) are needed, the perturbation method may not be applicable directly. Liquids, for examples, are also difficult to test under the perturbation conditions. For these types of materials, the approach is to include a much larger sample or, if appropriate, to fill the whole cavity. The perturbation method allows for this type of test as was indicated in the more general relations in (3.77), (3.78), and (3.80). These equations however are more difficult to apply since some of the approximations used in the perturbation method for small samples cannot be used. In particular, we cannot assume that the fields internal to the sample remain the same as the fields before the sample was introduced. Instead of using the approximations for the fields, we simply integrate the fields over the cavity volume for the empty cavity and over the material volume, which now may take a significant part of the cavity or the whole cavity. Assuming that the test material fills the whole cavity, we proceed with (3.105) and (3.107) modified as follows:

$$\omega_r - \omega_0 = -\frac{\omega_0 \int_V [(\varepsilon' - \varepsilon_0)\mathbf{E} \cdot \mathbf{E}_0] dv}{2\int_V \varepsilon_0 \mathbf{E} \cdot \mathbf{E}_0^2 dv} \tag{8.31}$$

$$\frac{1}{Q} = \frac{\int_V [\varepsilon'' \mathbf{E} \cdot \mathbf{E}_0] dv}{\int_V \varepsilon_0 \mathbf{E} \cdot \mathbf{E}_0 dv} \tag{8.32}$$

In these relations, E_0 is the field in the empty cavity, E the field in the perturbed cavity, and Q is found from

$$\frac{1}{Q} = \frac{1}{Q_s} - \frac{1}{Q_0} \tag{8.33}$$

where Q_0 is the Q-factor of the empty cavity and Q_s that of the perturbed cavity.

If the test material only fills a portion V_s of the cavity, then the integral in the nominator is over that volume only.

The disadvantage of this approach is simply in the difficulty of performing the integrals since the approximation we used previously to obtain (8.27)–(8.30), namely that $E \sim E_0$, cannot be used. That is, the formulas for $\Delta\omega/\omega_0$ and for $1/Q$ are more complex. However, from a testing point of view, we proceed with the same steps:

1. Measure the resonant frequency for the S_{11}-parameter of the empty cavity to obtain f_0.

2. Repeat for the cavity with the material to obtain f_r.
3. Measure Q_s and Q_0 as described above from the half-power points.

These provide the left-hand side of (8.32) and (8.33). The right-hand side is usually calculated analytically or even numerically if need be.

These measurements are easily done on a VNA and, unless the losses in the tested medium are comparable to the losses in the cavity walls, the results are accurate. Of all the methods for evaluation of the dielectric constant, the resonant methods are the most sensitive and often the most accurate. They are also relatively simple to perform once an appropriate cavity and sample have been prepared.

8.4.2 Transmission line methods

Perhaps, the most common method of evaluation of dielectric properties of materials is the transmission line (or waveguide) method described in Section 4.9. In most cases, the reflection-transmission method is used, which implies that the sample is in a transmission line and both the reflection and transmission coefficients are used for the measurement. The configuration was shown in Figure 4.29. The complex permittivity is calculated using (4.135) for nonmagnetic materials:

$$\varepsilon_r = \varepsilon r' + j\varepsilon r'' = \frac{\lambda_0^2}{\mu_r}\left(\frac{1}{\lambda_c^2} - \left[\frac{1}{2\pi l}\ln\left(\frac{1}{T}\right)\right]^2\right) \tag{8.34}$$

where T is found from (4.134):

$$T = \frac{S_{11} + S_{21} - \Gamma}{1 - (S_{11} + S_{21})\Gamma} \tag{8.35}$$

and Γ is given in (4.133):

$$\Gamma = \frac{S_{11}^2 - S_{21}^2 + 1}{2S_{11}} \pm \sqrt{\left(\frac{S_{11}^2 - S_{21}^2 + 1}{2S_{11}}\right)^2 - 1} \tag{8.36}$$

λ_0 is the wavelength in the empty transmission line or waveguide, and λ_c is the cutoff wavelength (equals infinity for transverse electromagnetic (TEM) waves in transmission lines but is finite in waveguides). Clearly then, the measurement is essentially that of measuring the S_{11}- and S_{21}-parameters using a VNA (requires both amplitude and phase of the S-parameters) and then calculating the permittivity.

If the material is magnetic, then the permeability in (4.135) must be found first from (4.136):

$$\mu_r = \frac{1 - \Gamma}{\Lambda(1 - \Gamma)\sqrt{(1/\lambda_0^2) - (1/\lambda_c^2)}} \tag{8.37}$$

where

$$\frac{1}{\Lambda^2} = -\left(\frac{1}{2\pi l}\ln\left(\frac{1}{T}\right)\right)^2 \tag{8.38}$$

As was indicated in Section 4.9, the solution for the imaginary part of permittivity is not unique and the correct solution must be found through additional analytic steps. These usually require the measurement of the group delay on the line, a quantity that can also be obtained using the network analyzer. The process described in Section 4.9 and in this section is often referred to the NRW (Nicholson–Ross–Weir) algorithm and is almost always available on network analyzers as a function.

It goes without saying that the normal steps in measurements such as calibration must precede any measurement and issues of connections and matching must be resolved. The preparation of samples must be carefully done and the fixture (transmission line or waveguide) must be a quality fixture with low wall losses. An example of such a fixture can be seen in Figure 8.19, and the S_{11}- and S_{21}-parameters one can obtain for various materials can be seen in Figures 8.20–8.26. Finally, it is worth mentioning as well that the method is well suited for materials with relatively high permittivities and losses and that it fails around the resonant frequencies. Low-loss materials are best tested in resonant cavities. The method is not suited to measurement of liquids simply because it is not possible to place a sample in the middle of a transmission line or waveguide.

A second option for testing in transmission lines and cavities is afforded by the so-called reflection method in which the sample is placed against the shorted end of the transmission line or waveguide as shown in Figure 4.28. There are two advantages to this method: first, only the reflection coefficient due to the sample is used and, second, liquid samples can be tested as well. The network analyzer only measures the S_{11}-parameter. Figure 4.33 shows how a liquid sample can be measured by simply setting the transmission line vertically so that the liquid collects at the bottom above the short. The measurement for a coaxial transmission line using the sample holder in Figure 8.19 is shown in Figure 8.28. With a fluid height t, and assuming the sample is placed against the short, the S_{11}-parameter, i.e., the slab reflection coefficient can be written as

$$S_{11} = \frac{-p + \tanh \gamma t}{p + \tanh \gamma t} \tag{8.39}$$

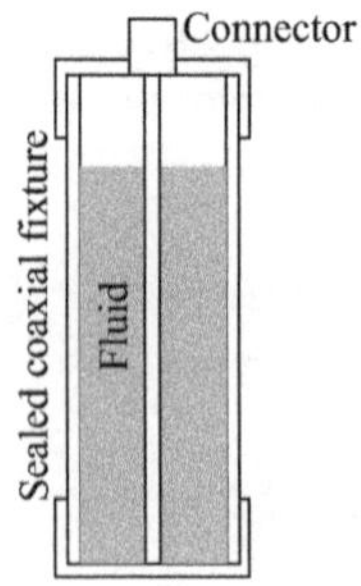

Figure 8.28 Measurement of the properties of a fluid in a shorted coaxial line

where

$$p = \frac{\gamma\mu_0}{\gamma_0\mu} \tag{8.40}$$

$$\gamma = \alpha + j\beta \tag{8.41}$$

α is the attenuation constant in the material under test (MUT), β the phase constant, and γ_0 is the propagation constant in the empty waveguide or line.

The propagation constants as well as the attenuation and phase constants depend on the type of transmission lines or waveguide and on the modes being propagated.

In a coaxial transmission line, only the TEM mode exists and there is no cutoff frequency; therefore, the propagation constant in the MUT is

$$\gamma_{\mathrm{TEM}} = j\omega\sqrt{\mu\varepsilon'}\sqrt{1 - j\frac{\sigma}{\omega\varepsilon}} = j\omega\sqrt{\mu}\sqrt{\varepsilon' - j\varepsilon''} \tag{8.42}$$

The attenuation and phase constants are obtained from γ:

$$\alpha = \omega\sqrt{\frac{\mu\varepsilon'}{2}\left[\sqrt{1 + \left(\frac{\sigma}{\omega\varepsilon'}\right)^2} - 1\right]} = \omega\sqrt{\frac{\mu}{2}\left[\sqrt{(\varepsilon')^2 + (\varepsilon'')^2} - 1\right]} \tag{8.43}$$

$$\beta = \omega\sqrt{\frac{\mu\varepsilon'}{2}\left[\sqrt{1 + \left(\frac{\sigma}{\omega\varepsilon'}\right)^2} + 1\right]} = \omega\sqrt{\frac{\mu}{2}\left[\sqrt{(\varepsilon')^2 + (\varepsilon'')^2} + 1\right]} \tag{8.44}$$

The propagation constant in the empty line is usually taken as that of lossless air:

$$\gamma_0 = j\omega\sqrt{\mu_0\varepsilon_0} \tag{8.45}$$

Once (8.42) and (8.45) are substituted into (8.40) and then (8.40) and (8.42) are substituted into (8.39), there is a direct relation between the permittivity and permeability and the terms of S_{11}. If the material is nonmagnetic, then this relation suffices to calculate ε' and ε'' from the real and imaginary parts of S_{11}. If both the permittivity and permeability are needed, one can either measure two samples, each of a different thickness or one can move the sample away from the short a certain distance Δt and obtain a second relation with this condition. However, since the interest here is only for nonmagnetic materials, we can set $\mu = \mu_0$ and proceed with the solution.

The measurement on a network analyzer consists of measuring the S_{11}-parameter followed by the calculations above. The algorithm necessary to do so may already exist in the network analyzer's computer or may be done separately.

The measurement in waveguides is identical, but the calculations are somewhat different because the propagation constant, and hence the phase and attenuation constants are different and mode dependent. In most cases, the lowest propagating mode is used, which, in a rectangular waveguide, is the TE_{10} mode. Under these conditions, the propagation constant in the waveguide is

$$\gamma_{TE} = \sqrt{k_c^2 - \omega^2\mu(\varepsilon' - j\varepsilon'')} \tag{8.46}$$

where k_c is the cutoff wavenumber for the particular waveguide and mode of propagation. For a rectangular waveguide with the larger dimension equal to a, the cutoff wavenumber is

$$k_c = \frac{\pi}{a} \tag{8.47}$$

We get

$$\gamma_{TE} = \sqrt{\left(\frac{\pi}{a}\right)^2 - \omega^2\mu(\varepsilon' - j\varepsilon'')} \tag{8.48}$$

The rest of the calculations are the same as for the coaxial line as long as the frequency is above the cutoff frequency for the mode and below the cutoff frequency of the next, higher mode.

8.4.3 Measurements in space

Many of the measurements made with network analyzers are in transmission lines and in waveguides including in sections of these that act as resonators. This is entirely understandable since under these conditions, the ports and the impedances are well defined and the fields generated and measured are contained within the structures. There are however other important uses for the network analyzer. One of these is the measurement of propagation properties in open space or in configurations in which the inputs of the network analyzer are not connected directly to transmission line ports. An example is shown in Figure 4.31. In this case, the purpose is to evaluate the material such as in trying to measure its complex permittivity or permeability or perhaps to detect inclusions in the material as part of a nondestructive testing regime.

Looking at Figure 4.31, we can view this as a radar system, whereby the source port transmits power, some of which is reflected back to the port, some of which is transmitted to the load port of the network analyzer. Therefore, the basic measurements of the network analyzer can be extended to this type of configuration with some modifications. The most important modification has to do with the fact that the DUT in Figure 4.31 is on a holder or support structure and that structure modifies the interaction of waves with the DUT. Similarly, one cannot assume that all power is either reflected from or transmitted into the DUT—some may be transmitted through air to the load port of the network analyzer, some may be lost by passing over the DUT into space. Another problem to be resolved is the interaction of waves with structures in the vicinity—walls, furniture, etc. Nevertheless, with careful setups, proper calibration, and test procedures, these measurements can be accurate and repeatable.

The process of obtaining the S-parameters under the conditions in Figure 4.31 is a two-step process:

1. The S-parameters of the holder without the DUT are obtained first.
2. The S-parameter with the holder and the DUT are then obtained separately.

These two sets of S-parameters are now used to deembed the holder effects from the S-parameter leading to a set of parameters that are entirely due to the DUT. The deembedding functions available on most network analyzers are often used to eliminate the effects of holders, connectors, cables, etc. The measurement then proceeds as for the transmission line measurement described above.

Although this type of measurement can be very accurate, some precautions are essential. Reflections from fixed surfaces and objects can be eliminated by the deembedding process, but transient effects such as moving objects and personnel moving in the vicinity may present a totally different signature during the two tests and hence lead to errors. It is also required that the transmission of power through the sample be done with a focused beam using directive antennas to ensure that all power either reflects from or transmits through the sample. That may require that the sample being tested be relatively large and/or that the transmitting and receiving antennas be placed at short distances from each other.

8.5 Integration of network analyzers in designs

The network analyzer is an expensive, accurate, measuring instrument. It is rare that one would consider using it for anything other than a laboratory instrument. Nevertheless, because it is so versatile and because it combines a source and analysis modules, it can be used as part of an industrial sensing system. Its advantage, in addition to its exceptional qualities as a measuring instrument, is that in certain cases, it can replace all the electronics, computation, and analysis requirements of the sensing system. An example is the production line sensing in a cavity resonator of, for example, plastic rods and profiles or fabrics for the purpose of integrity and quality control. In the present work, the thickness of rubber and the coating of fabrics are controlled by an open resonant sensor. In this type of application, the resonant cavity must be set into resonance and the resonant frequency measured accurately to obtain information for the purpose of control of the rubber thickness or the amount of coating material on the fabric. It is a simple matter of using the source port to drive the cavity and the load port to measure the resonant frequency. One can go beyond that and use the S-parameters to monitor the Q-factor as well. And all that over a wide range of frequencies and power levels, all accurately controlled. The computer in the network analyzer can then be employed to do additional tasks and to communicate data out of the system. The use of a dedicated network analyzer as part of the sensor system, if it can be justified in terms of cost, means that the design is limited to the sensor itself (the resonant cavity in this case) and its mechanical components. All other functions are taken care of by the network analyzer. This approach has additional advantages:

1. The single-module approach, whereby all electronics and computational functions are in a single instrument, has advantages in maintenance. In the case of malfunction, the network analyzer can be replaced and the faulty unit sent for repair. The interruption in production is minimal, and the need for skilled personnel is limited.
2. Development time is greatly reduced. In addition to having the hardware available, network analyzers also have powerful software tools that can be brought to bear on the design. The common necessary steps of prototyping the electronic circuits, testing, writing software, etc. are obviated, and the time required to integrate the system reduced to a minimum.
3. Future upgrades can be handled much easier. Replacement of the network analyzer or new software can be done independent of the rest of the system and vice versa.
4. Flexibility of the system is another important point. One can change parameters of the sensing system to match required conditions, and the use of the internal computer allows for additional tasks such as documenting, archiving, and communicating data.
5. The cost itself, even though it is high, may in the end be comparable or even lower than a design based on dedicated electronics, especially when one takes into account development time and possible future upgrades. This is particularly important in microwave systems in which hardware design requires specialty components and special skills.

There are however some limitations to this nonconventional use of the network analyzer:

1. The instrument is not an industrial instrument and may not be able to operate under the conditions often found on the factory floor without some means of protection. This may include protection from the elements, vibrations, excessive humidity and heat, and the like. It will almost certainly have to be placed in an air-conditioned environment, perhaps with some means of power conditioning. Electromagnetic compatibility and interference must also be addressed in high field environments.
2. The power levels available from the instrument are relatively low meaning that it may not be suitable in applications where high power is needed. Clearly, it is limited to low-power applications.

Overall, while there are certainly applications where this approach is not viable, there are many that can benefit from the use of the network analyzer. The designer should at the very least consider the possibility of utilizing a first rate, flexible instrument that can replace many of the functions in the design, and, in the process, reduce design and testing time and produce a better system.

Depending on the application, one may be able to use a simpler network analyzer (scalar or vector) that connects to a computer. These relatively simple instruments are available as front-ends to a computer and are driven by software on the computer. These two-port devices are usually limited in capabilities such as frequency range, power, scan time, and accuracy but are simple and simple to use and may be perfectly suited for many tasks at a small fraction of the cost of a laboratory network analyzer even when including the cost of a dedicated computer.

Further reading

There are many good references on network analyzers as well as well-written tutorials. As a whole, academic works on the subject are less important in the context of this chapter than tutorials and manufacturers' literature. There are also many publications and tips on methods of measurement, algorithms, measurement tips, and so on. There is much to be learned from these publications. The user should start with the guide that comes with the instrument and consult the application notes from the manufacturer. Because different manufacturers produce relatively similar instruments, it is common to find answers to questions on measurement from sources other than those specific to the instrument in use.

The following is a short list of publications that may be of interest, in additions to the references on measurements listed in the "Bibliography" section of Chapter 4. Description of the operation of the network analyzer can be found in the following publications available from various manufacturers of network analyzers.

[1] "Measurement Guide: Vector Network Analyzer for Anritsu RF and Microwave Handheld Instruments," Anritsu Corp., 2016.
[2] "Fundamentals of Vector Network Analysis, Version 1.1," Rohde & Schwarz USA, Inc.
[3] "Measurement Guide: Vector Network Analyzer for Anritsu RF and Microwave Handheld Instruments," 10580-00289, Anritsu Corp., 2016.
[4] "Introduction to Network Analyzer Measurements, Fundamentals and Background, National Instruments."
[5] "Network Analyzer Basics," Agilent Technologies 5965-7917E, 2004.
[6] Z. Wu, "Software VNA and Microwave Network Design and Characterisation," Wiley, New York, 2007.
[7] Hewlett Packard Product Note 8510-3, "Measuring Dielectric Constant with the HP 8510 Network Analyzer," 1985.

Some general interest publications related to network analyzer measurements (as well as spectrum analysis) are as follows:

[8] Rohde & Schwarz, "A test and measurement retrospective — 75 years of Rohde & Schwarz," Microwave Journal, 10 July 2008.
[9] G. Simpson, "Vector network analysis and ARFTG: a historical perspective," 50th ARFTG Conference Digest, 1997.
[10] J. Verspecht, "Large-signal network analysis," IEEE Microwave Magazine, Vol. 6, No. 4, December 2005, pp. 82–92.

[11] J. P. Dunsmore, "Handbook of Microwave Component Measurements: With Advanced VNA Techniques," Wiley, New York, 2012.
[12] N. Shoaib, "Vector Network Analyzer (VNA) Measurements and Uncertainty Assessment," Springer, 2016.
[13] M. Engelson and F. Telewski, "Spectrum Analyser Measurements," Artech House, Norwood, MA, 1974.
[14] P. Roblin, "Nonlinear RF Circuits and Nonlinear Vector Network Analyzers: Interactive Measurement and Design Techniques," Cambridge University Press, Cambridge, 2011.

The measurement of complex permittivity and permeability has received considerable attention in the literature because of its important in material characterization. In many ways, these measurements are associated with network analyzers since it is very difficult to obtain accurate data at high frequency otherwise. The following are in addition to [11–15] in the "Bibliography" section of Chapter 4. The references to the NRW method of computation are in [16,17]. These are listed here since the computational algorithms are usually integral to the software available on network analyzers.

[15] J. Baker-Jarvis, "Transmission/Reflection and Short-Circuit Line Permittivity Measurements," National Institute of Standards and Technology, July 1990.
[16] W. Weir, "Automatic measurement of complex dielectric constant and permeability at microwave frequencies," Proceedings of the IEEE, Vol. 62, No. 1, January 1974, pp. 33–36.
[17] A. M. Nicholson and G. F. Ross, "Measurement of the intrinsic properties of materials by time-domain technique," IEEE Transactions on Instrumentation and Measurements, Vol. IM-19, No. 4, 1970, pp. 377–382.
[18] R. N. Clarke (Ed.), "A Guide to the Characterisation of Dielectric Materials at RF and Microwave Frequencies," The Institute of Measurement & Control (UK) & NPL, London, 2003.
[19] J. Baker-Jarvis, M. D. Janezic, R. F. Riddle, *et al.*, "Measuring the Permittivity and Permeability of Lossy Materials: Solids, Liquids, Metals, Building Materials, and Negative-Index Materials," NIST Technical Note 15362005.
[20] ASTM Standard D2520, "Test Methods for Complex Permittivity (Dielectric Constant) of Solid Electrical Insulating Materials at Microwave Frequencies and Temperatures to 1650°C," ASTM Standard D2520, American Society for Testing and Materials.
[21] M. Janezic and J. Baker-Jarvis, "Full-wave analysis of a split-cylinder resonator for nondestructive permittivity measurements," IEEE Transactions on Microwave Theory and Techniques, Vol. 47, No. 10, October 1999, pp. 2014–2020.
[22] J. Krupka, A. P. Gregory, O. C. Rochard, R. N. Clarke, B. Riddle, and J. Baker-Jarvis, "Uncertainty of complex permittivity measurement by split-post dielectric resonator techniques," Journal of the European Ceramic Society, Vol. 21, No. 15, 2001, pp. 2673–2676.

[23] "Basics of Measuring the Dielectric Properties of Materials," Agilent Application Note. 5989-2589EN, April 28, 2005.
[24] "Electronic vs. Mechanical Calibration Kits: Calibration Methods and Accuracy," White Paper, Agilent Technologies 5988-9477EN, 2003.
[25] "In-Fixture Measurements Using Vector Network Analyzers," Application Note AN 1287-9, Agilent Technologies 5968-5329E, 2000.
[26] D. K. Godgaonkar, V. V. Varadan and V. K. Varadan, "Free space measurement of complex permittivity and complex permeability of magnetic materials at microwave frequencies," IEEE Transactions on Instrumentation and Measurements, Vol. 39, No. 2, 1990, pp. 387–394.
[27] P. K. Kadaba, "Simultaneous measurement of complex permittivity and permeability in the millimeter region by a frequency domain technique," IEEE Transactions on Instrumentation and Measurements, Vol. 33, 1984, pp. 336–340.

Appendix A
Electromagnetic radiation safety

A.1 Introduction

The sensors described in this work use an open stripline resonator driven by a network analyzer. As the sensors are open, it is important to ensure that the radiated fields are within safety standards, so that personnel are properly protected.

These sensors use frequencies in the range of 350–450 MHz generated by an Agilent model 8712ET vector network analyzer. The reference power of the analyzer is 1 mW, and the highest power used in any of the sensors was 16 dBm or 40 mW. Although the power is low, the fields in the resonant sensor can be rather high, and because personnel can get close to the open resonator, it is important to establish the risk associated with exposure to these fields.

The electromagnetic fields were measured using a Wayne Kerr spectrum analyzer and two types of receiving antennas, a hand-held monopole and a larger dipole receiving antenna. Measurements were taken at distances ranging from 50 mm to 10 m from the network analyzer's source antenna, which is located in the resonant cavity.

The measurement results were compared to the 1.25 mW/cm^2 power density exposure limit per published standards for the frequency range in which the sensor device will operate. The limits are those published in the American Conference of Governmental Industrial Hygienists worldwide publication, "2017 Threshold Limit Values (TLV) and Biological Exposure Indices."

In the United States, the limits defined by the Federal Communication Commission (FCC) are somewhat different and are given in Table A.1. These values are published in the "OET Bulletin No. 65 (August 1997), Evaluating Compliance With FCC Guidelines for Human Exposure to Radio Frequency Electromagnetic Fields." The FCC distinguishes between occupational/controlled and general population/uncontrolled exposures and, within these categories, allows larger exposures on the extremities—hands, wrists, feet, and ankles as shown. The limits for the general population is lower by a factor of 5 compared to the occupational limits, but in both cases, the limits allowed on the extremities are 50 times higher than on the body. Table A.1 also shows limits on specific absorption rate (SAR). These are two different ways of assessing the risk from exposure. Power density is specified on the surface of the body, whereas SAR is specified in the volume.

Table A.1 FCC limits of allowable exposure to electromagnetic fields in the very high frequency (VHF) and ultra high frequency (UHV) ranges

		Occupational	**General population**
Power density	30–300 MHz	1 mW/cm^2	0.2 mW/cm^2
	300–1,500 MHz	f/300 mW/cm^2	f/1,500 mW/cm^2
SAR		0.4 W/kg* (8 W/kg†)	0.08 W/kg* (1.6 W/kg†)

*Averaged over the whole body.
†Averaged over any 1 g of the body.

Although the limits in Table A.1 are somewhat different than those in the TLV document, they are close and below those in the TLV document. It should also be noted that other countries have their own limits and are often lower than those indicated here. Other bodies, including professional societies, trade organizations, national offices, and international agencies, publish limits and standards for exposure. However, for the sake of simplicity, the field levels measured as part of this work are compared to the FCC and TLV documents. It is shown that the fields produced by the sensors described here are orders of magnitude lower than these limits; hence, they also comply with any of the other sources available. In the interest of generality, some of the more important documents on the subject are listed in the "Bibliography" section.

A.2 Field measurements

The spectrum analyzer performs measurements in terms of power ratios (dBm—with reference to 1 mW) or in terms of voltage ratios (dBμV—with reference to 1 μV). These two are entirely equivalent since any reading in dBm can be converted into dBμV as follows:

$$\mathrm{dB\mu V} = \mathrm{dBm} + 107\ \mathrm{dB} \tag{A.1}$$

However, what we need here are the electric-field intensities at the location of the antenna rather than the dBm readings at the input to the spectrum analyzer. To convert the readings, the following is used:

1. The power at the input to the spectrum analyzer is

$$P_{\text{in}} = \frac{V_a^{\,2}}{2R_L} = \frac{V^2_{a(\text{rms})}}{R_L} \tag{A.2}$$

where R_L is the input impedance to the spectrum analyzer (50 Ω). Since values are typically measured as rms, the following will be done as rms values, but peak values obviously are obtained just as easily.

2. The power may equally well be written in terms of the incident electric-field intensity:

$$P_{\text{in}} = pqSA_e = pq\frac{E^2{}_{i(\text{rms})}}{\eta_0}A_e \tag{A.3}$$

where A_e is the antenna effective area and p and q are the polarization factors of the antenna. S is the time-averaged power density at the location of the antenna. The effective area of the antenna is given as

$$A_e = G\frac{\lambda^2}{4\pi} = G\frac{c^2}{4\pi f^2} \tag{A.4}$$

where G is the antenna gain, c the speed of light and f the frequency of the wave. The polarization factors p and q indicate the fraction of the maximum field that is coupled to the antenna. For simplicity, it was assumed that both p and q equal 1 (i.e., that the antenna is parallel to the maximum electric-field intensity). This is justified from the fact that both horizontal and vertical components of the field are roughly equal. The latter means the field is not polarized. The assumption also guarantees that worst case values are obtained.

Equating the input power in (A.3) and (A.4):

$$pq\frac{E^2_{i(\text{rms})}}{\eta_0}\,G\frac{c^2}{4\pi f^2} = \frac{V^2_{a(\text{rms})}}{R_L} \tag{A.5}$$

or:

$$E^2_{i(\text{rms})} = \frac{V^2_{a(\text{rms})}\,4\pi f^2\,\eta_0}{R_L\,Gc^2\,pq} \tag{A.6}$$

Now, to write this in dBμV, we take 10 $\log_{10}$ on both sides as follows:

$$10\log_{10} E^2_{i(\text{rms})} = 10\log_{10}\left(\frac{V^2_{a(\text{rms})}\,4\pi f^2\,\eta_0}{R_L\,Gc^2\,pq}\right) \tag{A.7}$$

or

$$\begin{aligned} &E_{i(\text{rms})}(\text{dBmV/m}) \\ &\quad = V_{a(\text{rms})}(\text{dB}\mu\text{V/m}) + 20\log_{10} f - 10\log_{10} G - 10\log_{10} R_L \\ &\qquad - 10\log_{10} p - 10\log_{10} q - 10\log_{10}\frac{c^2}{4\pi\eta_0} \end{aligned} \tag{A.8}$$

In this relation, f is measured in MHz and $c = 300$ to ensure proper units. With these, and taking $p = 1$, $q = 1$, $\eta_0 = 377$

$$\begin{aligned} E_{i(\text{rms})}(\text{dBmV/m}) &= V_{a(\text{rms})}(\text{dBmV/m}) + 20\log_{10} f - G(\text{dB}) \\ &\quad - 10\log_{10} R_L - 12.787 \end{aligned} \tag{A.9}$$

In this configuration, $R_L = 50\ \Omega$, $f = 364$ MHz. Also, the gain of the antenna can be approximated as $G = 1.642$. A half wavelength antenna has a gain of 2.15 dB. The antenna used for the measurements was smaller, but, since an infinitesimal dipole has a gain of $G = 1.5$ or 1.76 dB, a value of 2 is a good approximation. With the readings we have taken, which will be denoted here as M (dBm), the incident electric-field intensity can be written as

$$E_{i(\text{rms})}(\text{dB}\mu\text{V/m}) = M(\text{dBm}) + (107 + 51.22 - 2 - 16.99 - 12.787)\text{dB} \tag{A.10}$$

or:

$$E_{i(\text{rms})}(\text{dB}\mu\text{V/m}) = M(\text{dBm}) + 126.44(\text{dB}) = k \tag{A.11}$$

Finally, the electric field itself is calculated as

$$E_{i(\text{rms})} = 10^{0.05k} \times 10^{-6}\ (\text{V/m}) \tag{A.12}$$

Figure A.1 shows the lines along which the measurements were taken relative to the sensor, and Table A.1 shows the actual measured values in V/m. Additional measurements were taken, but those shown in Table A.2 are the largest values.

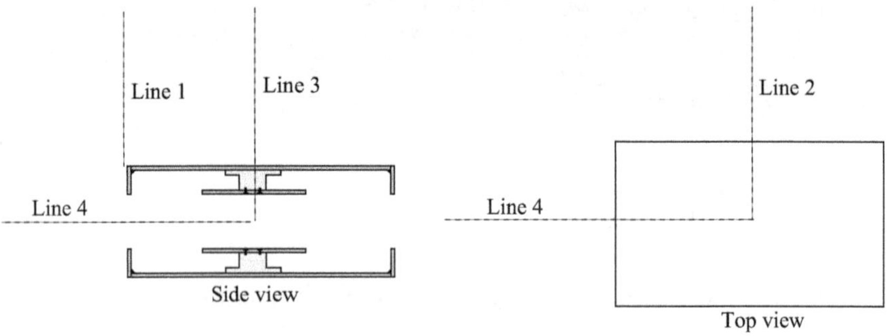

Figure A.1 Schematic showing the lines on which measurements of the electric-field intensity were taken. The values are shown in Table A.1

Table A.2 Electric-field intensities along the lines shown in Figure A.1

Distance (m)	Line 1 (μV/m)	Line 2 (μV/m)	Line 3 (μV/m)	Line 4 (μV/m)
0				
0.3	118.16	130.58	112.39	144.32
0.5	124.21	101.7	130.58	144.32
0.5	118.16		101.7	118.16
1	48.04	48.04	48.04	68.17
2	35.59	41.35	35.59	55.82
3	37.41	25.08	37.41	55.82
4	25.08	39.33	27.71	45.7
5	25.08	35.59	23.86	30.63
6	26.36	33.85	16.81	27.72

A.3 Conclusions

The largest field measured is less than 150 μV/m. This corresponds to a maximum power density of $(150\ \mu V/m)^2/377 = 59.7 \times 10^{-12}\ W/m^2$. The magnetic-field intensities were not measured separately, but rather it was assumed that the waves are plane waves.

Comparing these with the TLVs, the largest electric-field intensity is at least six orders of magnitude lower than the allowable TLVs. The same applies to the magnetic-field intensity. Allowable power density is $1.25\ mW/cm^2 = 12.5\ W/m^2$. The power radiated by any of the sensors is at least ten orders of magnitude lower than the allowable TLVs. Since we made some assumptions (such as plane wave propagation), the fields and power are well within the allowable exposure fields. This is not surprising since the total power to the sensor is 40 mW, and only a small part of this leaks out of the sensor.

Bibliography

The issue of exposure to electromagnetic fields has been taken up by many bodies, primarily as a potential health risk. Many of these bodies, especially scientific organizations, focus on methods of measurements and standards rather than the health effects. The references in [1–9] are a sampling of the literature available and the bodies behind it.

Health and safety is of course at the basis of any discussion on exposure to electromagnetic fields. Some of the documents on the relation between exposure and health both for occupational and general public are addressed in [10–15].

[1] D. Koradecka, "Handbook of Occupational Safety and Health," CRC Press, Boca Raton, 2010.

[2] R. Kitchen, "RF and Microwave Radiation Safety," Newness, Oxford, 2001.

[3] "Recommendation of occupational exposure limits (2016–2017): the Japan Society for occupational health," Journal of Occupational Health, Vol. 58, September 2016, pp. 489–518.

[4] G. Rosu, G. Samoilescu, M. C. Rau and O. Baltag, "Aspects Regarding the Occupational and Non-Occupational Exposure to Low Frequency and Radiofrequency Electromagnetic Fields," International Conference on Applied and Theoretical Electricity (ICATE), October 2016, pp. 1–6.

[5] International Commission on Non-Ionizing Radiation Protection, "Guidelines for limiting exposure to time-varying electric, magnetic, and electromagnetic fields (up to 300 GHz)," Health Physics, Vol. 74, 1998, pp. 494–522.

[6] "IEEE Standard for Safety Levels with Respect to Human Exposure to Radio Frequency Electromagnetic Fields, 3 kHz to 300 GHz," IEEE Std C95.1-2005 (Revision of IEEE Std C95.1-1991), April 2006, pp. 1–238.

[7] International Electrotechnical Commission, "Human Exposure to Radio Frequency Fields from Hand-Held and Body-Mounted Wireless Communication Devices-Human Models, Instrumentation, and Procedures—Part 2:

Procedure to Determine the Specific Absorption Rate (SAR) for Wireless Communication Devices Used in Close Proximity to the Human Body (Frequency Range of 30 MHz to 6 GHz)," IEC 62209-2:2010 ED1, 2010.
[8] "IEEE Recommended Practice for Measurements and Computations of Radio Frequency Electromagnetic Fields with Respect to Human Exposure to Such Fields, 100 kHz–300 GHz," IEEE Std C95.3-2002 (Revision of IEEE Std C95.3-1991), 2002, pp. i–126.
[9] Telecommunication Standardization Sector of ITU, "Guidance on Measurement and Numerical Prediction of Electromagnetic Fields for Compliance with Human Exposure Limits for Telecommunication Installations," Report, International Telecommunication Union, Geneva, Switzerland, 2008.
[10] "Electric and Magnetic Fields and Health: Review of the Scientific Research from March 1, 2012 to December 31, 2016," Report, Menlo Park, CA, 2017.
[11] "ICT Facts and Figures 2017," International Telecommunication Union (ITU), Geneva, CH, Report, July 2017, ICT Data and Statistics Division, Telecommunication Development Bureau, International Telecommunication Union.
[12] A. Jemal, E. M. Ward, C. J. Johnson, *et al.*, "Annual report to the nation on the status of cancer, 1975–2014, featuring survival," Journal of the National Cancer Institute, Vol. 109, 2017, pp. 1–22.
[13] Scientific Committee on Emerging and Newly Identified Health Risks, "Potential Health Effects of Exposure to Electromagnetic Fields (EMF)," Report, European Commission: DG Health and Food Safety, Luxembourg, January 2015.
[14] World Health Organization, "WHO Research Agenda for Radiofrequency Fields," Report, World Health Organization, Geneva, Switzerland, 2010.
[15] Commission in Science and Technology, "Potential Health Implications from Mobile Communication Systems," Report, European Commission: DG Health and Food Safety, Graz, Austria, October 2007.

Appendix B
Material properties

B.1 Introduction

The success of the sensors described in this work is due to the ability of the open stripline resonator to detect changes in the permittivity of the material being tested. In the case of the coated fabrics, the main material being sensed is the dip solution being picked up by the fabric. Although the solution contains various materials and upon drying it leaves a solid layer that is about 20% of the original solution, the sensing mechanism is based on sensing of water, which is about 80% of the solution. Since the fabric is very thin and the amount of solution is relatively small, the dielectric being sensed is a small fraction of the volume of the sensor. Because of that, what is really sensed is the effective permittivity of the space within the sensor. The same applies to the rubber thickness sensor. The rubber is only a few millimeter thick and again the effective permittivity is the sensed quantity. Nevertheless, the response of the sensor is linear with the effective permittivity because the amount of material involved is small. It is therefore obvious that the dielectric constants of the materials involved are important not the least because of the way calibration is performed.

This short appendix summarizes some of the properties used in this work as well as measurements of permittivity performed for the purpose of ascertaining published values. In some cases, the permittivity was calculated from simulations. Since the resonant frequency of the actual material was measured and given the dimensions of the medium, a simple process of iteration on the permittivity in the simulation until the simulated resonant frequency matched the measured resonant frequency provided the correct value for the material. In other cases, simulations were performed to ascertain the exact value of permittivity.

B.2 Measurements

Although permittivities of most materials we used are available in published documents, one cannot hope to find exact data at the frequencies of interest. In other cases, the relative permittivity is available in a range of values. In still other cases, such as the dip material, data are not available at all since the dip is a mixed composition of a number of substances. In such cases, the permittivity had to be measured.

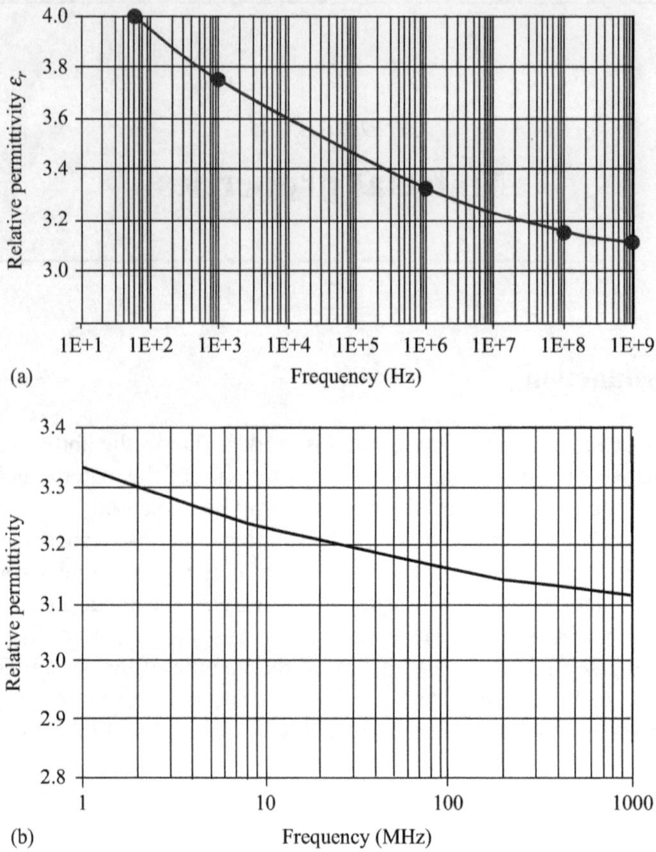

Figure B.1 Relative permittivity of nylon 6/6 with respect to frequency: (a) range between DC and 1 GHz and (b) range between 1 MHz and 1 GHz

Nylon and polyester. One of the fabrics used in this work is made of nylon. To obtain the relative permittivity, it was measured on sheets of Delrin of various thicknesses using a capacitive method in a holder, using the network analyzer. At 400 MHz, the relative permittivity obtained is 3.13, but it varies from 4.0 at very low frequencies—essentially DC to 3.11 at 1 GHz. The data are shown in Figure B.1. Similar measurements on polyester provide a value of 2.92 at 400 MHz with variations from 4.0 at DC to 2.9 at 1 GHz. Figure B.2 shows these data with a few points indicated. Unlike nylon, the scatter in the measurements on polyester is larger and the value of 2.92 is the best estimate that could have been obtained from the measurements. Aramid is also a common fabric. Its permittivity was obtained from published data (see the "Bibliography" section). Its permittivity at the frequencies of interest is between 4.5 and 5.3.

Dip material. As mentioned previously, the dip solution is composed of a number of ingredients. About 80% is water with additions that include epoxy,

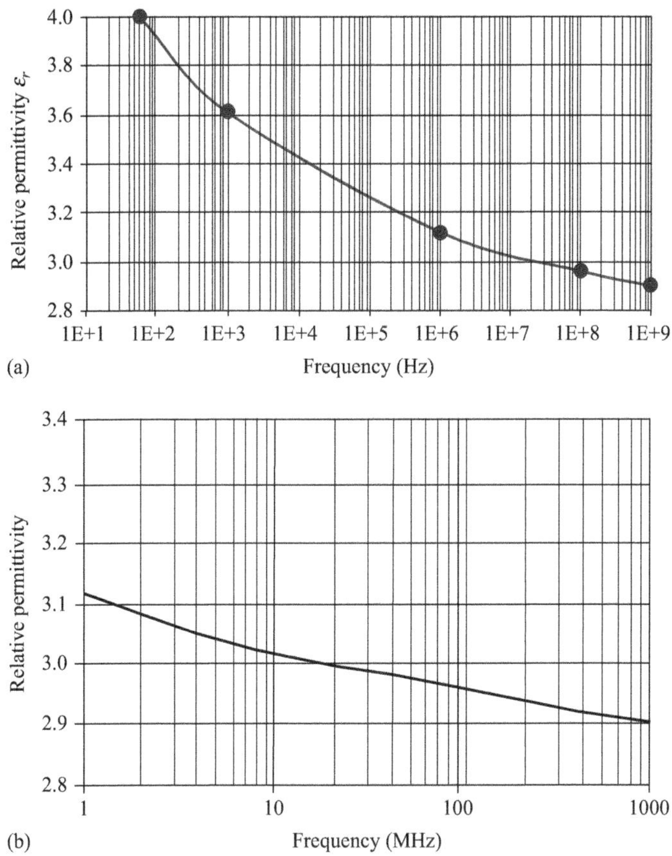

Figure B.2 Relative permittivity of polyester with respect to frequency: (a) range between DC and 1 GHz and (b) range between 1 MHz and 1 GHz

formalin, carbon black, and other additions that make up about 20% of the solution (by weight). Each of the ingredients has its own permittivity, available in published data. We have measured the permittivity of the dip solution using the method described in Chapter 4 in a coaxial resonator as well as used the mixing formula in Section 5.2 to calculate the relative permittivity of the mixed solution. The best estimate for the solution from both measurements and mixing formula is a relative permittivity of 62.9. This is reasonable considering the fact that 80% is water with a permittivity of 78.3 at 400 kHz. The mixing formula is as follows:

$$\sqrt{\varepsilon_{\mathrm{eq}}} = \frac{\sum_i \sqrt{\varepsilon_i} v_i}{\sum_i v_i} \tag{B.1}$$

Or, a simpler, less accurate formula may be used

$$\varepsilon_{\mathrm{eq}} = \frac{\sum_i \varepsilon_i v_i}{\sum_i v_i} \tag{B.2}$$

where ε_i is the relative permittivity of material i, and v_i its volume. The equivalent relative permittivity is that of the mixture. There are other formulas for mixing of dielectrics but all of them are based on certain assumptions including the shape of the composing media (sphere, cylinders, etc.) and hence each formula will produce different results. It should also be noted here that in the case of the fabric sensor, the fabric itself may either be modeled as a uniform mixture of the fabric material and the dip solution, whereas the solution itself is a mixture of material and again can be modeled as particles or as a uniform mixture of its components without any distinction between the constituents. In the cavity, the fabric constitutes a thin layer of dielectric with an effective permittivity based on its composition and dip pickup with the rest of the volume being air. The effective permittivity in the cavity now depends on these two dielectrics. As a simple estimate of the effective permittivity in the cavity, we have used (B.2) first to calculate the effective permittivity of the fabric with dip pickup and then the effective permittivity of the whole cavity volume, again using (B.2). These are very rough estimates and were only used as a guide in estimating the sensitivity of the system. In reality, the fields in various parts of the cavity are vastly different and depend on the mode of resonance. As was described in Sections 3.2.1 and 5.2, the fields of the even mode are highest at the edges of the center plates and in the fabric itself (and parallel to it), whereas the even mode fields are uniform across the fabric (and perpendicular to it). This means that the effect of the fabric on the even mode is larger than what a uniform permittivity throughout the cavity would imply, whereas the effect on the odd mode is smaller. For these reasons, pursuing more accurate methods of estimating the effective permittivity in the cavity was deemed unnecessary, whereas the use of (B.2) was seen as providing an underestimate of the permittivity and hence a lower limit on the sensitivity of the sensor.

In the case of rubber thickness sensing, the rubber itself was modeled as a uniform medium and the effective permittivity of the rubber layer and air was also estimated using (B.2), providing a lower limit on estimated sensitivity. Because in both cases the lower sensitivity limit was higher than the performance required of the sensors, no further refinement in these calculations was needed. Measurements on sample fabrics confirmed these considerations.

Fabric with dip material. The permittivities of the two common fabrics (nylon and polyester) were calculated from the relative permittivities of the base material and the dip pickup as they emerge from the dip bath and move into the sensor. The calculation is based on the mixing formula and the measured values of permittivity given above. The relative permittivity of the nylon fabric with dip is 15.59, whereas the relative permittivity of the polyester fabric is 17.79.

B.3 Effect of humidity and temperature

Humidity in air was a concern in these designs because the sensor is open and the dip solution as it passes through the sensor creates a moist environment close to 100% humidity under almost all conditions. The temperature is also likely to be

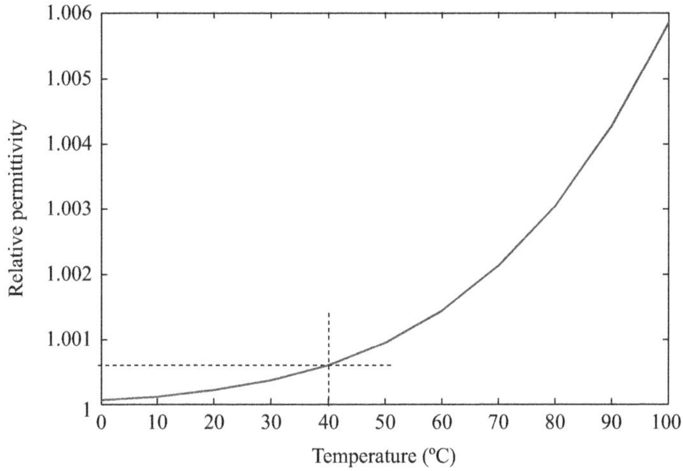

Figure B.3 Relative permittivity of saturated water vapor with respect to permittivity

Table B.1 Relative permittivity for saturated water vapor

Temperature	Relative permittivity
0	1.00007
10	1.00012
20	1.00022
30	1.00037
40	1.00060
50	1.00095
60	1.00144
70	1.00213
80	1.00305
90	1.00428
100	1.00587

high although it can vary considerably on the open factory floor. Direct measurement of the effect of humidity proved to be impractical primarily because of the difficulty of keeping a particular temperature and humidity constant for the measurement. Instead, we resorted to simulations based on published data for humidity. Such data are scarce, with the exception of saturated water vapor (100% humidity). The data for saturated water vapor are shown in Figure B.3 and tabulated in Table B.1.

To get an idea of the effect of humidity, we used these data to simulate the space in the sensor. At zero humidity, the resonant frequency of the empty was 379.3 MHz for the even mode and 424.3 MHz for the odd mode. At 100 °C and 100% humidity (relative permittivity of 1.00587), the resonant frequencies were

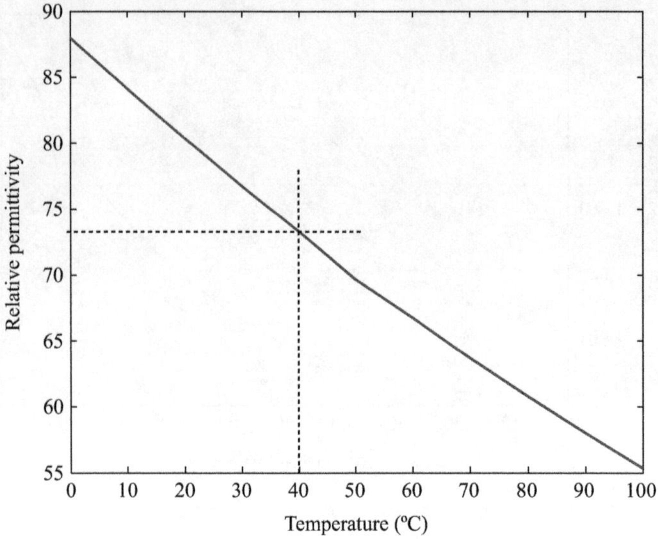

Figure B.4 Relative permittivity of water with respect to permittivity

378.9 MHz for the even mode and 423.7 for the odd mode. The changes seemed to be significant—0.4 MHz for the even mode and 0.6 MHz for the odd mode. However, these extreme conditions are not likely to ever be encountered. At a more reasonable maximum temperature of 40 °C at 100% humidity, the simulation shows a change below 100 kHz. In fact, assuming uniform fields in the resonator, the maximum change in the resonant frequency at 40° should be 120 kHz. This is on the very high end and is roughly twice the measured change. For these reasons, we decided to neglect the effects of humidity on the resonant frequency, a decision that was borne by the in-plant experiments reported in Chapter 6. Nevertheless, one can always opt for a compensation mechanism that uses the odd frequency to monitor humidity (and temperature) since the odd frequency is more sensitive to these effects. This method was used for the fabric sensor and is described in Section 7.5. It cannot be used with the rubber thickness sensor because the distinction between even and odd modes does not exist. Another simple method of compensation is to use a calibration curve using the simulated results and use these values to compensate for the changes in humidity for both sensors. That would require real-time measurements of temperature and humidity in the sensor, something that was not done in the implementations reported in this work.

Another concern related to material properties is the dependence of permittivity of the various constituents of the dip solution on temperature. Because water is the main constituent, the dependency of the permittivity of water on temperature was of primary interest. This dependency is shown in Figure B.4 and tabulated in Table B.2 for some temperatures. Indeed, the changes are significant. However, the temperature variations in plant turned out to be relatively small and hence, in the implementation described in this work, there was no attempt to compensate for any changes due to these effects. If

Table B.2 Relative permittivity of water as a function of temperature

Temperature	Relative permittivity
0	88
10	84.11
20	80.36
30	76.75
40	73.28
50	69.94
60	66.74
70	63.68
80	60.76
90	57.98
100	55.33

necessary, one can measure the temperature of the dip solution in real time and add a compensation scheme to the resonant frequency in the network analyzer itself.

Effect of frequency. The permittivity of most materials is frequency dependent as can be seen from Figures B.1 and B.2 for nylon and polyester. Clearly one has to use the correct permittivity. However, for other materials, the variations may not be as large as that indicated in these figures. The permittivity of water vapor and water also goes down with frequency, but, at the relatively low frequency used in this work, the permittivity is virtually the same as the static (or low frequency) permittivity.

Bibliography

The data for relative permittivity of saturated water vapor were taken from [1]. Permittivity of Aramid. Approximate relative permittivities of Aramid (Kevlar) can be found in various sources, and the values vary considerably depending on composition, frequency, and source. The range is 3.5–4.5 with some sources reporting values as high as 5.3. See for example [2].

[1] J.R. Rumble, ed., "CRC Handbook of Physics & Chemistry," 98th edition, CRC, Boca Raton, Fl, 2017.

[2] M.L. Minges, "Electronic Materials Handbook," ASM International, Materials Park, OH, 1989, p. 608.

Other sources for permittivity are as follows:

[3] W. L. Masterton and E. J. Slowinski, "Chemical Principles," W.B. Saunders, Philadelphia, 1973, p. 221.

[4] R.E. Bolz, "CRC Handbook of Tables for Applied Engineering Science," 2nd edition, CRC, Boca Raton, 1987.

[5] R.J. Lewis, "Sax's Dangerous Properties of Industrial Materials," 5th edition, Wiley, New York, 2012.

Because of the importance of mixed and composite dielectrics, considerable research into mixing formulas has been undertaken. A few references dealing with the issue are listed below.

[6] E. Tuncer, Y. V. Serdyuk and S. M. Gubanski, "Dielectric mixtures: electrical properties and modeling," IEEE Transactions on Dielectrics and Electrical Insulation, Vol. 9, No. 5, 2002, pp. 809–811.

[7] J. A. Reynolds and J. M. Hough, "Formulae for dielectric constant of mixtures," Proceedings of the Physical Society, Section B, Vol. 70, No. 8, 1957, pp. 769–775.

[8] Y. Wu, X. Zhao, F. Li and Z. Fan, "Evaluation of mixing rules for dielectric constants of composite dielectrics by MC-FEM calculation on 3D cubic lattice," Journal of Electroceramics, Vol. 11, 2003, pp. 227–239.

[9] K. K. Karkkainen, A. H. Shivola and K. I. Nikoskinen, "Effective permittivity of mixtures: numerical validation by the FDTD method," IEEE Transactions on Geoscience and Remote Sensing, Vol. 38, No. 3, May 2000, pp. 1303–1308.

[10] D. Polder and J. H. van Santen, "The effective permeability of mixtures of solids," Physica, Vol. XII, No. 5, 1946, pp. 257–271.

[11] H. Looyenga, "Dielectric Constants of Mixtures," Physica, Vol. 31, 1965, pp. 401–406.

In addition, there are many published material data on various websites. These are often published by manufacturers or by researchers. They are not listed here because websites tend to change with time, but a quick search should provide the needed data.

It should be noted as well that different sources list different values for material properties. While confusing, it should be remembered that the method of testing, testing conditions, instrumentation, and sample holders as well as frequency all affect material properties. Different batches of the "same" material will also exhibit variations in material properties because of small variations in composition in the production process. In most cases, published values should suffice, but, when accurate values are needed or when the composition or the structure of materials is unique, one needs to measure the permittivity directly unless the manufacturer has tested the same exact material and the data are available. A good example is the fabric referred to in this work. Although made of well-defined materials, its permittivity depends on the density of the fabric and this is not available in published data. In the fabrication process, the addition of oils, stretching, and normal variations in yarn thickness also produce variations in permittivity.

Appendix C

The finite-difference time-domain (FDTD) method

C.1 The finite difference time domain equations

The simulations used in this work were all done using the finite-difference time-domain (FDTD) method. One of a fairly large number of methods and variations on methods applicable to the simulation of electromagnetic fields, the FDTD is particularly simple and intuitive, and its application is well adapted to computation of high-frequency electromagnetic fields. The FDTD method approximates the electromagnetic-field equations directly without the need for intermediate steps such as the approximation and minimization processes that are so important in many numerical computation methods such as the finite-element methods. To understand how this is done, one can start with general Maxwell's equations in the time domain:

$$\nabla \times \mathbf{E} = -\frac{\partial \mathbf{B}}{\partial t} \tag{C.1}$$

$$\nabla \times \mathbf{H} = \mathbf{J} + \frac{\partial \mathbf{D}}{\partial t} \tag{C.2}$$

$$\nabla \cdot \mathbf{D} = \rho \tag{C.3}$$

$$\nabla \cdot \mathbf{B} = 0 \tag{C.4}$$

where **E** is the electric-field intensity, **B** the magnetic-flux density, **H** the magnetic-field intensity, **D** the electric-flux density, **J** the current density, and ρ the charge density in the volume of interest. All quantities are vectors, except the charge density, which is a scalar. One immediately recognizes (C.1) as Faraday's law, (C.2) as Ampere's law, (C.3) as Gauss's law, and (C.4) as the statement of non-existence of magnetic poles. Alternatively, the equations define the curl and divergence of the electric and magnetic fields to satisfy the Helmholtz theorem which requires that both the divergence and the curl of a vector field be specified for a vector to be uniquely defined. In addition, one has to take into account the material properties in which these fields occur through the constitutive relations:

$$\mathbf{B} = \mu \mathbf{H} \tag{C.5}$$

$$\mathbf{D} = \varepsilon \mathbf{E} \tag{C.6}$$

where μ is the permeability of the medium and ε its permittivity. Both μ and ε are, in general, complex values and are, for the most part, frequency dependent.

The current density $\mathbf{J}$ includes all possible current densities that may exist in the medium. These are source current densities $\mathbf{J}_s$ (such as those that may be applied in an antenna), induced current densities in conductors $\mathbf{J}_e$, and convection current densities $\mathbf{J}_v$ due to motion of charges in the medium. The second term on the right-hand side of (C.2) is also a current density called a displacement current density $\mathbf{J}_d$. The current density in conductors affords the third constitutive relation:

$$\mathbf{J}_e = \sigma \mathbf{E} \tag{C.7}$$

The first step is to remove the source current density from the equations since the sources will be applied to the computational model separately as "driving" functions, in the cases described in this work, through the probes connected to the network analyzer. Then, by using the constitutive relations, we rewrite (C.1) and (C.2) in terms of the electric- and magnetic-field intensity alone:

$$\frac{\partial \mathbf{H}}{\partial t} = -\frac{1}{\mu}(\nabla \times \mathbf{E}) \tag{C.8}$$

$$\frac{\partial \mathbf{E}}{\partial t} = \frac{1}{\varepsilon}(\nabla \times \mathbf{H} - \sigma \mathbf{E}) \tag{C.9}$$

The equations are then written explicitly as follows:

From (C.9) (Ampere's law)

$$\frac{\partial E_x}{\partial t} = \frac{1}{\varepsilon}\left(\frac{\partial H_z}{\partial y} - \frac{\partial H_y}{\partial z} - \sigma E_x\right) \tag{C.10}$$

$$\frac{\partial E_y}{\partial t} = \frac{1}{\varepsilon}\left(\frac{\partial H_x}{\partial z} - \frac{\partial H_z}{\partial x} - \sigma E_y\right) \tag{C.11}$$

$$\frac{\partial E_z}{\partial t} = \frac{1}{\varepsilon}\left(\frac{\partial H_y}{\partial x} - \frac{\partial H_x}{\partial y} - \sigma E_z\right) \tag{C.12}$$

From (C.8) (Faraday's law)

$$\frac{\partial H_x}{\partial t} = -\frac{1}{\mu}\left(\frac{\partial E_z}{\partial y} - \frac{\partial E_y}{\partial z}\right) \tag{C.13}$$

$$\frac{\partial H_y}{\partial t} = -\frac{1}{\mu}\left(\frac{\partial E_x}{\partial z} - \frac{\partial E_z}{\partial x}\right) \tag{C.14}$$

$$\frac{\partial H_z}{\partial t} = -\frac{1}{\mu}\left(\frac{\partial E_y}{\partial x} - \frac{\partial E_x}{\partial y}\right) \tag{C.15}$$

Note again that source-current densities are not included. These as well as boundary and initial conditions are specified separately although sources can be included directly in the formulation that follows.

The derivatives in (C.10)–(C.15) are approximated directly by finite differences in time and space. For this purpose, the partial derivatives are written using the common approximation for finite differences. One possibility is as follows:

$$f'(u) = \frac{df(u)}{du} = \frac{f(u + \Delta u) - f(u)}{\Delta u} \tag{C.16}$$

This is called a forward difference formula since it only uses terms at given points in space or time and points ahead of that. u represents any of the space variables or time and du is the distance between the two points used to approximate the derivative. A similar formula may be written as a backward difference formula:

$$f'(u) = \frac{df(u)}{du} = \frac{f(u) - f(u - \Delta u)}{\Delta u} \tag{C.17}$$

The average between the two provides a central difference formula in which points ahead and behind the center point are used:

$$f'(u) = \frac{df(u)}{du} = \frac{f(u + \Delta u) - f(u - \Delta u)}{2\Delta u} \tag{C.18}$$

One can obtain second-order derivatives as well but the approximation of the equations in (C.10)–(C.15) only require first-order derivatives.

The finite-difference formulas require that the space of interest be divided in space into cells that are Δx by Δy by Δz in size and by so doing define an assembly of points in space at which the values of **E** an **H** are calculated. In practice, one selects $\Delta x = \Delta y = \Delta z = h$, forming a regular mesh. The same must be done in time where, in principle, one would require a separate mesh for each time step as shown in Figure C.1.

In practice, the time approximation is done on the same mesh in the sense that a separate mesh is not required for each time step.

Because of the curl relations in (C.10)–(C.15), the electric-field intensity is calculated using the magnetic-field components in directions other than that of the

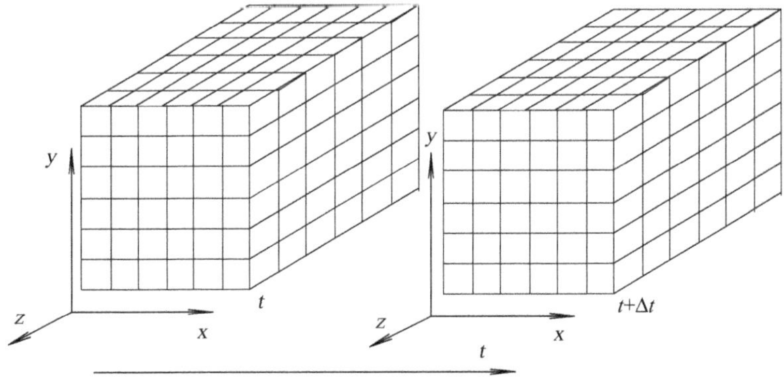

Figure C.1 A three-dimensional mesh for time-dependent applications

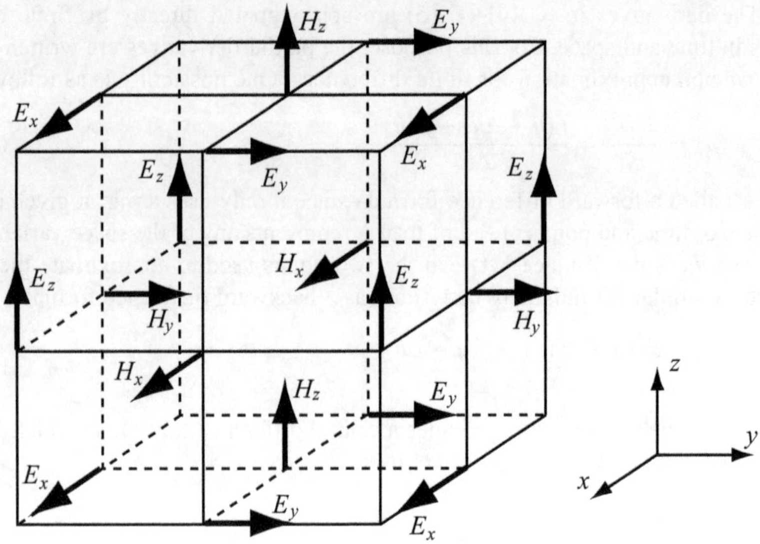

Figure C.2 Spatial relation between the magnetic and electric field. The electric and magnetic fields are calculated at different locations on the grid. The magnetic-field intensity is calculated a half grid away from the electric-field intensity. The center of the cell is at (i,j,k)

electric-field component. For example, to calculate the x-component of the electric-field intensity requires the y- and z-components of the magnetic-field intensity and vice versa. For this reason, each cell in the mesh is subdivided into 2 in each direction as shown in Figure C.2, and the electric and magnetic fields are calculated at half-cell distances away from each other both in time and space.

Because the formulation links the magnetic- and electric-field components in each of (C.10)–(C.15), the calculation of the electric-field components in the first three equations requires the calculation of the magnetic field from the second set of three equations and vice versa. For this reason, the calculation proceeds in a leapfrog fashion by which the electric field is calculated at a certain time step assuming the magnetic field is known. Then, the magnetic field at the same time step but different location in space is calculated on the basis of the now-known electric-field components.

To do so, a grid is defined over the space of the solution domain as shown in Figures C.1 and C.2. The spatial distribution in each dimension is assumed to be the same, denoted by h but that is only a convenience to simplify the relations. Based on this basic cell, (C.10)–(C.15) are discretized as follows. First, a finite-difference scheme is defined for space and for time discretization:

$$\frac{\partial E^m_{i,j,k}}{\partial x} = \frac{\partial E^m_{i+1/2,j,k} - \partial E^m_{i-1/2,j,k}}{h} \tag{C.19}$$

$$\frac{\partial E^{m}_{i,j,k}}{\partial t} = \frac{\partial E^{m+1/2}_{i,j,k} - \partial E^{m-1/2}_{i,j,k}}{\Delta t} \tag{C.20}$$

where m indicates a point in time and i,j,k a point in space, h is the cell size, and Δt is the time step. The notation 1/2 indicates a half step in time and/or space. Applying these approximations to (C.10)–(C.15) on the grid in Figure C.2:

$$\begin{aligned} &H_x^{n+1/2}(i,j+1/2,k+1/2) \\ &\quad = H_x^{n-1/2}(i,j+1/2,k+1/2) \\ &\quad - \frac{\Delta t}{\mu h}\Big[E_z^n(i,j+1,k+1/2) - E_z^n(i,j,k+1/2) - E_y^n(i,j+1/2,k+1) \\ &\quad + E_y^n(i,j+1/2,k)\Big] \end{aligned} \tag{C.21}$$

$$\begin{aligned} &H_y^{n+1/2}(i+1/2,j,k+1/2) \\ &\quad = H_y^{n-1/2}(i+1/2,j,k+1/2) \\ &\quad - \frac{\Delta t}{\mu h}\left[E_x^n(i+1/2,j,k+1) - E_x^n(i+1/2,j,k) - E_z^n(i+1,j,k+1/2) \right. \\ &\quad \left. + E_z^n(i,j,k+1/2)\right] \end{aligned} \tag{C.22}$$

$$\begin{aligned} &H_z^{n+1/2}(i+1/2,j+1/2,k) \\ &\quad = H_z^{n-1/2}(i+1/2,j+1/2,k) \\ &\quad - \frac{\Delta t}{\mu h}\Big[E_y^n(i+1,j+1/2,k) - E_y^n(i,j+1/2,k) - E_x^n(i+1/2,j+1,k) \\ &\quad + E_x^n(i+1/2,j,k)\Big] \end{aligned} \tag{C.23}$$

$$\begin{aligned} E_x^{n+1/2}(i+1/2,j,k) &= \frac{1}{(\varepsilon/\Delta t + \sigma/2)_{i+1/2,j,k}} \\ &\times \left\{ \begin{aligned} &(\varepsilon/\Delta t - \sigma/2)_{i+1/2,j,k} E_x^{n+1/2}(i+1/2,j,k) \\ &\quad + \frac{1}{h}\Big[H_z^{n+1/2}(i+1/2,j+1/2,k) - H_z^{n+1/2}(i+1/2,j-1/2,k)\Big] \\ &\quad - \frac{1}{h}\Big[H_y^{n+1/2}(i+1/2,j,k+1/2) - H_y^{n+1/2}(i+1/2,j,k-1/2)\Big] \end{aligned} \right\} \end{aligned} \tag{C.24}$$

$$E_y^{n+1/2}(i,j+1/2,k) = \frac{1}{(\varepsilon/\Delta t + \sigma/2)_{i,j+1/2,k}} \times \left\{ \begin{array}{l} (\varepsilon/\Delta t - \sigma/2)_{i,j+1/2,k} E_y^n(i,j+1/2,k) \\ \quad + \frac{1}{h}\left[H_x^{n+1/2}(i,j+1/2,k+1/2) - H_x^{n+1/2}(i,j+1/2,k-1/2) \right] \\ \quad - \frac{1}{h}\left[H_z^{n+1/2}(i+1/2,j+1/2,k) - H_z^{n+1/2}(i-1/2,j+1/2,k) \right] \end{array} \right\} \tag{C.25}$$

$$E_z^{n+1/2}(i,j,k+1/2) = \frac{1}{(\varepsilon/\Delta t + \sigma/2)_{i,j,k+1/2}} \times \left\{ \begin{array}{l} (\varepsilon/\Delta t - \sigma/2)_{i,j,k+1/2} E_z^n(i,j,k+1/2) \\ \quad + \frac{1}{h}\left[H_y^{n+1/2}(i+1/2,j,k+1/2) - H_y^{n+1/2}(i-1/2,j,k+1/2) \right] \\ \quad - \frac{1}{h}\left[H_x^{n+1/2}(i,j+1/2,k+1/2) - H_x^{n+1/2}(i,j-1/2,k+1/2) \right] \end{array} \right\} \tag{C.26}$$

The most important observation here is that the electric and magnetic fields are interlaced and calculation of one field can only occur after the second has been evaluated. Note also that material properties ε and σ are specified for each mesh point at which E is evaluated. That means that they can vary from point to point and hence one can easily model material variations within the solution domain. On the other hand, it is assumed here that the permeability is constant throughout the solution space. Although the permeability can also be specified at each point, this is not usually necessary, especially in the context of the present work. Assuming that initial values for the electric- and magnetic-field intensities are specified, the magnetic field components in (C.21)–(C.23) are first evaluated. Then one evaluates the electric field intensity components in (C.24)–(C.26). Note that these values, both for the electric and magnetic fields, are at a time step $n + 1/2$ and are based on the previous time step n. Once the fields are evaluated at all points of the mesh at the current time step, the time step advances by Δt, and the process repeats until an error criterion has been satisfied.

A couple of additional observations are in order here. First, the mesh cell size was assumed to be equal in the three dimensions ($\Delta x = \Delta y = \Delta z = h$) in each cell and uniform throughout the mesh. This is not a necessary condition, and, in fact, one can use different values for Δx, Δy, and Δz by replacing h in (C.21)–(C.26) with the appropriate value. In principle, each cell can be of different size although that would complicate the algorithms needed for computation considerably. Some of this is done in adaptive algorithms, but in FDTD, it is most common to use a constant uniform value. The second observation is that the conductivity in the space is assumed to be low, that is, that if conductivity exists, it is due to a lossy dielectric rather than a conductor. In conductors, the displacement current is negligible and hence Maxwell's equations reduce to a diffusion process rather than a wave process. In such cases, the FDTD method is not the best approach for solution.

The cell spacing and time step used for the geometrical mesh and the time discretization must satisfy the following stability criterion:

$$\Delta t \leq \frac{1}{v_p\sqrt{(1/\Delta x)^2 + (1/\Delta y)^2 + (1/\Delta z)^2}} \quad \text{or}$$
$$\Delta t \leq \frac{h}{v_p\sqrt{3}} \qquad \text{for} \qquad \Delta x = \Delta y = \Delta z = h \tag{C.27}$$

where v_p is the maximum phase velocity in the discretization space. From a practical point of view, one would prefer that Δt be as large as possible and hence, h should be as large as possible so that solution can be fast. The latter however is constrained by the accuracy required and is usually taken as $h < \lambda/10$ where λ is the wavelength in the medium being discretized. In many cases, even smaller spatial steps may be needed to simulate features that may be smaller than $\lambda/10$. That then constrains the time step to very small values. For this reason, an FDTD model requires a very large mesh. As an example, consider a 1-m^3 space in which the fields are to be simulated at 1 GHz. The wavelength is 0.03 m, requiring a mesh of at least $33\times33\times33$ or about 36,000 points. The time step that corresponds to the stability criterion is 5×10^{-11} s. This would require 20 time steps per cycle of a sinusoidal source and may require thousands of cycle to obtain a solution with the required accuracy. In practice, a 30-mm special spacing may be too large to account for features in the solution space. A 1-mm resolution would require 10^6 nodes and a time step of 10^{-12} s. Although the model may be vast, the calculations are rather simple, and in general, FDTD solutions are rather fast.

C.2 Boundary conditions

Any numerical model must be bounded in space and limited in time if one hopes to obtain a valid solution in finite solution time. In cases where the boundaries coincide with conducting surfaces, the reflecting condition on the boundary may be used, that is, any wave impinging on a (perfect) conductor is reflected, simply modifying the difference formulae at the boundary to take into account the behavior of the fields at the conducting boundary. On the other hand, unbounded space requires special attention. To bound an unbounded space, one can introduce an artificial boundary that encloses the region of interest. For this boundary to be valid, it must be "transparent" to ongoing waves. That means it should not reflect any of the waves back into the solution domain. This can be accomplished in a number of ways. One is to define an absorbing boundary condition (ABC), which modifies the difference formulae to ensure there are no reflections. These boundary conditions are mathematical in nature. An important aspect of absorbing boundary conditions is that they are only approximate. One can improve their performance but one cannot entirely eliminate reflections.

Another approach is to actually introduce an artificial medium with material properties at the boundary, again, to ensure there are no reflections from that medium. These boundaries are called perfectly matched layers (PMLs). PMLs are designed to be frequency and angle of incidence independent and to be as close as possible to true matched layers.

Both the ABC and PML approaches are rather complex mathematically and hence will not be described here. These are usually incorporated in commercial implementation of the FDTD method requiring little intervention from the user.

C.3 Near-to-far-field transformation

In many cases, the fields in the vicinity of sources such as antennas are the quantities of interest. In this work, we are particularly interested in the fields within and in the vicinity of open resonant cavities. However, if one requires the fields at large distances from the sources such as the far fields of antennas or the effect of the open cavity fields on personnel, one must find a way of deriving these fields without modeling the very large spaces that would otherwise be needed. To do so, one can use a near-to-far-field transformation. The far-field transformation arises directly from Huygens principle. In the case here that means that the electric and magnetic fields on any surface enclosing the sources can be viewed as the sources for the far fields. Thus, the near-to-far-field transformation starts with a convenient artificial surface enclosing the sources. Then the equivalent electric- and magnetic-current densities are defined as $\hat{\mathbf{n}} \times \mathbf{H}$ and $\mathbf{E} \times \hat{\mathbf{n}}$. These equivalent current densities then serve to calculate the far fields. Again, as with boundary conditions, the transformation is usually incorporated into commercial FDTD software.

C.4 Modeling material interfaces

The approximations in (C.21)–(C.26) were defined on a Cartesian grid, and material properties are also inserted into the equations based on this grid. As long as geometry boundaries and material interfaces comply with this structure, the application of these equations for calculation presents no issues. However, any boundary or interface that is not rectangular in nature must somehow adapt to this structure. Figure C.3(a) shows a curved interface on a finite-difference grid. Clearly, most nodes do not fall on the interface. The simplest solution to this problem is to approximate the interface using the staircase shape that passes through nodes of the mesh as shown in Figure C.3(b). The advantage of this approach is that the mesh does not need to be modified and, if the steps are small compared to the wavelength, the solution is usually acceptable. If necessary, one can decrease the step size (h) to accommodate these types of interfaces or boundary, but this increases the mesh size and decreases the time step resulting in larger and slower models. Figure C.4 shows this aspect of modeling and its consequences.

There are however alternative methods that are much more effective in modeling curved boundaries and interfaces, without the need for excessive refinement

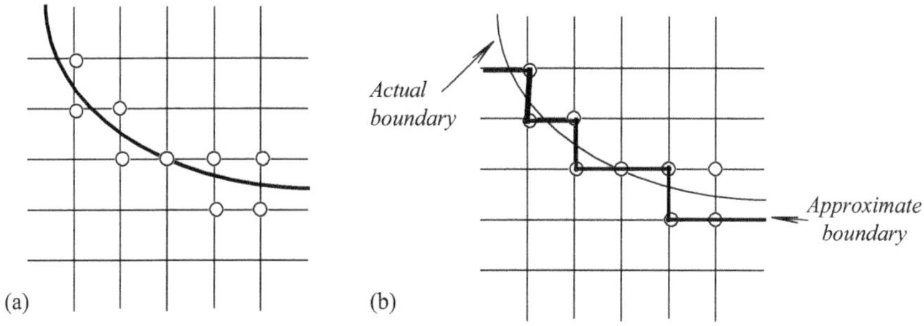

Figure C.3 (a) Curved boundary in a finite-difference grid and (b) approximation of a curved boundary by nearest nodes

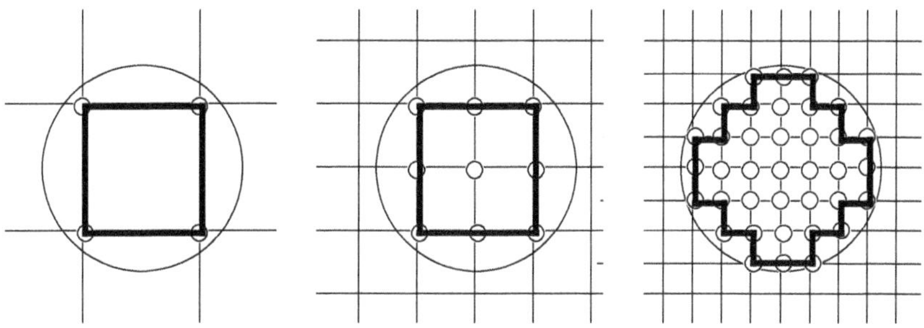

Figure C.4 Improving curved boundary representation by mesh refinement

of the mesh. These are available under the generic name of conformal FDTD cells or algorithms. The methods can treat curved conducting boundaries or dielectric interfaces without the staircasing effects that would otherwise reduce accuracy. Conformal FDTD grids are nonorthogonal and can follow the shape of a boundary accurately. They rely on modifications of the grid locally without affecting the grid elsewhere. There are many methods of doing so but perhaps the simplest is based on keeping the grid itself unchanged but modifying the permittivity at the appropriate nodes based on their position with respect to the interface.

Conducting boundaries can also be treated using an interpolation technique as shown in Figure C.5. In this method, any grid line (lines in a grid are artifices: only the nodes actually exist and are used) that cuts the boundary creates a node on the boundary at the point of intersection. The result is a grid, with nodes on the boundary, but the uniformity of the mesh has been lost. To use this grid, we perform interpolation between the available points and find a modified expression for the nodes at the boundary. By using Figure C.5, the two dark points on the boundaries can be approximated by linear interpolation as

$$V_{11} = V_5 \frac{\Delta x + a}{\Delta x} - V_4 \frac{a}{\Delta x} \tag{C.28}$$

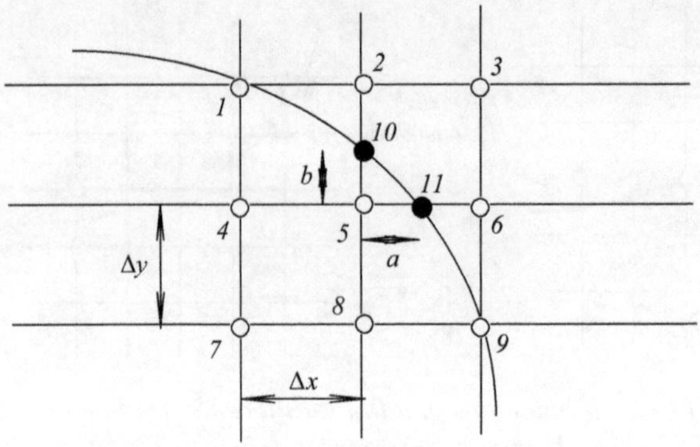

Figure C.5 Use of interpolation to approximate nodes on curved boundaries

$$V_{10} = V_5 \frac{\Delta y + b}{\Delta y} - V_8 \frac{b}{\Delta x} \tag{C.29}$$

These equations may be used, for example, to calculate the value at node 5 in terms of values at nodes 4, 8, and the two boundary values V_{10} and V_{11}. (V is used here as a generic variable.)

C.5 Inclusion of sources

As was mentioned above, the sources were so far excluded from the FDTD approximation. Of course, they must be part of the solution and hence they must be reintroduced. There are a number of ways this can be done. The most obvious method is to keep the sources in (C.8) and hence to modify (C.10)–(C.12) as well as (C.24)–(C.26) to include the sources. All that is required is to add the term $-J_x^{n+1/2}(i+1/2, j, k)$ in the curly brackets in (C.24), the term $-J_y^{n+1/2}(i, j+1/2, k)$ in the curly brackets in (C.25), and $-J_z^{n+1/2}(i, j, k+1/2)$ in the curly brackets in (C.26). If the components of the current density J_x, J_y, and J_z at the nodes of the mesh are known (many of which would normally be zero), that specifies the sources directly into the approximations and drives the solution.

In many cases, however, the driving function is not a current density but, perhaps, the electric-field intensity produced by an antenna on part of the boundary of the solution domain, or at a point (or points) within it. In such cases, these points are held at the required electric-field intensity as boundary values and serve as the driving functions for the solution. For example, in the simulations in Chapter 5, the transmission line resonator is driven by a probe, which in turn is supplied by the network analyzer. Figure C.6 shows the basic model used for these simulations. The source probe is replaced by a port at which the electric field (or magnetic field) is specified and that serves as a boundary condition at this location (it can be as simple as a single node or, perhaps a few nodes). The load is similarly modeled so that resonance can be detected. The metallic shield and the center plates are

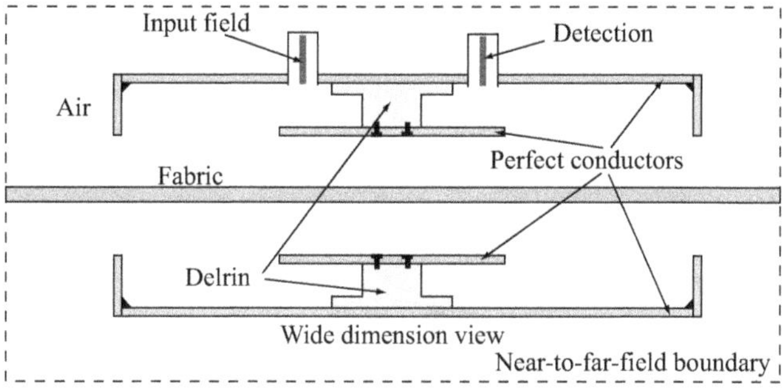

Figure C.6 Geometry model of the resonator with boundary conditions, source, load, and the near-to-far-field transformation

modeled as perfect conductors (perfectly reflecting surfaces), and the open space is modeled with a near-to-far-field boundary on which the near-to-far-field transformation is applied. The Delrin supports are implemented with the dielectric property of Delrin and the fabric with its own properties including the effect of dip pickup on permittivity. Some details such as the penetration of the bolts that hold the center plates into the Delrin supports or the channel in which the supports sit are neglected. In this particular application, the magnitudes of the fields are not important since the quantity of interest is the resonant frequency. That is detected by monitoring the fields in the cavity or at the load port, where, in the actual device, the load probe would be located.

The simulation starts by defining the geometry of the sensor. This includes the folded ground planes and the center plates as shown in the center of Figure 5.8. This is done through the human–machine interface (HMI) of the software, and, while the HMI for various software programs may be different, it usually involves either drawing the object or specifying its coordinates and dimensions as well as the type of materials. The ground planes and center planes are modeled as perfect conductors. In addition, one has to specify the driving source for the simulation. This is the source (or feed) probe in Figure 5.3 or Figure 5.10. In the FDTD program used here, the probe is replaced with an input plane, shown as the rectangular structure (port) at the left side, below the lower center plane in Figure C.7. The second, identical structure is the load port at which the fields are monitored to detect resonance (or, in other terms, to compute the S_{21}-parameter).

These structures, which look like small square-based towers, are designed as well to present the proper impedance to the transmission line connecting to them (not part.of the simulation). In most cases, this would be 50 Ω and would represent the connectors to the transmission lines. Because the resonator is open, one must assume that the electric and magnetic fields outside the resonator extend to infinity, something that computer models cannot simulate. To take this into account, artificial boundaries are created to enclose the geometry at some reasonable distance away (shown in Figure C.7 with triangular markings). These boundaries are used for the near-to-far-field transformation so that the net effect

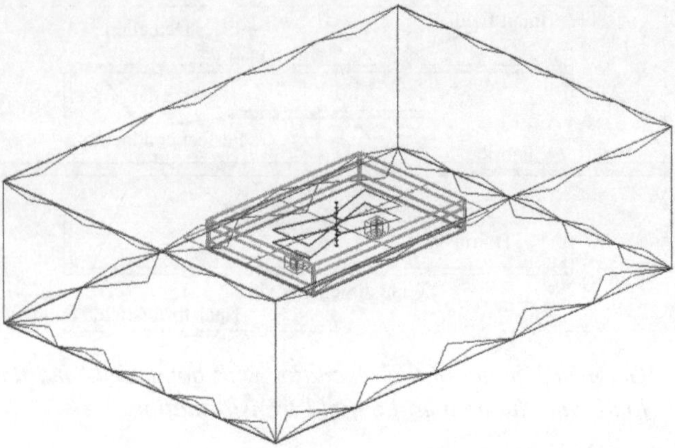

Figure C.7 The input model as displayed by the FDTD program. The outer surfaces indicated by the triangular frill are the near-to-far-field transformation. The center plates in this model are butterfly shaped

is that fields at any distance can be calculated without the need to model large volumes.

As with any numerical method, many more variations and modifications are possible, including definition of grids on cylindrical or spherical geometries, inclusion of anisotropic materials, and others. In all cases, however, it is important to note that for the solution to start, the initial conditions for the fields must be available before one can start (often zero values are used for this purpose). In other cases, perhaps a wave front is available, which can serve as initial condition. Also to be noted again is that material properties are defined point by point. Therefore, anisotropic media are easily taken into account. Finally, the simplicity of this formulation should be again indicated.

To obtain a solution, it is also necessary to properly terminate the space discretization through imposition of boundary conditions and termination in time based on some appropriate criterion. Sometimes boundary conditions are imposed by the geometry. A perfect electric conductor (PEC) is an example of this type of boundary condition. More often, especially with propagating waves, some type of radiation or absorption boundary condition is necessary. These allow close truncation of the solution domain so that only regions of interest are included thereby reducing the time needed for solution. Here again the FDTD method holds a clear advantage in the simplicity with which boundary conditions are taken into account. These boundary conditions are incorporated directly into the formulation. Because of this ease of implementation, some very useful radiation boundary conditions have been introduced allowing the FDTD to work effectively in open geometry configurations. Because of this, it has been the method of choice in high-frequency applications for many years. Nevertheless, it can also be used at low frequencies, either through use of modified radiation boundary conditions or, indeed, through the use of surface impedance boundary conditions (SIBCs).

In addition to boundary conditions, the initial conditions needed to start the solution must be specified. This may take the form of sources at nodes or may take the form of, say, a plane wave, propagating into the solution domain.

There are however some limitation in the use of the FDTD. A basic one, which is shared by other methods such as the transmission line method, is discretization of nonrectangular boundaries. The basic solution to this difficulty is the staircase discretization of such boundaries, that is, one tries to fit the geometry on a rectangular grid. If the staircase is done properly (i.e., sufficiently fine), then this method is acceptable. Others have adapted the FDTD for use in non-Cartesian grids and for nonuniform grids, all in attempts to extend the applicability of the method. Another limitation is the stability criterion defined in (C.27). It is always possible to find a combination of time/space steps that will satisfy these conditions, but the time or space steps may have to be so small as to make the solution either slow and difficult or, indeed, impractical.

Of course, there is much more to the FDTD method than this short appendix implies. The purpose here is merely to indicate some basics of the method rather than to serve as a tutorial.

The software used for the simulations in this work is a commercial one called Concerto. It has many facilities that have not been included in the basics of the method above. First, it features a user interface or editor that allows the user to input a geometry with its parameters and material properties. The issue of stability is handled internally so that the user does not have to worry about it. Boundary conditions may be specified as conducting (PEC boundaries), magnetic boundaries (perfect magnetic conductor boundaries), symmetry planes, and so on. Curved geometries are treated equally well and infinite geometries are handled with special projection methods. From the basic data, the software calculates fields, scans over frequency ranges, calculates resonant frequencies, the Q-factors, and so on.

Bibliography

The FDTD method is usually considered to have started with a paper by Yee [1], but it actually can be traced back to work by Courant and Lewy in 1928 [2]. The texts listed in [3–5] are accessible presentations of the method and discuss a variety of aspects associated with its theory and implementation. Developments, variations, and applications of the FDTD method abound. A small selection of publications is listed in [6–9].

[1] K. Yee, "Numerical solution of initial boundary value problems involving Maxwell's equations in isotropic media," IEEE Transactions on Antennas and Propagation, Vol. 14, No. 3, May 1966, pp. 302–307.

[2] R. Courant, K. O. Friedrichs and H. Lewy, "Über die partiellen Differenzengleichungen der mathematischen Physik," Mathematische Annalen, Vol. 100, No. 1, 1928, pp. 32–74.

[3] A. Taflove and S. Hagness, "Computational Electrodynamics: The Finite-Difference Time-Domain Method," 3rd edition, Artech House, Boston, MA, 2005.
[4] K. S. Kunz and R. J. Luebbers, "The Finite Difference Time Domain Method for Electromagnetics," CRC Press, Boca Raton, FL, 1993.
[5] J. M. Jin, "Theory and Computation of Electromagnetic Fields," Wiley-IEEE Press, New York, 2010.
[6] D. S. Katz, A. Taflove, M. J. Piket-May and K. R. Umashankar, "FDTD analysis of electromagnetic wave radiation from systems containing horn antennas," IEEE Transactions on Antennas and Propagation, Vol. 39, No. 8, 1991, pp. 1203–1212. Bibcode:1991ITAP...39.1203K. doi:10.1109/8.97356.
[7] D. M. Sullivan, O. P. Gandhi and A. Taflove, "Use of the finite-difference time-domain method in calculating EM absorption in man models," IEEE Transactions on Biomedical Engineering, Vol. 35, No. 3, 1988, pp. 179–186. PMID 3350546. doi:10.1109/10.1360.
[8] D. H. Choi and W. J. Hoefer, "The finite-difference time-domain method and its application to eigenvalue problems," IEEE Transactions on Microwave Theory and Techniques, Vol. 34, No. 12, 1986, pp. 1464–1470. Bibcode:1986ITMTT..34.1464C. doi:10.1109/TMTT.1986.1133564.
[9] A. Taflove, "Application of the finite-difference time-domain method to sinusoidal steady state electromagnetic penetration problems," IEEE Transactions on Electromagnetic Compatibility, Vol. 22, No. 3, 1980, pp. 191–202. doi:10.1109/TEMC.1980.303879.

Boundary conditions have received considerable attention, with efforts on accuracy and efficiency and with attempts to improve the performance of absorbing boundary conditions and perfectly matched layers. The following is a selection of publications on the subject (References [10–19]).

[10] G. Mur, "The modeling of singularities in the finite-difference approximation of the time-domain electromagnetic-field equations," IEEE Transactions on Microwave Theory and Techniques, Vol. MTT-29, No. 10, October 1981, pp. 1073–1077.
[11] G. Mur, "Absorbing boundary conditions for the finite-difference approximation of the time-domain electromagnetic-field equations," IEEE Transactions on Electromagnetic Compatibility, Vol. EMC-23, No. 4, November 1981, pp. 377–382.
[12] D. M. Sullivan, O. P. Gandhi and A. Taflove, "Use of the finite-difference time-domain method in calculating EM absorption in man models," IEEE Transactions on Biomedical Engineering, Vol. 35, No. 3, 1988, pp. 179–186. PMID 3350546. doi:10.1109/10.1360.
[13] J. Anderson, M. Okoniewski and S. S. Stuchly, "Practical 3-D contour/staircase treatment of metals in FDTD," IEEE Microwave Guided Wave Letters, Vol. 6, No. 3, March 1996, pp. 146–148.
[14] A. C. Cangellaris and D. B. Wright, "Analysis of the numerical error caused by the stair-stepped approximation of a conducting boundary in FDTD

simulations of electromagnetic phenomena," IEEE Transactions on Antennas and Propagation, Vol. 39, No. 10, October 1991, pp. 1518–1525.

[15] R. Holland, "Pitfalls of Staircase Meshing," IEEE Transactions on Electromagnetic Compatibility, Vol. 35, No. 4, November 1993, pp. 434–439.

[16] J. P. Berenger, "A perfectly matched layer for the absorption of electromagnetic waves," Journal of Computational Physics, Vol. 114, No. 2, 1994, pp. 185–200.

[17] A. P. Zhao, "Uniaxial perfectly matched layer media for an unconditionally stable 3-D ADI–FD-TD method," IEEE Microwave and Wireless Components Letters, Vol. 12, No. 12, December 2002, pp. 497–499.

[18] D. Sullivan and J. L. Young, "Far-field time-domain calculation from aperture radiators using the FDTD method," IEEE Transactions on Antennas and Propagation, Vol. 49, No. 3, March 2001, pp. 464–469.

[19] X. Li, A. Taflove and V. Backman, "Modified FDTD near-to-far-field transformation for improved backscattering calculation of strongly forward-scattering objects," IEEE Antennas and Wireless Propagation Letters, Vol. 4, 2005, pp. 35–38.

[20] T. G. Jurgens, A. Taflove, K. R. Umashankar and T. G. Moore, "Finite-difference time-domain modeling of curved surfaces," IEEE Transactions on Antennas and Propagation, Vol. 40, No. 4, 1992, pp. 357–366. Bibcode:1992ITAP...40..357J. doi:10.1109/8.138836.

[21] S. Dey and R. Mittra, "A locally conformal finite-difference time-domain (FDTD) algorithm for modeling three-dimensional perfectly conducting objects," IEEE Microwave Guided Wave Letters, Vol. 7, No. 9, September 1997, pp. 273–275.

[22] S. Benkler, N. Chavannes and N. Kuster, "A new 3-D conformal PEC FDTD scheme with user defined geometric precision and derived stability criterion," IEEE Transactions on Antennas and Propagation, Vol. 54, No. 6, June 2006, pp. 1843–1849.

[23] T. Su, Y. Liu, W. Yu and R. Mittra, "A conformal meshgenerating technique for the conformal finite-difference time domain (CFDTD) method," IEEE Antennas and Propagation Magazine, Vol. 46, No. 1, February 2004, pp. 37–49.

[24] S. Dey and R. Mittra, "A locally conformal finite-difference time domain (FDTD) algorithm for modeling three-dimensional perfectly conducting objects," IEEE Microwave Guided Wave Letters, Vol. 7, No. 9, September 1997, pp. 273–275.

[25] W. Yu and R. Mittra, "A conformal FDTD algorithm for modeling perfectly conducting objects with curve-shaped surfaces and edges," Microwave and Optical Technology Letters, Vol. 27, No. 2, October 2000, pp. 136–138.

Appendix D
Selected elements of electromagnetics

D.1 Maxwell's equations

The purpose of this appendix is to collect some of the quantities related to propagation of electromagnetic waves and provide definitions to the important quantities that are used in this work. It is of course not possible to deal with the whole theory of electromagnetics nor there is a need for that. It is assumed the reader is familiar with most of the concepts, but there is still value in collecting together these concepts. The reader will, of course, have to refer to more extensive exposition of electromagnetics for the details of the relations given here and for any extension of the concepts beyond the scope of the present work.

As is well known, all aspects of electromagnetics are governed by Maxwell's equations. These form a set that define the divergence and the curl of the electric and magnetic field intensities as required by the Helmholtz theorem, that is, a vector field is defined within an additive constant by its curl and divergence. In addition, one needs to define the relations of the fields with material properties. Maxwell's equations and the constitutive relations are shown in Table D.1 in differential and integral forms. **E** is called the electric field intensity (V/m), **H** the magnetic field intensity (A/m), **D** the electric-flux density (C/m^2), and **B** the magnetic-flux density (T). μ is the permeability of the medium (H/m) and ε the permittivity (F/m). In these relations, **E**, **H**, **B**, **J**, and **D** are complex vector, ρ_v is the volume charge density in the region of interest, Q the total charge (if any), $\boldsymbol{\Phi}$ the magnetic flux, and both ε and μ are, in general, complex tensors. In addition, one defines the relation between current density and the electric field intensity in conducting media and in lossy dielectrics as $\mathbf{J} = \sigma\mathbf{E}$. The Lorenz force equation is usually added to the system of equation and defines the electric and magnetic forces on charges. In (D.11), **v** refers to the velocity of moving charge.

It should be also clarified that these are what one might call macroscopic equations and in most cases we will assume that media are linear, isotropic, and homogeneous although Maxwell's equations do not require that. Under these assumptions, both μ and ε are single valued.

The equations, either in differential form in (D.1)–(D.4) or in integral form in (D.5)–(D.8), contain four vector variables **E**, **D**, **B**, and **H**, and two sources **J** (or I)

Table D.1 Summary of the electromagnetic field equations in differential and integral forms

Maxwell's equations	Differential form		Integral form	
Faraday's law	$\nabla \times \mathbf{E} = -\dfrac{\partial \mathbf{B}}{\partial t}$	(D.1)	$\oint_C \mathbf{E} \cdot d\mathbf{l} = -\dfrac{d\Phi}{dt}$	(D.5)
Ampere's law	$\nabla \times \mathbf{H} = \mathbf{J} + \dfrac{\partial \mathbf{D}}{\partial t}$	(D.2)	$\oint_C \mathbf{H} \cdot d\mathbf{l} = \int_s \left(\mathbf{J} + \dfrac{\partial \mathbf{D}}{\partial t}\right) \cdot d\mathbf{s}$	(D.6)
Gauss's law	$\nabla \cdot \mathbf{D} = \rho_v$	(D.3)	$\oint_s \mathbf{D} \cdot d\mathbf{s} = Q$	(D.7)
No monopoles	$\nabla \cdot \mathbf{B} = 0$	(D.4)	$\oint_s \mathbf{B} \cdot d\mathbf{s} = 0$	(D.8)
Constitutive relations			$\mathbf{B} = \mu \mathbf{H}$	(D.9)
			$\mathbf{D} = \varepsilon \mathbf{E}$	(D.10)
The Lorentz force equation			$\mathbf{F} = q(\mathbf{E} + \mathbf{v} \times \mathbf{B})$	(D.11)

and ρ_v (or Q). The first is a vector source, whereas the second is a scalar source. Each vector variable has three components in space, and, therefore, the equations contain 12 unknown values for the 12 components of the fields. Since the first two equations are vector equations, they are equivalent to six scalar equations. The last two equations [(D.3) and (D.4) or (D.7) and (D.8)] are scalar equations. Thus, we have 8 scalar equations in 12 unknowns.

In fact, the last two equations in each set are not independent of the first two and can be derived from the first two with the aid of the continuity equation. The latter is a statement of conservation of charge and is usually stated as follows:

$$\nabla \cdot \mathbf{J} = -\frac{\partial \rho_v}{\partial t} \tag{D.12}$$

Using this, one can derive (D.3) from (D.2) and (D.4) from (D.1). Therefore, there are really only two independent vector equations for a total of six unknowns. The constitutive equations in (D.9) and (D.10) are used as the remaining six scalar equations to ensure proper solution of Maxwell's equations. That this must be so can also be seen from the fact that Maxwell's equations as written in (D.1)–(D.4) or (D.5)–(D.8) do not refer to material properties at all. On the other hand, we know that fields are very much dependent on materials. This dependency is expressed by the constitutive relations.

D.1.1 Maxwell's equations: the time-harmonic form

Maxwell's equations are often written in terms of phasors, again, assuming linearity in material properties and, often, monochromatic (sinusoidal) excitation. This is the case in this work as well and the use of phasors simplifies both discussion and solution. The time-harmonic differential and integral forms of Maxwell's equations together with the constitutive relations and the Lorentz force are summarized in Table D.2 where all quantities are phasors.

Table D.2 Summary of the time-harmonic electromagnetic field equations

Maxwell's equations	Differential form		Integral form	
	$\nabla \times \mathbf{E} = -j\omega\mathbf{B}$	(D.13)	$\oint_C \mathbf{E} \cdot d\mathbf{l} = -j\omega \int_s \mathbf{B} \cdot d\mathbf{s}$	(D.17)
	$\nabla \times \mathbf{H} = \mathbf{J} + j\omega\mathbf{D}$	(D.14)	$\oint_C \mathbf{H} \cdot d\mathbf{l} = \int_s (\mathbf{J} + j\omega\mathbf{D}) \cdot d\mathbf{s}$	(D.18)
	$\nabla \cdot \mathbf{D} = \rho_v$	(D.15)	$\oint_s \mathbf{D} \cdot d\mathbf{s} = Q$	(D.19)
	$\nabla \cdot \mathbf{B} = 0$	(D.16)	$\oint_s \mathbf{B} \cdot d\mathbf{s} = 0$	(D.20)
Constitutive relations		$\mathbf{B} = \mu\mathbf{H}$		(D.21)
		$\mathbf{D} = \varepsilon\mathbf{E}$		(D.22)
The Lorentz force equation		$\mathbf{F} = q(\mathbf{E} + \mathbf{v} \times \mathbf{B})$		(D.23)

Table D.3 The source-free time-dependent Maxwell's equations

	Differential		Integral	
Faraday's law	$\nabla \times \mathbf{E} = -\frac{\partial \mathbf{B}}{\partial t}$	(D.24)	$\oint_C \mathbf{E} \cdot d\mathbf{l} = -\frac{d\Phi}{dt}$	(D.28)
Ampere's law	$\nabla \times \mathbf{H} = -\frac{\partial \mathbf{D}}{\partial t}$	(D.25)	$\oint_C \mathbf{H} \cdot d\mathbf{l} = \int_s \frac{\partial \mathbf{D}}{dt} \cdot d\mathbf{s}$	(D.29)
Gauss's law	$\nabla \cdot \mathbf{D} = 0$	(D.26)	$\oint_s \mathbf{D} \cdot d\mathbf{s} = 0$	(D.30)
No monopoles	$\nabla \cdot \mathbf{B} = 0$	(D.27)	$\oint_s \mathbf{B} \cdot d\mathbf{s} = 0$	(D.31)

Note that the constitutive relations and the Lorentz force equations have not changed although all vector quantities are now assumed to be phasors. Of course, velocity is still a real number. ε and μ remain unaffected by the phasor notation but they are, in general, complex values. The charge Q or the charge density ρ_v may, in some cases, be time dependent, in which case they also become phasors.

D.1.2 Source-free equations

The general forms of Maxwell's equations can sometimes be simplified if the sources do not need to be taken into account. Under these conditions, the current density $\mathbf{J}$, the charge density ρ_v, or both are removed from the equations, and a much simpler form of the equations is obtained. This is true in the time-dependent or phasor forms of the equations. The time-dependent and time-harmonic source-free Maxwell's equations are summarized in Tables D.3 and D.4. This is the form adopted in the present work since we are primarily interested in resonant frequencies without regard to amplitudes or, for that matter, how the fields in cavities

Table D.4 The source-free time-harmonic Maxwell's equations

Faraday's law	$\nabla \times \mathbf{E} = -j\omega\mathbf{B}$	(D.32)	$\oint_C \mathbf{E} \cdot d\mathbf{l} = -j\omega\oint_s \mathbf{B} \cdot d\mathbf{s}$	(D.36)
Ampere's law	$\nabla \times \mathbf{H} = -j\omega\mathbf{D}$	(D.33)	$\oint_C \mathbf{H} \cdot d\mathbf{l} = j\omega\int_s \mathbf{D} \cdot d\mathbf{s}$	(D.37)
Gauss's law	$\nabla \cdot \mathbf{D} = 0$	(D.34)	$\oint_s \mathbf{D} \cdot d\mathbf{s} = 0$	(D.38)
No monopoles	$\nabla \cdot \mathbf{B} = 0$	(D.35)	$\oint_s \mathbf{B} \cdot d\mathbf{s} = 0$	(D.39)

Table D.5 Electromagnetic interface conditions for general materials

	Electric field	Magnetic field
Tangential components	$E_{1t} = E_{2t}$	$\hat{\mathbf{n}} \times (\mathbf{H}_1 - \mathbf{H}_2) = \mathbf{J}_s$ (A/m) or: $H_{1t} - H_{2t} = J_s^*$ (A/m)
	$D_{1t}/\varepsilon_1 = D_{2t}/\varepsilon_2$	$\hat{\mathbf{n}} \times ((\mathbf{B}_1/\mu_1) - (\mathbf{B}_2/\mu_2)) = \mathbf{J}_s$ (A/m) or: $(B_{1t}/\mu_1) - (B_{2t}/\mu_2) = J_s$ (A/m)
Normal components	$\hat{\mathbf{n}} \cdot (\mathbf{D}_1 - \mathbf{D}_2) = \rho_s$ (C/m^2) or: $D_{1n} - D_{2n} = \rho_s$ (C/m^2)	$B_{1n} = B_{2n}$
	$\hat{\mathbf{n}} \cdot (\varepsilon_1\mathbf{E}_1 - \varepsilon_2\mathbf{E}_2) = \rho_s$ (C/m^2) or: $\varepsilon_1 E_{1n} - \varepsilon_2 E_{2n} = \rho_s$ (C/m^2)	$\mu_1 H_{1n} = \mu_2 H_{2n}$

are generated since these are supplied by the network analyzer in the form of pure sinusoidal signals.

D.1.3 Interface conditions

The behavior of electric and magnetic fields as defined by Maxwell's equations depend on material properties. At the interface between different materials, the fields must also satisfy Maxwell's equations. This imposes relations between the fields on the two sides of any interface. These relations are called interface conditions.

In these relations it is assumed that the normal to the interface points into medium 1. ρ_s is the surface charge density at the interface given in C/m^2, and $\mathbf{J}_s$ is the current density on the interface given in A/m. The tangential components of the electric field intensity and the normal components of the magnetic-flux density are always continuous across interfaces but the magnetic field intensity and the electric-flux density are not, unless the surface charge density and the surface current density vanish. The interface conditions are summarized in Table D.5.

D.2 The electromagnetic wave equation and its solution

Except for low frequency applications, in which the displacement current density in Maxwell's equations [second term in (D.2) or (D.6) or in (D.14) or (D.18)] can be neglected, one usually deals with propagation of waves. At low frequencies, the existence of waves is neglected and only the distribution of fields due to sources is of interest. To obtain the appropriate wave equations and, consequently, their solutions, one can start either with the time-domain equations in Table D.1 or the frequency domain equations in Table D.2 if the sources must be included or those in Table D.3 or Table D.4 if sources are not included. In the present work, we are not concerned with sources but we are concerned with losses due to induced currents in lossy dielectrics, especially in the way these manifest themselves in resonators. Therefore we will look at wave propagation in lossy media without external sources, and in the frequency domain, that is, we will use the equations in Table D.2 and replace the term $\mathbf{J}$ by the induced current density $\mathbf{J}_e = \sigma\mathbf{E}$.

D.2.1 Time-harmonic wave equations

The time-harmonic wave equation is obtained either by starting with the time-harmonic Maxwell's equations and following steps similar to those in the previous section or with the time-dependent equation and then transforming the resulting time-dependent wave equations to time-harmonic wave equations.

To obtain the time-harmonic wave equation in terms of the electric field intensity $\mathbf{E}$, we start with Maxwell's equations in time-harmonic form [see (D.13)–(D.16)], but written in terms of $\mathbf{E}$ and $\mathbf{H}$. Assuming linear, isotropic, homogeneous materials, these are

$$\nabla \times \mathbf{E} = -j\omega\mathbf{B} = -j\omega\mu\mathbf{H} \tag{D.40}$$

$$\nabla \cdot \mathbf{H} = 0 \tag{D.41}$$

$$\nabla \times \mathbf{H} = \mathbf{J} + j\omega\mathbf{D} = \sigma\mathbf{E} + j\omega\varepsilon\mathbf{E} \tag{D.42}$$

$$\nabla \cdot \varepsilon\mathbf{E} = 0 \tag{D.43}$$

We start by taking the curl on both sides of Faraday's law [see (D.40)]:

$$\nabla \times (\nabla \times \mathbf{E}) = -j\omega\mu(\nabla \times \mathbf{H}) \tag{D.44}$$

Substituting for $\nabla \times \mathbf{H}$ from Ampere's law [see (D.42)]

$$\nabla \times (\nabla \times \mathbf{E}) = -j\omega\mu(\sigma + j\omega\varepsilon)\mathbf{E} \tag{D.45}$$

Using the identity $\nabla \times (\nabla \times \mathbf{E}) = -\nabla^2\mathbf{E} + \nabla(\nabla \cdot \mathbf{E})$

$$-\nabla^2\mathbf{E} + \nabla(\nabla \cdot \mathbf{E}) = -j\omega\mu(\sigma + j\omega\varepsilon)\mathbf{E} \tag{D.46}$$

The divergence of $\mathbf{E}$ is given in (D.43) and happens to be zero for source-free environments. The source-free wave equation in lossy media under source-free conditions is

$$\nabla^2\mathbf{E} = j\omega\mu(\sigma\mathbf{E} + j\omega\varepsilon\mathbf{E}) \tag{D.47}$$

If, in addition, the losses are zero ($\sigma = 0$), the source-free, lossless wave equation is obtained:

$$\nabla^2 \mathbf{E} = j\omega\mu(j\omega\varepsilon)\mathbf{E} \tag{D.48}$$

Multiplying the terms on the right-hand side of (D.47) and rearranging them gives

$$\nabla^2 \mathbf{E} + \omega^2\mu\varepsilon\mathbf{E} - j\omega\mu\sigma\mathbf{E} = 0 \tag{D.49}$$

If the medium is lossless, this reduces to

$$\nabla^2 \mathbf{E} + \omega^2\mu\varepsilon\mathbf{E} = 0 \tag{D.50}$$

One can easily obtain the wave equation for the magnetic field intensity in a similar manner.

D.2.2 Solution of the wave equation

Now that the wave equation has been obtained, it is time to solve it. First, we must decide which wave equation to solve and under what conditions. To observe the behavior of fields and define the important aspects of propagation, we use a one-dimensional wave equation; that is, we assume that the electric field intensity **E** or the magnetic field intensity **H** has a single component in space. This is by no means a restriction since we can do the same for each component of the electric and magnetic field intensities to obtain a general solution. We start by solving the equation under the following conditions:

1. Fields are time harmonic.
2. The electric field intensity is directed in the x direction but varies in the z direction; that is, the field is perpendicular to the direction of propagation.
3. The medium in which the wave propagates is lossless ($\sigma = 0$).
4. The wave equation is source free ($\mathbf{J}_s = 0$, $\rho_v = 0$).

This set of assumptions seems to be rather restrictive. In fact, it is not. Although the direction in space is fixed, we are free to choose this direction and we can repeat the solution with a field in any other direction in space. Also, and perhaps more importantly, many of the above assumptions are actually satisfied, at least partially in practical waves. Similarly, propagation in general media, although not identical to propagation in lossless media, is quite similar in many cases. The benefit of this approach is in keeping the solution simple while still capturing all the important properties of the wave.

In fact, the conditions stated in this section specify what is called a uniform plane wave.

D.2.3 Solution for uniform plane waves in lossless media

A uniform plane wave is a wave (i.e., a solution to the wave equation) in which the electric and magnetic field intensities are directed in fixed directions in space and are *constant in magnitude and phase on planes perpendicular to the direction of propagation*.

Clearly, for a field to be constant in amplitude and phase on infinite planes, the source must also be infinite in extent. In this sense, a plane wave cannot be generated in practice. However, many practical situations can approximate plane waves to such an extent that plane waves are actually more common than one might think. Even when fields are not plane waves, one can show that they are superpositions of plane waves. Hence, the idea of plane waves is extremely useful for analysis.

D.2.3.1 The one-dimensional wave equation in free space and perfect dielectrics

With the assumptions in Section D.2.2, the electric field intensity is

$$\mathbf{E} = \hat{\mathbf{x}} E_x(z) \quad (\mathrm{V/m}) \tag{D.51}$$

where $\mathbf{E}$ is a phasor (i.e., $e^{j\omega t}$ is implied). These assumptions imply the following conditions:

$$E_y = E_z = 0 \quad \text{and} \quad \frac{\partial E_*}{\partial x} = \frac{\partial E_*}{\partial y} = 0 \tag{D.52}$$

where * denotes any component of $\mathbf{E}$. Substitution of these into (D.50) results in

$$\frac{d^2 E_x}{dz^2} + \omega^2 \mu \varepsilon E_x = 0 \tag{D.53}$$

where the partial derivative was replaced with the ordinary derivative because of the field dependence on z alone. Also, since the electric field is directed in a fixed direction in space, a scalar equation is sufficient. We denote

$$k = \sqrt{\omega^2 \mu \varepsilon} = \omega \sqrt{\mu \varepsilon} \quad (\mathrm{rad/m}) \tag{D.54}$$

Since (D.53) describes simple harmonic motion, it has a solution

$$E_x(z) = E_0^+ e^{-jkz} + E_0^- e^{jkz} \quad (\mathrm{V/m}) \tag{D.55}$$

where E_0^+ and E_0^- are constants (amplitudes) to be determined from the boundary conditions of the problem. The notations (+) and (−) indicate that the first term is a propagating wave in the positive z direction called a *forward-propagating wave* and the second a propagating wave in the negative z direction called a *backward-propagating wave*, as in Figure D.1(a) (horizontal arrows indicate the direction of propagation, the electric field intensity components are vertical). The amplitudes E_0^+ and E_0^- are real (but they may, in general, be complex) and are arbitrary. This solution can be verified by direct substitution into (D.53).

Using the phasor transformation, we can write the solution in the time domain as

$$E_x(z,t) = \mathrm{Re}\{E_x(z) e^{j\omega t}\} = E_0^+ \cos(\omega t - kz + \phi) + E_0^- \cos(\omega t + kz + \phi) \quad (\mathrm{V/m}) \tag{D.56}$$

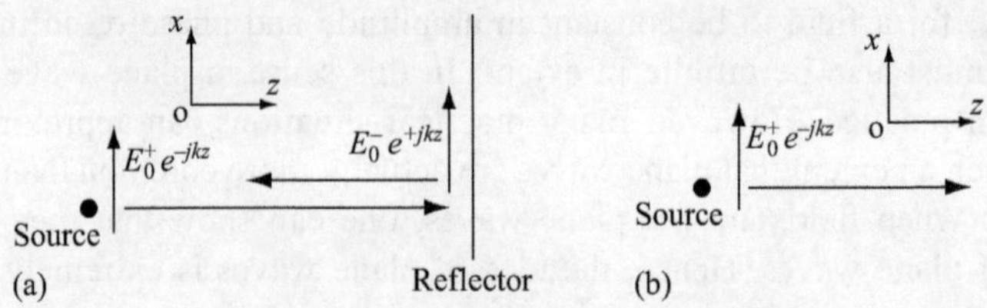

Figure D.1 (a) Forward and backward-propagating waves in bounded space and (b) forward-propagating wave in unbounded space (the horizontal arrows show the direction of propagation)

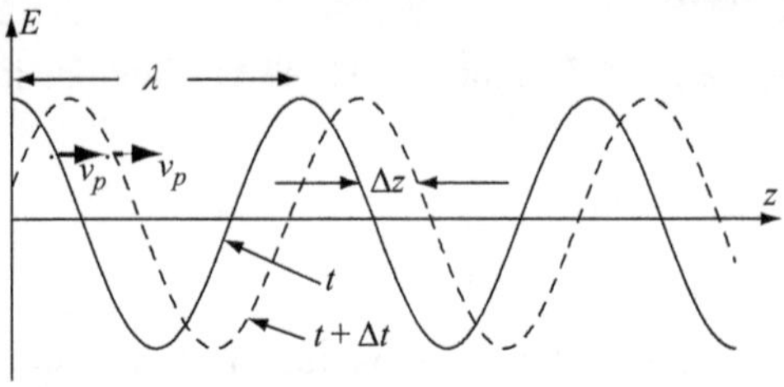

Figure D.2 Definition of wavelength and calculation of phase velocity

where the initial (arbitrary) phase angle ϕ was added for completeness.

If the wave propagates in boundless space, only an outward wave exists and E_0^- is zero (all power propagates away from the source and there can be no backward-propagating waves, such as in a transmitting antenna). If the forward-propagating wave is reflected without losses, the amplitudes of the two waves are equal.

Assuming only a forward-propagating wave, the solution is

$$E_x(z) = E_0^+ \, e^{-jkz} e^{j\phi} \quad \text{or} \quad E_x(z,t) = E_0^+ \cos(\omega t - kz + \phi) \quad (\text{V/m}) \tag{D.57}$$

Examining these expressions, it becomes apparent that what changes with time is the phase of the wave. In other words, the phase of the wave "travels" at a certain velocity. To see what this velocity is, we use Figure D.2 and follow a fixed point on the wave, for which the phase of the field is $\omega t - kz + \phi = \text{constant}$:

$$z = \frac{\omega t}{k} + \frac{\phi}{k} - \text{constant} \tag{D.58}$$

The speed of propagation of the phase is

$$v_p = \frac{dz}{dt} = \frac{\omega}{k} = \frac{1}{\sqrt{\mu\varepsilon}} \quad (\text{m/s}) \tag{D.59}$$

where $k = \omega\sqrt{\mu\varepsilon}$ was used [see (D.54)]. v_p is called the *phase velocity* of the wave.

The phase velocity of electromagnetic waves is material dependent. In particular, in free space,

$$v_p = \frac{1}{\sqrt{\mu_0 \varepsilon_0}} = \frac{1}{\sqrt{4 \times \pi \times 10^{-7} \times 8.8541853 \times 10^{-12}}} = 3 \times 10^8 \quad (\mathrm{m/s}) \qquad \text{(D.60)}$$

The phase velocity of electromagnetic waves in free space is the speed of light.

The phase velocity in all materials is lower than c since μ_r, $\varepsilon_r \geq 1$.

As the wave propagates, the distance between two successive crests of the wave depends both on the frequency of the wave and its phase velocity. We define the *wavelength* λ (in meters) as that distance a wave front (a front of constant phase) travels in one cycle:

$$\lambda = \frac{v_p}{f} = \frac{2\pi}{k} \quad (\mathrm{m}) \qquad \text{(D.61)}$$

In the electromagnetic case, the wavelength can be very short or very long, depending on frequency and phase velocity. For example, the wavelength in free space for a wave at 50 Hz is 6,000 km. At 30 GHz (a frequency used to communicate with satellites), the wavelength is 10 mm. From the definition of the wavelength in (D.61), we can write k as

$$k = \frac{2\pi}{\lambda} \quad (\mathrm{rad/m}) \qquad \text{(D.62)}$$

k is called the *wave number*. If the wavelength in free space is given, then k is called the *free-space wave number.*

D.3 Propagation of plane waves in materials

That waves are affected by the material in which they propagate has been shown in Section D.2.3.1, where propagation in lossless dielectrics, including free space, was discussed. The phase velocity, wavelength, wave number, and intrinsic impedance are material dependent. The behavior of waves in the presence of materials and in particular of lossy dielectrics is discussed next. In the process, we define the important parameters of propagating waves, which, in addition to those defined in Section D.2, describe an electromagnetic wave. These parameters include the propagation, phase, and attenuation constants, as well as the skin depth and the complex permittivity.

D.3.1 Propagation of plane waves in lossy dielectrics

The source-free wave equation with losses was written in (D.47) for the electric field intensity as

$$\nabla^2 \mathbf{E} = j\omega\mu(\sigma + j\omega\varepsilon)\mathbf{E} \qquad \text{(D.63)}$$

If the term $\sigma + j\omega\varepsilon$ in (D.63) is replaced with a complex term $j\omega\varepsilon_c$, that is,

$$j\omega\varepsilon_c = \sigma + j\omega\varepsilon \tag{D.64}$$

Equation (D.63) becomes identical in form to (D.50). The only difference is that the permittivity is replaced with the complex permittivity. The term ε_c can be written as

$$\varepsilon_c = \frac{\sigma + j\omega\varepsilon}{j\omega} = \varepsilon - j\frac{\sigma}{\omega} = \varepsilon\left[1 - j\frac{\sigma}{\omega\varepsilon}\right] = \varepsilon + j\varepsilon'' \quad (\text{F/m}) \tag{D.65}$$

This is called the *complex permittivity* and, in general, replaces the permittivity ε in the field equations. The imaginary part of the complex permittivity is associated with losses. Now, the term lossless dielectric becomes obvious: these are dielectrics in which $\sigma = 0$ and ε_c is real and equal to ε. The definition of complex permittivity is not merely a mathematical nicety: it is an accurate model of material behavior. The real and imaginary parts of the complex permittivity are measurable.

The ratio between the imaginary and real parts of the complex permittivity is called the *loss tangent* of the material and is a common measure of how lossy materials are:

$$\tan\theta_{\text{loss}} = \frac{\sigma}{\omega\varepsilon} = \frac{\varepsilon''}{\varepsilon'} \quad (\text{dimensionless}) \tag{D.66}$$

Since the loss tangent may be viewed as the ratio between induced and displacement current densities, we will use it to define approximation limits to the complex permittivity. A very low conductivity means that the permittivity is real, whereas a high conductivity means that the imaginary part of the complex permittivity dominates and the real part may be neglected.

To obtain a solution to the wave equation in lossy media, we will rely on the solution we already obtained for the lossless equation. Since the two are identical in form if the permittivity in the lossless equation is replaced with the complex permittivity, we can write the wave equation in lossy dielectrics as

$$\nabla^2\mathbf{E} = j\omega\mu(j\omega\varepsilon_c)\mathbf{E} = j\omega\mu\left(j\omega\varepsilon\left[1 - j\frac{\sigma}{\omega\varepsilon}\right]\right)\mathbf{E} \tag{D.67}$$

or writing this in the form of the Helmholtz equation in (D.50)

$$\nabla^2\mathbf{E} - j\omega\mu\left(j\omega\varepsilon\left[1 - j\frac{\sigma}{\omega\varepsilon}\right]\right)\mathbf{E} = 0 \tag{D.68}$$

Comparing this with the source-free (Helmholtz) equation and denoting

$$\gamma = j\omega\sqrt{\mu\varepsilon}\sqrt{\left[1 - j\frac{\sigma}{\omega\varepsilon}\right]} \tag{D.69}$$

Equation (D.68) can be written as

$$\nabla^2\mathbf{E} - \gamma^2\mathbf{E} = 0 \tag{D.70}$$

The quantity γ is called the *propagation constant* and is, in general, a complex number. The propagation constant can also be written directly from (D.63) by comparison with (D.70) as

$$\gamma = \sqrt{j\omega\mu(\sigma + j\omega\varepsilon)} \tag{D.71}$$

Equation (D.70) for lossy materials is similar to (D.50) for lossless materials. We can put them in exactly the same form if we write

$$\gamma = jk_c \tag{D.72}$$

where

$$k_c = \omega\sqrt{\mu\varepsilon}\sqrt{\left[1 - j\frac{\sigma}{\omega\varepsilon}\right]} \quad (\text{rad/m}) \tag{D.73}$$

The importance of this is that now we can use all the relations obtained for the lossless propagation of waves by replacing the term jk in (D.55) and (D.57) by γ.

The general solution for propagation in a lossy dielectric has the same two wave components as in (D.55): one traveling in the positive z direction, the other in the negative z direction

$$E_x(z) = E_0^+ e^{-\gamma z} + E_0^- e^{+\gamma z} \quad (\text{V/m}) \tag{D.74}$$

Similarly, assuming only an outgoing wave, we have from (D.57):

$$E_x(z) = E_0^+ e^{-\gamma z} \quad (\text{V/m}) \tag{D.75}$$

Since the propagation constant is a complex number, it can also be written as

$$\gamma = \alpha + j\beta \tag{D.76}$$

This gives for the general solution

$$E_x(z) = E_0^+ e^{-\alpha z} e^{-j\beta z} + E_0^- e^{+\alpha z} e^{+j\beta z} \quad (\text{V/m}) \tag{D.77}$$

Similarly, in the case of forward propagation only

$$E_x(z) = E_0^+ e^{-(\alpha + j\beta)z} = E_0^+ e^{-\alpha z} e^{-j\beta z} \quad (\text{V/m}) \tag{D.78}$$

The general solution in the time domain may be written as

$$E_x(z,t) = E_0^+ e^{-\alpha z}\cos(\omega t - \beta z) + E_0^- e^{+\alpha z}\cos(\omega t + \beta z) \quad (\text{V/m}) \tag{D.79}$$

For a wave propagating in the positive z direction only, this reduces to the first term of (D.79):

$$E_x(z,t) = E_0^+ e^{-\alpha z}\cos(\omega t - \beta z) \quad (\text{V/m}) \tag{D.80}$$

In this form, the propagating wave has the same form as (D.57) where β has replaced k and the exponential term $e^{-\alpha z}$ multiplies the amplitude. This is

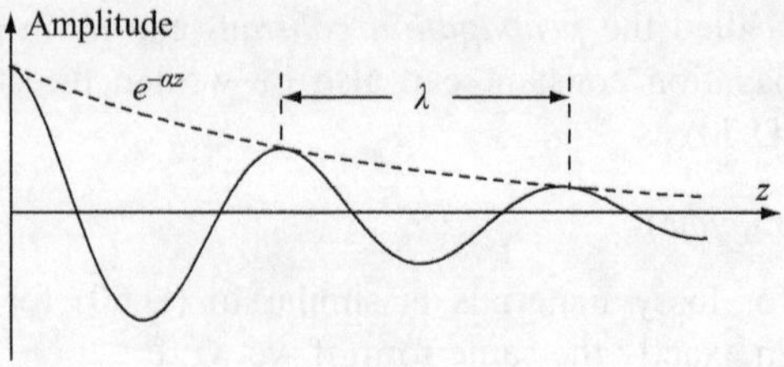

Figure D.3 Propagation of a wave in a lossy material showing exponential attenuation

therefore a wave, propagating in the positive z direction, with a phase velocity v_p and with an exponentially decaying amplitude. Thus, unlike the lossless case, in which the amplitude remained constant as the wave propagated, this time the amplitude changes as the wave propagates (Figure D.3). Much more will be said about this decay in amplitude in future sections. Perhaps, the most important general comment is that the decay can be quite rapid and that it depends on conductivity. If $\sigma = 0$, $\alpha = 0$, $e^{-\alpha z} = 1$, and the amplitude does not decay as the wave propagates.

We note that α causes an attenuation of the amplitude of the wave and is called the *attenuation constant*. The attenuation constant α is measured in nepers/meter. The *neper* defines the fraction of the attenuation the wave undergoes in 1 m. Attenuation of 1 Np/m reduces the wave amplitude to $1/e$ as it propagates a distance of 1 m. Therefore, it is equivalent to 8.69 dB/m ($20 \log_{10} e = 8.69$), that is, 1 Np/m = 8.69 dB/m.

The imaginary part, β, only affects the phase of the wave and is called the *phase constant.* The phase constant for lossless materials is identical to the wave number k as defined in (D.54). However, we will use k as notation for wave number and use the phase constant β for all media, including lossless dielectrics.

A propagating wave in a lossy material is shown schematically in Figure D.3. As the wave propagates in space, its amplitude is reduced exponentially. All aspects of propagation presented in the previous section remain the same except for replacing k by β and including the exponential decay in the amplitude.

The attenuation and phase constants for a general, lossy material are found by separating the real and imaginary parts of γ in (D.71). These are

$$\alpha = \omega\sqrt{\frac{\mu\varepsilon}{2}\left[\sqrt{1+\left(\frac{\sigma}{\omega\varepsilon}\right)^2} - 1\right]} \quad \text{(Np/m)} \tag{D.81}$$

$$\beta = \omega\sqrt{\frac{\mu\varepsilon}{2}\left[\sqrt{1+\left(\frac{\sigma}{\omega\varepsilon}\right)^2} + 1\right]} \quad \text{(rad/m)} \tag{D.82}$$

The other parameters required for description of the wave in general lossy media are the phase velocity, wavelength, and intrinsic impedance. The phase velocity and wavelength are now

$$v_p = \frac{\omega}{\beta} = \frac{1}{\sqrt{\mu\varepsilon/2\left[\sqrt{1+(\sigma/\omega\varepsilon)^2}+1\right]}} \quad (\text{m/s}) \tag{D.83}$$

$$\lambda = \frac{2\pi}{\beta} = \frac{2\pi}{\omega\sqrt{\mu\varepsilon/2\left[\sqrt{1+(\sigma/\omega\varepsilon)^2}+1\right]}} \quad (\text{m}) \tag{D.84}$$

Thus, both phase velocity and wavelength are smaller in lossy dielectrics, depending on conductivity. For lossless materials ($\sigma = 0$), (D.83) and (D.84) reduce to those for lossless materials given in (D.59) and (D.61). To find the intrinsic impedance, we return to (D.75):

$$\frac{\partial E_x^+}{\partial z} = \frac{\partial}{\partial z}\left(E_0^+ e^{-\gamma z}\right) = -\gamma E_x^+(z) \tag{D.85}$$

Substituting this in (D.32) gives

$$-\gamma E_x^+(z) = -j\omega\mu H_y \tag{D.86}$$

The intrinsic impedance is now written as

$$\eta = \frac{E_x^+(z)}{H_y^+(z)} = \frac{j\omega\mu}{\gamma} \quad (\Omega) \tag{D.87}$$

In the case considered here, the intrinsic impedance becomes

$$\eta = \frac{j\omega\mu}{\gamma} = \frac{j\omega\mu}{\sqrt{j\omega\mu(\sigma+j\omega\varepsilon)}} = \sqrt{\frac{j\omega\mu}{\sigma+j\omega\varepsilon}} \quad (\Omega) \tag{D.88}$$

The intrinsic impedance (also called the wave impedance) is now a complex number. It has both a resistive and a reactive part. In practical terms, this means that **E** and **H** are out of time phase in all but lossless materials and are out of phase in space for all materials (i.e., for plane waves, they are perpendicular to each other).

Some important general observations are appropriate here:

1. The phase velocity in lossy dielectrics is lower than in perfect dielectrics. This can be seen from (D.83) since β for lossy materials is larger than k for perfect dielectrics. Thus, the speed of propagation of electromagnetic waves is lower in lossy dielectrics (for the same ε and μ). The larger the losses, the lower the speed.
2. The intrinsic impedance (wave impedance) in lossy dielectrics is complex, indicating a phase difference between the electric and magnetic field intensity

in the same way as the phase difference between voltage and current in a circuit which contains reactive components. The magnitude of the intrinsic impedance is lower in conductive media. The higher the conductivity (losses), the lower the magnitude of the impedance.

3. The electric and magnetic field intensity remain perpendicular to each other and to the direction of propagation regardless of losses. This is a property of the uniform plane waves we assumed.
4. Attenuation of the wave in lossy media is exponential. This means that in materials with high conductivity, the attenuation is rapid. These materials will be called high-loss materials. Low-loss materials are materials with low conductivity.

D.3.2 Propagation of plane waves in low-loss dielectrics

We define low-loss materials as those materials in which the loss tangent is small: $\sigma/\omega\varepsilon \ll 1$ [or, equivalently, that the imaginary part of the complex permittivity in (D.65) is small compared to the real part]. This relation also indicates that a material may be considered to be low loss at a given frequency range, whereas in another frequency range, this assumption may not hold. Thus, the classification of materials changes, depending on frequencies. In practice, the permittivity of the material also changes with frequency, changing the range in which a material may be considered to be a low-loss material.

All properties of the wave propagating in low-loss dielectrics remain the same as for any lossy material. But the above condition for low-loss materials simplifies some of these relations, allowing easier application and better understanding of behavior of waves propagating in these materials. The propagation constant now can be approximated using the binomial expansion (because $\sigma/\omega\varepsilon < 1$) as

$$\gamma = j\omega\sqrt{\mu\varepsilon}\sqrt{\left(1 - \frac{j\sigma}{\omega\varepsilon}\right)} = j\omega\sqrt{\mu\varepsilon}\left[1 - \frac{j\sigma}{2\omega\varepsilon} + \frac{1}{8}\left(\frac{\sigma}{\omega\varepsilon}\right)^2 + H.O.T.\right] \quad \text{(D.89)}$$

Deciding, somewhat arbitrarily, to neglect all but the first three terms in the expansion, the attenuation constant is approximated by the second (real) term in (D.89):

$$\alpha \approx \left(\frac{\sigma}{2}\right)\sqrt{\frac{\mu}{\varepsilon}} = \frac{\sigma}{2}\eta_n \quad \text{(Np/m)} \quad \text{(D.90)}$$

where η_n is the no-loss intrinsic impedance (i.e., the intrinsic impedance of a material with the same μ and ε but in which $\sigma = 0$), and the phase constant is

$$\beta \approx \omega\sqrt{\mu\varepsilon}\left(1 + \frac{1}{8}\left(\frac{\sigma}{\omega\varepsilon}\right)^2\right) \quad \text{(rad/m)} \quad \text{(D.91)}$$

In very low-loss cases, the second term in (D.91) may also be neglected and the phase constant may often be approximated as

$$\beta \approx \omega\sqrt{\mu\varepsilon} \quad \text{(rad/m)} \quad \text{(D.92)}$$

The phase constant and, therefore, phase velocity and wavelength for low-loss dielectrics are essentially unchanged from those for the lossless dielectric because the second term in (D.91) is small, but the attenuation constant can be quite significant. Thus, the phase velocity and wavelength are

$$\begin{aligned} v_p &\approx \frac{1}{\sqrt{\mu\varepsilon}\left(1+(1/8)(\sigma/\omega\varepsilon)^2\right)} \quad \text{(m/s)} \\ \lambda &= \frac{2\pi}{\beta} = \frac{v_p}{f} = \frac{1}{f\sqrt{\mu\varepsilon}\left(1+(1/8)(\sigma/\omega\varepsilon)^2\right)} \quad \text{(m)} \end{aligned} \tag{D.93}$$

for general low-loss materials and

$$v_p \approx \frac{1}{\sqrt{\mu\varepsilon}} \quad \text{(m/s)} \qquad \lambda = \frac{2\pi}{\beta} = \frac{v_p}{f} \quad \text{(m)} \tag{D.94}$$

for very low-loss materials. The intrinsic impedance in low-loss dielectrics is still a complex number. Substituting the value of γ from (D.89) in (D.87), and using the expansion again, η can be approximated as

$$\eta = \frac{j\omega\mu}{\gamma} = \sqrt{\frac{\mu}{\varepsilon}}\frac{1}{\sqrt{(1-(j\sigma/\omega\varepsilon))}} \approx \sqrt{\frac{\mu}{\varepsilon}}(1+(j\sigma/2\omega\varepsilon)) = \eta_n\left(1+\frac{j\sigma}{2\omega\varepsilon}\right) \quad (\Omega) \tag{D.95}$$

where η_n is the no-loss intrinsic impedance for the same material. The reactive part of the intrinsic impedance is quite small since $\sigma/\omega\varepsilon \ll 1$. Thus, for many practical applications, the intrinsic impedance of the lossless material may be used with little error.

D.3.3 Propagation of plane waves in conductors or high-loss dielectrics

In highly conductive materials, the losses are high, and we can assume that $\sigma \gg \omega\varepsilon$ or that the imaginary part of the complex permittivity is not negligible compared to the real part (i.e., conduction currents dominate). Under this condition, the complex propagation constant can be approximated from (D.89) as

$$\gamma \approx j\omega\sqrt{\mu\varepsilon}\sqrt{-\frac{j\sigma}{\omega\varepsilon}} = \sqrt{\frac{j\omega\mu\varepsilon\sigma}{\varepsilon}} = (1+j)\sqrt{\frac{\omega\mu\sigma}{2}} \tag{D.96}$$

by neglecting 1 compared to $j\sigma/\omega\varepsilon$ and using $\sqrt{j} = (1+j)/\sqrt{2}$. From this, we get

$$\alpha = \sqrt{\frac{\omega\mu\sigma}{2}} = \sqrt{\pi f\mu\sigma} \quad \text{(Np/m)}, \quad \beta = \sqrt{\frac{\omega\mu\sigma}{2}} = \sqrt{\pi f\mu\sigma} \quad \text{(rad/m)} \tag{D.97}$$

The attenuation and phase constants are equal and are very large. The wave is attenuated rapidly to the point where propagation in conducting media can only exist within short distances. The propagating wave can now be written as

$$E_x(z) = E_0^+ e^{-z/\delta} e^{-jz/\delta} \quad \text{(V/m)} \tag{D.98}$$

where the term

$$\delta = \sqrt{\frac{2}{\omega\mu\sigma}} = \sqrt{\frac{1}{\pi f \mu\sigma}} = \frac{1}{\alpha} \quad \text{(m)} \tag{D.99}$$

is known as the *skin depth* or *depth of penetration* of the wave. It is defined as that distance in which the amplitude of a plane wave is attenuated to $1/e$ of its original amplitude. The skin depth in conductors is small. In the microwave range, it can be of the order of a few microns (depending on material and frequency). Because waves at these high frequencies penetrate very little in conductors, it is quite common to use the perfect conductor approximation for conducting materials.

The phase velocity in good conductors is [from (D.83) and (D.97)]:

$$v = \frac{\omega}{\beta} = \omega\delta = \sqrt{\frac{2\omega}{\mu\sigma}} \quad \text{(m/s)} \tag{D.100}$$

and is obviously small compared to the phase velocity in dielectrics or free space, because δ is small.

The wavelength also changes drastically compared to free space or lossless dielectrics. It is very short and given by

$$\lambda = \frac{2\pi}{\beta} = 2\pi\delta \quad \text{(m)} \tag{D.101}$$

The intrinsic impedance is [using (D.88)]:

$$\eta \approx \frac{j\omega\mu}{\gamma} = \frac{j\omega\mu}{(1+j)\sqrt{\omega\mu\sigma/2}} = (1+j)\sqrt{\frac{\omega\mu}{2\sigma}} = (1+j)\frac{1}{\sigma\delta} = (1+j)\frac{\omega\mu\delta}{2} \quad (\Omega) \tag{D.102}$$

where $j/(j+1) = (j+1)/2$ was used. The phase angle of the intrinsic impedance is, therefore, 45°. This is characteristic of good conductors for which the magnetic field intensity lags behind the electric field intensity by 45°. The intrinsic impedance of conductors can be very low and is much lower than the intrinsic impedance of free space. For example, the intrinsic impedance in copper at 1 GHz is $(1+j) \times 8.3 \times 10^{-3}$ Ω compared to 377 Ω in free space.

D.4 The Poynting theorem and electromagnetic power

One of the most important characteristics of waves is their ability to transport energy and the power associated with the process. To examine power and energy relations in the electromagnetic wave, it is convenient to look first at the general time-dependent expression for the rate of energy transfer that includes time-rate of change in stored magnetic and stored electric energy, and dissipated power. As always, the starting point must be with Maxwell's equations.

D.4.1 The Poynting theorem in the time domain

Before formalizing the expressions for energy transfer, first consider Ampere's law [see (D.2)]:

$$\nabla \times \mathbf{H} = \mathbf{J} + \frac{\partial \mathbf{D}}{\partial t} \quad (\mathrm{A/m^2}) \tag{D.103}$$

where $\mathbf{J}$ includes all possible current densities as follows

$$\mathbf{J} = \mathbf{J}_0 + \mathbf{J}_e = \mathbf{J}_0 + \sigma\mathbf{E} \quad \left(\mathrm{A/m^2}\right) \tag{D.104}$$

where $\mathbf{J}_0$ indicates source current densities and $\mathbf{J}_e$ indicates induced current densities in conducting media. Now, suppose we take the scalar product of (D.103) with the electric field intensity $\mathbf{E}$:

$$\mathbf{E} \cdot (\nabla \times \mathbf{H}) = \mathbf{E} \cdot \mathbf{J} + \mathbf{E} \cdot \frac{\partial \mathbf{D}}{\partial t} \tag{D.105}$$

The term $\mathbf{E} \cdot \mathbf{J}$ is the volume power density associated with the general current density in (D.104):

$$\frac{dP}{dv} = \mathbf{E} \cdot \mathbf{J} \quad \left(\mathrm{W/m^3}\right) \tag{D.106}$$

The terms on the right-hand side of (D.105) are electric volume power densities since they are both associated with the electric field intensity.

The following vector identity can be written for any two vectors:

$$\nabla \cdot (\mathbf{E} \times \mathbf{H}) = \mathbf{H} \cdot (\nabla \times \mathbf{E}) - \mathbf{E} \cdot (\nabla \times \mathbf{H}) \tag{D.107}$$

The second term on the right-hand side is (D.105), and, therefore, all three terms in (D.107) represent power densities. The first term on the right-hand side comes from taking the scalar product of Faraday's law [see (D.1)] and the magnetic field intensity $\mathbf{H}$:

$$\mathbf{H} \cdot (\nabla \times \mathbf{E}) = -\mathbf{H} \cdot \left(\frac{\partial \mathbf{B}}{\partial t}\right) \tag{D.108}$$

The right-hand side of (D.108) is a magnetic power density as it is associated with the magnetic field intensity. From (D.107) we have

$$\nabla \cdot (\mathbf{E} \times \mathbf{H}) = -\mathbf{H} \cdot \frac{\partial \mathbf{B}}{\partial t} - \mathbf{E} \cdot \left(\mathbf{J} + \frac{\partial \mathbf{D}}{\partial t} \right) = -\mathbf{H} \cdot \frac{\partial \mathbf{B}}{\partial t} - \mathbf{E} \cdot \frac{\partial \mathbf{D}}{\partial t} - \mathbf{E} \cdot \mathbf{J} \quad (\mathrm{W/m^3}) \tag{D.109}$$

Assuming that we consider the power relations in a volume v, bounded by an area s, the total power in the volume is obtained by integrating over the volume:

$$\int_v \nabla \cdot (\mathbf{E} \times \mathbf{H}) dv = -\int_v \left(\mathbf{H} \cdot \frac{\partial \mathbf{B}}{\partial t} + \mathbf{E} \cdot \frac{\partial \mathbf{D}}{\partial t} \right) dv - \int_v \mathbf{E} \cdot \mathbf{J} dv \quad (\mathrm{W}) \tag{D.110}$$

The left-hand side is transformed from a volume integral to a surface integral using the divergence theorem. We also use the following identities:

$$\mathbf{E} \cdot \frac{\partial \mathbf{D}}{\partial t} = \frac{\partial}{\partial t} \left(\frac{\mathbf{E} \cdot \mathbf{D}}{2} \right), \quad \mathbf{H} \cdot \frac{\partial \mathbf{B}}{\partial t} = \frac{\partial}{\partial t} \left(\frac{\mathbf{H} \cdot \mathbf{B}}{2} \right) \tag{D.111}$$

With these, (D.110) becomes

$$\oint_s (\mathbf{E} \times \mathbf{H}) \cdot d\mathbf{s} = -\frac{\partial}{\partial t} \int_v \left(\frac{\mathbf{H} \cdot \mathbf{B}}{2} + \frac{\mathbf{E} \cdot \mathbf{D}}{2} \right) dv - \int_v \mathbf{E} \cdot \mathbf{J} \, dv \quad (\mathrm{W}) \tag{D.112}$$

or performing the scalar products

$$\mathbf{H} \cdot \mathbf{B} = \mu \mathbf{H} \cdot \mathbf{H} = \mu H^2, \quad \mathbf{E} \cdot \mathbf{D} = \varepsilon \mathbf{E} \cdot \mathbf{E} = \varepsilon E^2 \tag{D.113}$$

we get

$$\oint_s (\mathbf{E} \times \mathbf{H}) \cdot d\mathbf{s} = -\frac{\partial}{\partial t} \int_v \left(\frac{\mu H^2}{2} + \frac{\varepsilon E^2}{2} \right) dv - \int_v \mathbf{E} \cdot \mathbf{J} \, dv \quad (\mathrm{W}) \tag{D.114}$$

The left-hand side of (D.114) represents the total outward flow of power through the area s bounding the volume v or, alternately, the energy per unit time crossing the surface s. If this flow is inwards, it is a negative flow; if outward, it is positive (because $d\mathbf{s}$ is always positive pointing out of the volume). The expression $\mathbf{E} \times \mathbf{H}$ has units of (V/m)×(A/m) = (W/m^2) and is therefore a surface power density. This power density is called the ***Poynting vector*** $\mathcal{P}$:

$$\mathcal{P} = \mathbf{E} \times \mathbf{H} \quad (\mathrm{W/m^2}) \tag{D.115}$$

The advantage of this expression is that it also indicates the direction of the power flow, information that is important for wave propagation calculations. Thus, power flows in the direction perpendicular to both **E** and **H**, according to the right-hand rule.

The first term on the right-hand side of (D.114) represents the time-rate of decrease in the potential or stored energy in the system. It has two components: one is the time-rate of change of the stored electric energy and the other is time-rate of change of the stored magnetic energy.

The second term is due to any sources that may exist in the volume. There are two possibilities that must be considered here. One is that the current density is a source current density such as produced by a battery or a generator inside volume v. The second is a current density provided by external sources (outside the volume v).

The expressions in (D.114) and (D.115) are instantaneous quantities. For practical purposes, a time-averaged quantity is sometimes more useful. For a periodic time variation of fields, this can be obtained by averaging over a time T (usually a cycle of the field), giving the time-averaged Poynting vector:

$$\mathcal{P}_{av} = \frac{1}{T}\int_0^T \mathcal{P}(t)dt \quad \left(\mathrm{W/m^2}\right) \tag{D.116}$$

The time-averaged Poynting vector is a time-averaged power density. To calculate the total power, either instantaneous or time averaged, the Poynting vector must be integrated over the surface through which the power crosses. This usually means a closed surface enclosing a volume, but not always. The instantaneous power is given as

$$P(t) = \oint_s \mathcal{P}(t) \cdot d\mathbf{s} = \oint_s (\mathbf{E}(t) \times \mathbf{H}(t)) \cdot d\mathbf{s} \quad (\mathrm{W}) \tag{D.117}$$

while the time-averaged power through a closed surface s is

$$P_{av} = \oint_s \mathcal{P}_{av} \cdot d\mathbf{s} \quad (\mathrm{W}) \tag{D.118}$$

The important properties of the Poynting theorem and the Poynting vector are as follows:

1. The Poynting theorem gives the power relations of the fields in any volume.
2. The Poynting vector is the power density on the surface of a volume. The direction of the Poynting vector is the direction of flow of power.
3. The Poynting vector gives the direction of propagation of electromagnetic power.
4. The Poynting theorem gives the net flow of power out of a given volume through its surface.

D.4.2 The complex Poynting vector

As pointed out earlier, most electromagnetic relations encountered here, including most applications, are handled in the frequency domain, assuming sinusoidal excitation. Thus, it often becomes necessary to define the Poynting vector in the frequency domain. This definition also shows the relation between real and reactive power and is closely related to time-averaged power.

In analogy with time-averaged power in circuits using the phasor notation, the time-averaged Poynting vector in a general field under sinusoidal conditions may be written as

$$\mathcal{P}_{av} = \frac{1}{2}\mathrm{Re}\{\mathbf{E} \times \mathbf{H}^*\} \quad \left(\mathrm{W/m^2}\right) \tag{D.119}$$

where * indicates the complex conjugate form. Comparing this to (D.115), we are led to define a complex Poynting vector as

$$\mathcal{P}_c = \mathbf{E} \times \mathbf{H}^* = \mathbf{E}^* \times \mathbf{H} \quad (\mathrm{W/m^2}) \tag{D.120}$$

where **E** and **H** are phasors. Clearly, the value of the complex Poynting vector is the ease with which the time-averaged power density and, therefore, time-averaged power are evaluated.

A formal derivation of the complex Poynting vector starts with Maxwell's first two equations in the frequency domain [see (D.13) and (D.14)]:

$$\nabla \times \mathbf{E} = -j\omega\mu\mathbf{H} \tag{D.121}$$

$$\nabla \times \mathbf{H} = \mathbf{J} + j\omega\varepsilon\mathbf{E} \tag{D.122}$$

The conjugates of (D.121) and (D.122) are

$$\nabla \times \mathbf{E}^* = j\omega\mu\mathbf{H}^* \tag{D.123}$$

$$\nabla \times \mathbf{H}^* = \mathbf{J}^* - j\omega\varepsilon\mathbf{E}^* \tag{D.124}$$

The current density $\mathbf{J}^*$ includes source and induced current densities [see (D.104)]:

$$\mathbf{J}^* = \mathbf{J}_0^* + \mathbf{J}_e^* = \mathbf{J}_0^* + \sigma\mathbf{E}^* \tag{D.125}$$

First, we write the scalar product between $\mathbf{H}^*$ and (D.121) as

$$\mathbf{H}^* \cdot (\nabla \times \mathbf{E}) = -j\omega\mu\mathbf{H} \cdot \mathbf{H}^* \tag{D.126}$$

Next, we write the scalar product between **E** and (D.124) as

$$\mathbf{E} \cdot (\nabla \times \mathbf{H}^*) = (\mathbf{J}^* - j\omega\varepsilon\mathbf{E}^*) \cdot \mathbf{E} \tag{D.127}$$

Equations (D.126) and (D.127) may be combined using the following vector identity:

$$\mathbf{H}^* \cdot (\nabla \times \mathbf{E}) - \mathbf{E} \cdot (\nabla \times \mathbf{H}^*) = \nabla \cdot (\mathbf{E} \times \mathbf{H}^*) \tag{D.128}$$

Substituting for $\mathbf{H}^* \cdot (\nabla \times \mathbf{E})$ from (D.126) and for $\mathbf{E} \cdot (\nabla \times \mathbf{H}^*)$ from (D.127) and rearranging terms gives

$$\nabla \cdot (\mathbf{E} \times \mathbf{H}^*) = j\omega(\varepsilon\mathbf{E} \cdot \mathbf{E}^* - \mu\mathbf{H} \cdot \mathbf{H}^*) - \mathbf{E} \cdot \mathbf{J}^* \tag{D.129}$$

The first two terms on the right-hand side represent the electric and magnetic power densities. The third term represents the input and dissipated power densities.

Using the ideas of the transmitter and receiver cases discussed in the previous section, the term $\mathbf{E} \cdot \mathbf{J}^*$ is replaced by $\sigma\mathbf{E} \cdot \mathbf{E}^*$ for the receiver case and by $-\mathbf{E} \cdot \mathbf{J}^*$ for the transmitter case, as was done earlier. To write this in terms of power rather than power density, we integrate (D.129) over an arbitrary volume v:

$$\int_v \nabla \cdot (\mathbf{E} \times \mathbf{H}^*) dv = j\omega \int_v (\varepsilon\mathbf{E} \cdot \mathbf{E}^* - \mu\mathbf{H} \cdot \mathbf{H}^*) dv - \int_v \mathbf{E} \cdot \mathbf{J}_0^* dv - \int_v \sigma\mathbf{E} \cdot \mathbf{E}^* dv \quad (\mathrm{W}) \tag{D.130}$$

and using the divergence theorem on the left-hand side, we get

$$\oint_s (\mathbf{E}\times\mathbf{H}^*)\cdot d\mathbf{s} = j\omega\int_v (\varepsilon\mathbf{E}\cdot\mathbf{E}^* - \mu\mathbf{H}\cdot\mathbf{H}^*)dv - \int_v \mathbf{E}\cdot\mathbf{J}_0^* dv - \int_v \sigma\mathbf{E}\cdot\mathbf{E}^* dv \quad \text{(W)} \tag{D.131}$$

where $\mathbf{J}_0$ indicates a source current density. The left-hand side is the complex power flow through the surface s enclosing the volume v. The first term on the right-hand side is the reactive power in the volume, the second term on the right-hand side is the complex source power (either positive or negative depending on the location of the source), and the last term is the dissipated power in the volume if dissipation occurs (in conducting media).

Equation (D.131) is the complex Poynting theorem. As mentioned at the beginning of this section when using the complex Poynting vector, it is for the purpose of calculating time-averaged quantities. It is therefore more useful to write this relation as two terms as follows [using the notation in (D.119)]:

$$\frac{1}{2}\text{Re}\left\{\oint_s \mathcal{P}_c\cdot d\mathbf{s}\right\} = -\frac{1}{2}\int_v \mathbf{E}\cdot\mathbf{J}_0^* dv - \frac{1}{2}\int_v \sigma\mathbf{E}\cdot\mathbf{E}^* dv \quad \text{(W)} \tag{D.132}$$

$$\frac{1}{2}\text{Im}\left\{\oint_s \mathcal{P}_c\cdot d\mathbf{s}\right\} = \omega\int_v \left(\frac{\varepsilon\mathbf{E}\cdot\mathbf{E}^*}{2} - \frac{\mu\mathbf{H}\cdot\mathbf{H}^*}{2}\right)dv \quad \text{(W)} \tag{D.133}$$

Equation (D.132) gives the real power balance in the volume. The left-hand side is the net outward flow of power through the surface enclosing the volume. The first term on the right-hand side is the net source power (in this case, the source is outside the volume hence the negative sign) and the last term is the dissipated power in the volume. $\mathbf{E}\cdot\mathbf{J}^*$ is positive for the receiver case (power propagates into the volume) and negative for the transmitter case (power propagates out of the volume). Therefore, the second term on the right-hand side of (D.132) is negative for the receiver case and positive for the transmitter case.

Usually, in the transmitter case, we will assume there are no losses in the volume, whereas in the receiver case, there are no sources in the volume. If this is so, the corresponding terms are deleted from (D.132).

Equation (D.133) is the balance of reactive power. It shows it is the rate of flow of reactive power across the surface. The right-hand side gives the time-averaged reactive power. The second term on the right-hand side is the reactive power due to the source. The time-averaged magnetic and electric energy densities can be shown to be

$$w_{m(av)} = \frac{1}{4}\mu\mathbf{H}\cdot\mathbf{H}^* \quad w_{e(av)} = \frac{1}{4}\mu\mathbf{E}\cdot\mathbf{E}^* \quad (\text{J/m}^3) \tag{D.134}$$

To emphasize the time-averaged energy densities, (D.133) may be written as

$$\frac{1}{2}\text{Im}\left\{\oint_s \mathcal{P}_c\cdot d\mathbf{s}\right\} = 2\omega\int_v \left(\frac{\varepsilon\mathbf{E}\cdot\mathbf{E}^*}{4} - \frac{\mu\mathbf{H}\cdot\mathbf{H}^*}{4}\right)dv \quad \text{(W)} \tag{D.135}$$

D.5 Reflection, transmission, and refraction of plane waves

In addition to propagation of waves in unbounded space, we also need to look at their behavior at interfaces between different media. At these interfaces, which can be between lossless or lossy dielectrics or between dielectrics and conductors, the waves are modified by reflections from and transmission through the interface. This interaction gives rise to additional properties of waves that need to be defined. In fact, the ideas of reflection and transmission are fundamental in many applications including those described in Chapters 3 and 4. We will develop here the relations for transmission and reflection coefficients for the most important configurations including lossy and lossless dielectrics and conductors.

The basic principle involved in describing the behavior of a wave at the interface between materials is to write the waves on both sides of the interface and to match the components of the electric and magnetic fields at the interface. In general, this means applying the interface conditions of the fields, from which the fields on both sides of the interface are found. The general conditions are shown in Figure D.4. There are two separate configurations that need to be considered. One is the case of perpendicular polarization shown in Figure D.4(a). The electric field intensity is perpendicular to the plane formed by the direction of propagation of the incident wave and the normal to the interface. This plane is called the *plane of incidence*. The second case is shown in Figure D.4(b) in which the electric field intensity is polarized parallel to the plane of incidence. The more general case of arbitrary polarization is handled by superposition of perpendicular and parallel polarization cases. The wave in material (1) propagates at an angle to the normal to the interface. This angle is called the *incidence angle*, θ_i. The tangential components of the electric field intensity are continuous at the interface (see Table D.5). The normal components are discontinuous. In general, we will assume that there are no charge densities or current densities at the interface (except for conducting interfaces). The wave is partly transmitted into material (2) and partly reflected at an angle θ_r called the *reflection angle*. We will also show that the incidence and reflection angles are equal. The *transmission angle* (or *refraction angle*) θ_t is

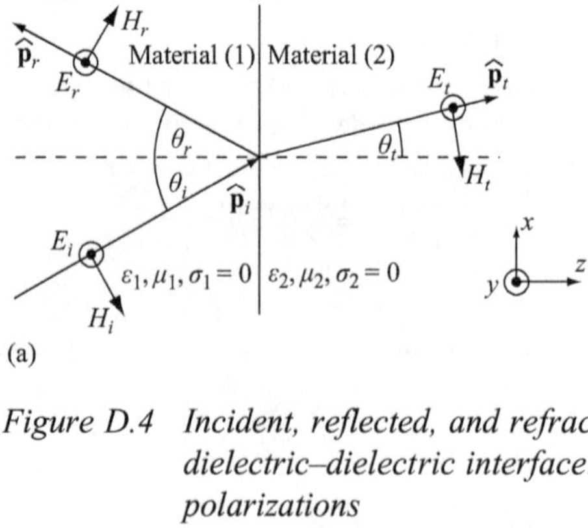

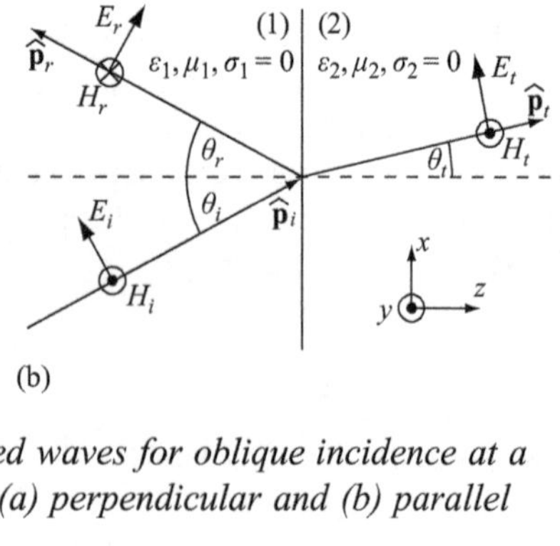

Figure D.4 Incident, reflected, and refracted waves for oblique incidence at a dielectric–dielectric interface: (a) perpendicular and (b) parallel polarizations

different than the incidence angle θ_i, as one would expect from two different materials. Note however that the discussion of waves at interfaces goes beyond simple interface conditions. The propagation properties of the wave must also be taken into account. This means that we must consider such properties as polarization of the wave and its speed of propagation.

As we discuss the behavior of waves at interfaces, it is useful to recall the behavior of light waves. We expect similar behavior, including reflection, transmission, and refraction of waves at the interface. We start with a wave incident at an angle on a dielectric interface and find the reflection and transmission coefficients. Then, from these relations, we find the conditions for transmission and reflection on conductors and the conditions for normal incidence on both conductors and dielectrics.

We start with a general, lossy dielectric interface and then proceed to discuss lossless and low-loss dielectrics and conductors. Normal incidence is treated as a special case of the general oblique incidence, and incidence on conductors is a limiting case of incidence on lossy media.

D.5.1 Oblique incidence on a dielectric interface: perpendicular polarization

To define the conditions for the reflection and transmitted waves, we use Figure D.4, write the electric and magnetic field intensities on both sides, and apply the boundary conditions on the interface for the tangential components of the electric field intensity. We assume here lossless propagation for simplicity. Losses will be discussed separately after the lossless reflection and transmission is discussed.

The direction of propagation can be written directly from Figure D.5(b). For the incident wave, the unit vector in the direction of propagation is

$$\hat{\mathbf{p}}_i = \hat{\mathbf{x}} \sin\,\theta_i + \hat{\mathbf{z}} \cos\theta_i \tag{D.136}$$

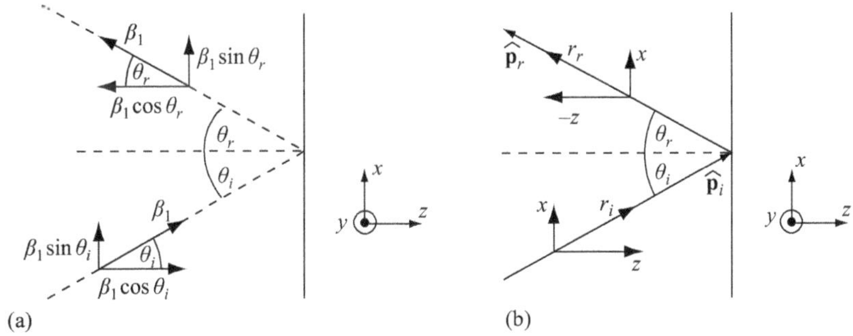

Figure D.5 Direction of propagation and phase constants at an interface: (a) relation between phase constants and (b) distances traveled by the incident and reflected waves

For the reflected wave, the direction of propagation is

$$\widehat{\mathbf{p}}_r = \widehat{\mathbf{x}} \sin\theta_r - \widehat{\mathbf{z}} \cos\theta_r \tag{D.137}$$

where the unit vectors $\widehat{\mathbf{p}}_i$ and $\widehat{\mathbf{p}}_r$ refer to the direction of the Poynting vector for the incident and reflected waves, respectively.

Inspection of the electric field intensity in (D.77) shows that the variable z in the exponent is a distance the wave propagates from some reference point, whereas β is the phase constant of the wave. The product βz is the phase of the wave after it has propagated a distance z from some reference point. Since the phase constant β_1 is given for the wave propagating in the $\widehat{\mathbf{p}}_i$ direction, we can view the wave in Figure D.5(b) as two components, one propagating in the positive x direction with phase constant β_{1x} and one in the positive z direction with phase constant β_{1z} [see Figure D.5(a)], where

$$\beta_{1x} = \beta_1 \sin\theta_i, \quad \beta_{1z} = \beta_1 \cos\theta_i \quad \text{(rad/m)} \tag{D.138}$$

When the wave propagates a distance r_i along $\widehat{\mathbf{p}}_i$ [see Figure D.5(b)], the vertical component of the wave propagates a distance x with phase constant β_{1x} and the horizontal component propagates a distance z with a phase constant β_{1z}. The phase of the incident wave is therefore

$$\beta_1 r_i = \beta_1 x \sin\theta_i + \beta_1 z \cos\theta_i \quad \text{(rad)} \tag{D.139}$$

Similarly, for the backward-propagating wave, the components of β are as in (D.138), but now the wave travels a distance x in the vertical direction and a distance $-z$ in the horizontal direction and the angle is θ_r. The phase of the reflected wave is

$$\beta_1 r_r = \beta_1 x \sin\theta_r - \beta_1 z \cos\theta_r \quad \text{(rad)} \tag{D.140}$$

The incident electric and magnetic fields can now be written directly from Figure D.4(a) as follows:

$$\mathbf{E}_i(x,z) = \widehat{\mathbf{y}} E_{i1} e^{-j\beta_1(x\sin\theta_i + z\cos\theta_i)} \quad \text{(V/m)} \tag{D.141}$$

$$\mathbf{H}_i(x,z) = \frac{E_{i1}}{\eta_1}(-\widehat{\mathbf{x}}\cos\theta_i + \widehat{\mathbf{z}}\sin\theta_i) e^{-j\beta_1(x\sin\theta_i + z\cos\theta_i)} \quad \text{(A/m)} \tag{D.142}$$

where, again, the direction of propagation of the incident wave is given by $\widehat{\mathbf{p}}_i = \widehat{\mathbf{x}} \sin\theta_i + \widehat{\mathbf{z}}\cos\theta_i$. The reflected electric and magnetic fields are

$$\mathbf{E}_r(x,z) = \widehat{\mathbf{y}} E_{r1} e^{-j\beta_1(x\sin\theta_i - z\cos\theta_i)} \quad \text{(V/m)} \tag{D.143}$$

$$\mathbf{H}_r(x,z) = \frac{E_{r1}}{\eta_1}(\widehat{\mathbf{x}}\cos\theta_i + \widehat{\mathbf{z}}\sin\theta_i) e^{-j\beta_1(x\sin\theta_i - z\cos\theta_i)} \quad \text{(A/m)} \tag{D.144}$$

Similarly, the transmitted electric and magnetic fields have the same form as the incident wave but with different amplitude and propagate at a different angle [see Figure D.4(a)]:

$$\mathbf{E}_t(x,z) = \hat{\mathbf{y}} E_{t2} e^{-j\beta_2(x \sin\theta_t + z\cos\theta_t)} \quad (\mathrm{V/m}) \tag{D.145}$$

$$\mathbf{H}_t(x,z) = \frac{E_{t2}}{\eta_2}(-\hat{\mathbf{x}}\cos\theta_t + \hat{\mathbf{z}}\sin\theta_t) e^{-j\beta_2(x\sin\theta_t + z\cos\theta_t)} \quad (\mathrm{A/m}) \tag{D.146}$$

To determine the transmission and reflection coefficients, the tangential components of the electric field intensity and those of the magnetic field intensity on both sides of the interface (i.e., at $z = 0$) are equated. From Figure D.4(a) and (D.141)–(D.146), and taking only the tangential components (y component for $\mathbf{E}$ and x component for $\mathbf{H}$) at $z = 0$, we have

$$(E_{i1} + E_{r1})e^{-j\beta_1 x\sin\theta_i} = E_{t2}e^{-j\beta_2 x\sin\theta_t} \quad \text{and}$$

$$\left(\frac{E_{i1}}{\eta_1} - \frac{E_{r1}}{\eta_1}\right)\cos\theta_i e^{-j\beta_1 x\sin\theta_i} = \frac{E_{t2}}{\eta_2}\cos\theta_t e^{-j\beta_2 x\sin\theta_t} \tag{D.147}$$

There are three relations that must be satisfied:

$$e^{-j\beta_1 x\sin\theta_i} = e^{-j\beta_2 x\sin\theta_t} \quad \text{or} \quad \beta_1\sin\theta_i = \beta_2\sin\theta_t \tag{D.148}$$

and

$$E_{i1} + E_{r1} = E_{t2} \quad \text{and} \quad \frac{E_{i1}}{\eta_1}\cos\theta_i - \frac{E_{r1}}{\eta_1}\cos\theta_i = \frac{E_{t2}}{\eta_2}\cos\theta_t \tag{D.149}$$

From (D.148), we get

$$\omega\sqrt{\varepsilon_1\mu_1}\sin\theta_i = \omega\sqrt{\varepsilon_2\mu_2}\sin\theta_t \tag{D.150}$$

or

$$\sin\theta_t = \frac{\sqrt{\varepsilon_1\mu_1}}{\sqrt{\varepsilon_2\mu_2}}\sin\theta_i \tag{D.151}$$

This relation between the incident and refraction angle is Snell's law of refraction. Since $\varepsilon_1 = \varepsilon_0\varepsilon_{r1}$, $\varepsilon_2 = \varepsilon_0\varepsilon_{r2}$, $\mu_1 = \mu_0\mu_{r1}$, and $\mu_2 = \mu_0\mu_{r2}$ (where $\varepsilon_{r1}, \varepsilon_{r2}, \mu_{r1}, \mu_{r2}$ $\varepsilon_{r1}, \varepsilon_{r2}, \mu_{r1}, \mu_{r2}$ are the relative permittivities and relative permeabilities of the two media), and since the phase velocities in medium (1) and (2) are $v_{p1} = 1/\sqrt{\varepsilon_1\mu_1}$ and $v_{p2} = 1/\sqrt{\varepsilon_2\mu_2}$, respectively, we can also write Snell's law of refraction as

$$\frac{\sin\theta_t}{\sin\theta_i} = \frac{n_1}{n_2} = \frac{v_{p2}}{v_{p1}} \tag{D.152}$$

where $n_1 = \sqrt{\varepsilon_{r1}\mu_{r1}}$ is the (optical) index of refraction in medium (1) and $n_2 = \sqrt{\varepsilon_{r2}\mu_{r2}}$ is the (optical) index of refraction in medium (2).

Now returning to (D.149), the solution of the two relations for E_{r1} and E_{t2} gives

$$E_{r1} = E_{i1}\frac{\eta_2\cos\theta_i - \eta_1\cos\theta_t}{\eta_2\cos\theta_i + \eta_1\cos\theta_t}, \quad E_{t2} = E_{i1}\frac{2\eta_2\cos\theta_i}{\eta_2\cos\theta_i + \eta_1\cos\theta_t} \tag{D.153}$$

Because E_{r1} and E_{i1} are in the same direction, the reflection coefficient may be written as $\Gamma_\perp = E_{r1}/E_{i1}$ and the transmission coefficient as $T_\perp = E_{t1}/E_{i1}$:

$$\Gamma_\perp = \frac{E_{r1}}{E_{i1}} = \frac{\eta_2\cos\theta_i - \eta_1\cos\theta_t}{\eta_2\cos\theta_i + \eta_1\cos\theta_t} \quad \text{(dimensionless)} \tag{D.154}$$

$$T_\perp = \frac{E_{t2}}{E_{i1}} = \frac{2\eta_2\cos\theta_i}{\eta_2\cos\theta_i + \eta_1\cos\theta_t} \quad \text{(dimensionless)} \tag{D.155}$$

The notation $\perp$ indicates these are the reflection and transmission coefficients for perpendicular polarization, because, as we shall see, the coefficients for parallel polarization differ.

Now, the total fields in each material can be written directly. In material (1), the fields are the sum of the incident and reflected waves [from (D.141) and (D.142) for E_1 and from (D.142) and (D.144) for H_1]:

$$\mathbf{E}_1(x,z) = \hat{\mathbf{y}}E_{i1}\left[e^{-j\beta_1 z\cos\theta_i} + \Gamma_\perp e^{j\beta_1 z\cos\theta_i}\right]e^{-j\beta_1 x\sin\theta_i} \quad (\text{V/m}) \tag{D.156}$$

$$\mathbf{H}_1(x,z) = \hat{\mathbf{x}}\frac{E_{i1}\cos\theta_i}{\eta_1}\left[\Gamma_\perp e^{j\beta_1 z\cos\theta_i} - e^{-j\beta_1 z\cos\theta_i}\right]e^{-j\beta_1 x\sin\theta_i} + \hat{\mathbf{z}}\frac{E_{i1}\sin\theta_i}{\eta_1}\left[e^{-j\beta_1 z\cos\theta_i} + \Gamma_\perp e^{j\beta_1 z\cos\theta_i}\right]e^{-j\beta_1 x\sin\theta_i} \quad (\text{A/m}) \tag{D.157}$$

In medium (2), where the only wave is the transmitted wave, (D.145) and (D.146) describe the wave. Using the transmission coefficient, we can write

$$\mathbf{E}_t(x,z) = \hat{\mathbf{y}}T_\perp E_{i1}e^{-j\beta_2(x\sin\theta_t + z\cos\theta_t)} \quad (\text{V/m}) \tag{D.158}$$

$$\mathbf{H}_t(x,z) = \frac{T_\perp E_{i1}}{\eta_2}(-\hat{\mathbf{x}}\cos\theta_t + \hat{\mathbf{z}}\sin\theta_t)^{-j\beta_2(x\sin\theta_t + z\cos\theta_t)} \quad (\text{A/m}) \tag{D.159}$$

In all these relations, we could also use (D.151) to write the refraction angle θ_t in terms of the incident angle θ_i. However, this would complicate the expressions considerably.

The electric field intensity $\mathbf{E}_1$ is in the y direction, but $\mathbf{H}_1$ has a component in the x and z directions. This indicates a propagating wave in the x direction and a standing wave in the z direction.

D.5.2 Oblique incidence on a dielectric interface: parallel polarization

The situation considered here is shown in Figure D.4(b). The incident electric field is parallel to the plane of incidence and the magnetic field is perpendicular in the y direction so that the incident wave propagates toward the interface. The directions of the reflected fields in Figure D.4(b) are assumed. The correct directions are found from the interface conditions that must be satisfied.

The incident and reflected electric and magnetic field intensities for the configuration in Figure D.4(b) are written directly from the figure:

$$\mathbf{E}_i(x,z) = E_{i1}(\hat{\mathbf{x}}\cos\theta_i - \hat{\mathbf{z}}\sin\theta_i)e^{-j\beta_1(x\sin\theta_i + z\cos\theta_i)} \quad (\text{V/m}) \tag{D.160}$$

$$\mathbf{H}_i(x,z) = \hat{\mathbf{y}}\frac{E_{i1}}{\eta_1}e^{-j\beta_1(x\sin\theta_i + z\cos\theta_i)} \quad (\text{A/m}) \tag{D.161}$$

$$\mathbf{E}_r(x,z) = E_{r1}(\hat{\mathbf{x}}\cos\theta_i + \hat{\mathbf{z}}\sin\theta_i)e^{-j\beta_1(x\sin\theta_i - z\cos\theta_i)} \quad (\text{V/m}) \tag{D.162}$$

$$\mathbf{H}_r(x,z) = -\hat{\mathbf{y}}\frac{E_{r1}}{\eta_1}e^{-j\beta_1(x\sin\theta_i - z\cos\theta_i)} \quad (\text{A/m}) \tag{D.163}$$

The transmitted wave into material (2) can be written directly from Figure D.4(b):

$$\mathbf{E}_t(x,z) = E_{t1}(\hat{\mathbf{x}}\cos\theta_t - \hat{\mathbf{z}}\sin\theta_t)e^{-j\beta_2(x\sin\theta_t + z\cos\theta_t)} \quad (\text{V/m}) \tag{D.164}$$

$$\mathbf{H}_t(x,z) = \hat{\mathbf{y}}\frac{E_{t2}}{\eta_2}e^{-j\beta_2(x\sin\theta_t + z\cos\theta_t)} \quad (\text{A/m}) \tag{D.165}$$

At the interface between the two media (at $z = 0$), the continuity condition on the tangential components of the electric and magnetic field intensities are

$$E_{i1}\cos\theta_t + E_{r1}\cos\theta_i = E_{t2}\cos\theta_t \quad \text{and} \quad \frac{E_{i1}}{\eta_1} - \frac{E_{r1}}{\eta_1} + \frac{E_{t2}}{\eta_2} \tag{D.166}$$

Solving for E_{r1} and E_{t2}, we get

$$E_{r1} = E_{i1}\frac{\eta_2\cos\theta_t - \eta_1\cos\theta_i}{\eta_2\cos\theta_t + \eta_1\cos\theta_i}, \quad E_{t2} = E_{i1}\frac{2\eta_2\cos\theta_i}{\eta_2\cos\theta_t + \eta_1\cos\theta_i} \tag{D.167}$$

Therefore, the reflection coefficient for parallel polarization is defined as

$$\Gamma_{\|} = \frac{E_{r1}}{E_{i1}} = \frac{\eta_2\cos\theta_t - \eta_1\cos\theta_i}{\eta_2\cos\theta_t + \eta_1\cos 0_i} \tag{D.168}$$

The transmission coefficient is

$$T_{\|} = \frac{E_{t2}}{E_{i1}} = \frac{2\eta_2\cos\theta_i}{\eta_2\cos\theta_t + \eta_1\cos\theta_i} \tag{D.169}$$

The total fields in medium (1) are calculated by summing the incident and reflected waves. With the use of the reflection coefficient (i.e., using $E_{r1} = \Gamma_{\|}E_{i1}$), these become

$$\begin{aligned}\mathbf{E}_1(x,z) = {} & \hat{\mathbf{x}}E_{i1}\cos\theta_i\left(\Gamma_{\|}e^{j\beta_1 z\cos\theta_i} + e^{-j\beta_1 z\cos\theta_i}\right)e^{-j\beta_1 x\sin\theta_i} \\ & + \hat{\mathbf{z}}E_{i1}\sin\theta_i\left(\Gamma_{\|}e^{j\beta_1 z\cos\theta_i} - e^{-j\beta_1 z\cos\theta_i}\right)e^{-j\beta_1 x\sin\theta_i} \quad (\text{V/m})\end{aligned} \tag{D.170}$$

$$\mathbf{H}_1(x,z) = \hat{\mathbf{y}}\frac{E_{i1}}{\eta_1}\left(\Gamma_{\|}e^{j\beta_1 z\cos\theta_i} - e^{-j\beta_1 z\cos\theta_i}\right)e^{-j\beta_1 x\sin\theta_i} \quad (\text{A/m}) \tag{D.171}$$

Using $E_{t2} = T_{\|}E_{i1}$ in (D.164) and (D.165), we get the fields in medium (2):

$$\mathbf{E}_t(x,z) = T_{\|}E_{i1}(\widehat{\mathbf{x}}\cos\theta_t - \widehat{\mathbf{z}}\sin\theta_t)e^{-j\beta_2(x\sin\theta_t+z\cos\theta_t)} \quad (\text{V/m}) \qquad \text{(D.172)}$$

$$\mathbf{H}_t(x,z) = \widehat{\mathbf{y}}\frac{T_{\|}E_{i1}}{\eta_2}e^{-j\beta_2(x\sin\theta_t+z\cos\theta_t)} \quad (\text{A/m}) \qquad \text{(D.173)}$$

D.5.3 Reflection and transmission on dielectric interfaces: normal incidence

The reflection coefficients in (D.154) and (D.155) and the transmission coefficients in (D.168) and (D.169) apply to normal incidence as well by simply setting the incidence angle to 0°. The transmission and reflection coefficients become (regardless of polarization):

$$\Gamma = \frac{\eta_2 - \eta_1}{\eta_2 + \eta_1} \quad \text{(dimensionless)} \qquad \text{(D.174)}$$

$$T = \frac{2\eta_2}{\eta_2 + \eta_1} \quad \text{(dimensionless)} \qquad \text{(D.175)}$$

Similarly, the electric and magnetic field intensities on either side of the interface have only components tangential to the interface as can be seen in either (D.141)–(D.146) and (D.156)–(D.159) or (D.160)–(D.165) and (D.170)–(D.173).

D.5.4 Reflection and transmission on perfect conductors

The transmission and reflection coefficients also apply to conducting media. In perfect conductors, the transmission coefficient is zero and the reflection coefficient is -1 as can be seen by direct substitution (the intrinsic impedance of perfect conductors is zero).

Real conductors can be treated as lossy dielectrics, that is, we simply insert the intrinsic impedance of the conductor and the propagation constant, both of which are dependent on conductivity as

$$\eta_1 = \sqrt{\frac{j\omega\mu_1}{\sigma_1 + j\omega\varepsilon_1}} \quad (\Omega), \quad \gamma_1 = \sqrt{j\omega\mu_1(\sigma_1 + j\omega\varepsilon_1)} \qquad \text{(D.176)}$$

This also means that in lossy dielectrics the reflection and transmission coefficients must be complex since the intrinsic impedances in (D.154), (D.155), (D.168), (D.169), (D.174), and (D.175) are complex.

The select elements of electromagnetics given in this appendix represent the fundamentals of wave propagation. There are of course many more and important aspects of electromagnetics that were not and could not be covered. Even within the given subjects, there are issues and details that were not addressed either because they are not relevant to this work or because their exposition would take too much space and detract from the idea of writing a short appendix. Some additional aspects of electromagnetics are discussed in Chapter 3 (waveguides and cavity resonators), Chapter 5 (reflection and transmission sensors), Chapter 8 (resonant methods of measurement), and the whole of Chapter 2, which discusses transmission lines. Any additional material that may be needed can be found in textbooks of various levels. Some are listed in the following short bibliography.

Further reading

[1] N. Ida, "Engineering Electromagnetics," 3rd edition, Springer, NY, 2015.
[2] D. K. Cheng, "Field and Wave Electromagnetics," 2nd edition, Addison-Wesley, Reading, MA, 1992.
[3] C. A. Balanis, "Advanced Engineering Electromagnetics," 2nd edition, Wiley, New York, 2012.

Index